SYNTHETIC ORGANIC ELECTROCHEMISTRY

HARPER'S CHEMISTRY SERIES
Under the Editorship of Stuart Alan Rice

SYNTHETIC ORGANIC ELECTROCHEMISTRY

Albert J. Fry
WESLEYAN UNIVERSITY

HARPER & ROW, PUBLISHERS
New York, Evanston, San Francisco, London

SYNTHETIC ORGANIC ELECTROCHEMISTRY

Standard Book Number: 06-042215-7
LIBRARY OF CONGRESS CATALOG CARD NUMBER: 72-82902

To Missy

CONTENTS

PREFACE

Organic chemists interested in electrochemical methods have long been aware of a sharp contrast between the rich synthetic potentialities in the field of organic electrochemistry and the relatively slight acceptance that electrochemical techniques have actually gained in organic synthesis. There is no doubt that one of the principal reasons for this unfortunate disparity is the lack of a published treatment of electrochemical principles as they relate specifically to organic synthesis. The present volume attempts to fill this gap. It is designed for the synthetic organic chemist who may be unfamiliar with electrochemical methods, and hopefully it will be evaluated in this light. No attempt has been made to cover the vast organic electrochemical literature exhaustively, nor was it considered pertinent to discuss the fascinating area of applications of electrochemical techniques in physical organic chemistry. Rather, there has been presented, as far as permitted by limitations of space, a reasonably full discussion of electrochemical principles and experimental techniques, along with a survey of the various types of organic electrode reactions.

It may be helpful for the reader to have an overall view of the organization of the book before beginning. Chapter 1 contains a brief introduction to electrochemical experimentation. Chapter 2 consists of a discussion of electrochemical principles. Although rather detailed for an introductory book of this sort, it is far from comprehensive. The aim is to bring the reader up to a level where he has an understanding of the basic principles underlying the various kinds of electrochemical techniques. Chapter 3 discusses a number of common electrochemical techniques and shows how each can be used to obtain information about the mechanism of a new organic electrode reaction. Chapter 4 illustrates the variety of ways in which the overall course of an electrochemical reaction can be influenced by changes in chemical parameters—the intent is to suggest ways in which electrode reactions can be manipulated to obtain improved yields or even new products. Chapters 5 to 8 summarize the electrochemical behavior of a wide variety of organic compounds, with emphasis on the mechanistic features common to various types of functional groups. Finally, Chapter 9 provides a brief introduction to experimental methods. Since this book is not meant to be a manual for laboratory operations (and indeed could not be so without a considerable increase in length), the principal aim in this chapter is to point out the kinds of problems one has to face when selecting equipment for electrochemical experimentation; references are supplied there to the ample literature on this subject.

Notice that the book is organized such that a discussion of electrochemical principles and techniques precedes the description of the application of these to organic chemical systems. This order of topics appears to me to be the most logical organization. It is no doubt true that many readers will wish first to know something about the kinds of organic reaction that can be carried out electrochemically, before delving into the fundamental electrochemical principles underlying these reactions. It is entirely possible for such readers, after reading the introduction, to skip to Chapter 4, using the earlier chapters as a reference where necessary.

One further point should be made concerning the general orientation of the book. Emphasis is placed throughout on mechanisms of organic electrode processes, since one must understand the mechanism of a reaction in order to carry it out most efficiently, to divert its course into new channels, or to design new chemistry based on it. Thus, by the word "synthetic" in the title of this book I imply a characteristic element of creativity. Unfortunately, not all organic electrode reactions and phenomena are well enough understood at present to permit manipulation in a rational manner. Much more needs to be learned about why certain reactions proceed as they do. This is not entirely bad, of course: there would be scant intellectual attraction to a research area with no controversies or unanswered questions. At a number of places in the manuscript, I have pointed out mechanistic questions that remain unclear, and on occasion I have indulged in speculation concerning such points.

I would like to take this opportunity to acknowledge the assistance of a number of individuals in the preparation of this book. I owe special thanks to my colleague, Professor John Sease of Wesleyan University, who introduced me to organic electrochemistry and thus was forced to suffer patiently my many questions. Mr. (now Dr.) Richard Reed, Mrs. (now Dr.) Roberta Gable Reed, and Mr. Wayne Britton—all Wesleyan University graduate students—have contributed to many helpful discussions, as well as providing some of the illustrative material used herein. Unpublished preprints, information, and advice have been supplied by Professor Lennart Eberson, Professor Ronald Erickson, Dr. Palle Iversen, Professor Harry Mark, Professsor Louis Meites, Dr. Klas Nyberg, Dr. M. Rifi, and Professor Irving Shain. The hospitality of the staff of the Institute for Organic Chemistry of the University at Heidelberg, West Germany, during a sabbatical leave during which most of this book was written, is gratefully acknowledged.

Professors Dennis Evans, Leo Paquette, and Larry Anderson read and commented upon the entire manuscript. They, especially Professor Evans, are thanked for their many helpful suggestions.

Mrs. Betsy Shiels, Mrs. Jan Virginski, and Mrs. Lila Scoville typed the final draft of the manuscript. My wife Melissa typed a previous draft, did many of the drawings and formulas, and not least of all, provided the constant encouragement and good humor without which this work would never have been completed.

ALBERT J. FRY

Middletown, Connecticut
July, 1972

SYNTHETIC ORGANIC ELECTROCHEMISTRY

1 INTRODUCTION

Let us state the odious truth at the outset: electrochemical techniques are generally ignored at present in synthetic organic chemistry, even in comprehensive textbooks and review series devoted to synthesis. This is both surprising and unfortunate in view of the fact that an extremely wide variety of electrochemical conversions can be carried out using relatively simple and inexpensive equipment, usually with the further advantages of speed, good yields, and simplicity of workup. It is possible to carry out many redox processes, that is, reductions and oxidations, far more efficiently electrochemically than with the aid of conventional reducing or oxidizing agents; furthermore, there are many other reactions that can only be carried out electrochemically, for example,

O, Cl, Br ⟶ O, Cl (96%) (*Ref. 1*)

$$RCO_2^- \xrightarrow[\text{step}]{\text{one}} R(CH_2)_nCO_2R \qquad (\textit{Ref. 2})$$

n = any integer; yields 50–90%

O–CH=CHCO$_2$Et, O–CH=CHCO$_2$Et ⟶ O, CH_2CO_2Et, O, CH_2CO_2Et (89%) (*Ref. 3*)

OCH_3, OCH_3 ⟶ CH_3O OCH_3, CH_3O OCH_3 (88%) (*Ref. 4*)

⟶ X (*Ref. 5*)

X = OH, OAc, NHAc; yields ~ 90%

The mere fact that interesting chemistry can be carried out electrochemically is not sufficient, however. To be of any real use to the nonspecialist, organic electrode reactions must be susceptible to rational analysis, at least sufficiently so that one can predict with reasonable confidence the course of a previously untried electrochemical reaction under a given set of conditions. For this reason, succeeding chapters do not consist simply of lists of preferred conditions for electrochemical interconversion of functional groups (although in fact advantageous experimental conditions will be *deduced* for a number of electrode reactions by consideration of chemical and electrochemical principles). Instead there is presented in the first part of the book as extensive a discussion of electrochemical principles and techniques as is feasible within the limitations of space and the synthetic orientation of the book (Chapters 2–4). Only then, after methods of analysis and modification of organic electrode reactions have been discussed, is the electrochemical behavior of specific functional groups examined (Chapters 5–8). The objective in these sections is to extract *generalizations* concerning various modes of organic electrochemical behavior. The mechanistic discussions are meant to focus attention on similarities in the electrochemical behavior of the functional groups in a given section. In keeping with the philosophy of the book, examples have been selected for their variety and possible implications for the development of new chemistry. A final chapter discusses some practical aspects of electrochemical experimentation.

This introduction would be incomplete without a passing reference to one of the most valuable features of organic electrochemical experimentation. Oxidations and reductions, particularly fast rates of formation and reaction of intermediates in redox processes, can be studied in considerably greater detail electrochemically than is usually possible with chemical oxidants and reductants and conventional kinetic methods. Detailed mechanistic and kinetic studies of this sort are more properly within the province of physical organic than synthetic organic chemistry, however. They have therefore been neglected here, although they are of great potential importance in physical organic studies.

1.1 DEFINITIONS

As with all branches of science, electrochemistry has its own terminology, some of which must be learned at the outset. Fortunately, the number of terms is fewer, and they are rather better standardized than is true of many other branches of science. Those terms that require definition at this time may be grouped accordingly as they relate to *electrolysis*, *potential*, and *vector representation*.

1.1.1 Electrolysis

Examine the simple electrochemical cell illustrated in Figure 1.1. Each terminal of a source of direct current *B* is connected to an inert metallic electrode, and both electrodes are immersed in a conducting salt solution. In this way an

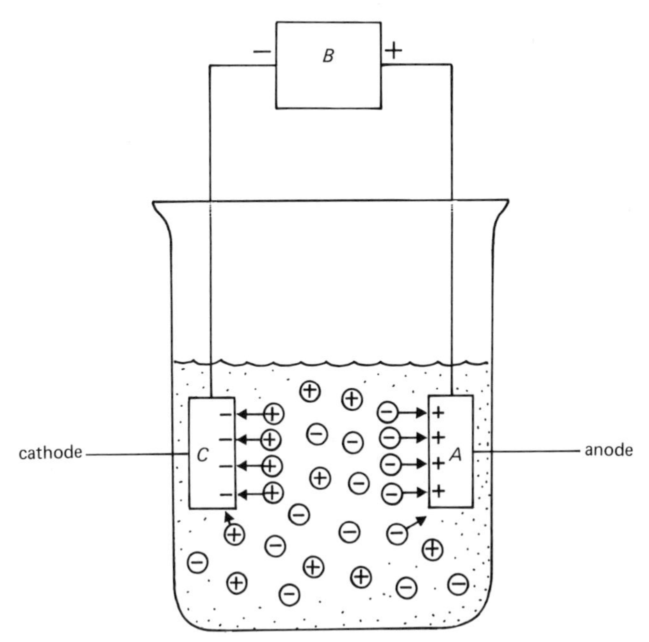

Figure 1.1 Undivided electrochemical cell. B is a source of direct current.

electrical circuit is completed: current flows as positive ions migrate to the negative electrode (*cathode*) and are reduced, while negative ions migrate to the positive *anode* and are oxidized. Obviously, every cell must contain both an anode and a cathode. Usually, however, one is interested in effecting either oxidation or reduction of a given substance, not both. Furthermore, species reduced at the cathode in a cell such as that in Figure 1.1 could later migrate to, and be oxidized at, the anode, or vice versa. Both to define more precisely the experimental conditions in the desired electrode process and to prevent further electrochemical destruction of products, the anode and cathode are usually placed in separate compartments, separated by a barrier that is porous enough to permit the passage of ions necessary for conduction of current but that prevents mixing of the solutions in the two compartments (Figure 1.2). The solution in the cathode compartment is known as the *catholyte* and that in the anode side is the *anolyte*; they may and often do differ considerably in composition.

The substance whose electrochemical behavior is to be investigated is known as the *electroactive substance* or *substrate* (or, in the older literature, the *depolarizer*). The quantity of electricity necessary to change the oxidation state of 1 gram-ionic weight of an inorganic ion or 1 gram-molecular weight of an organic substance by one unit is called the faraday, and is given the symbol $\mathscr{F}$.

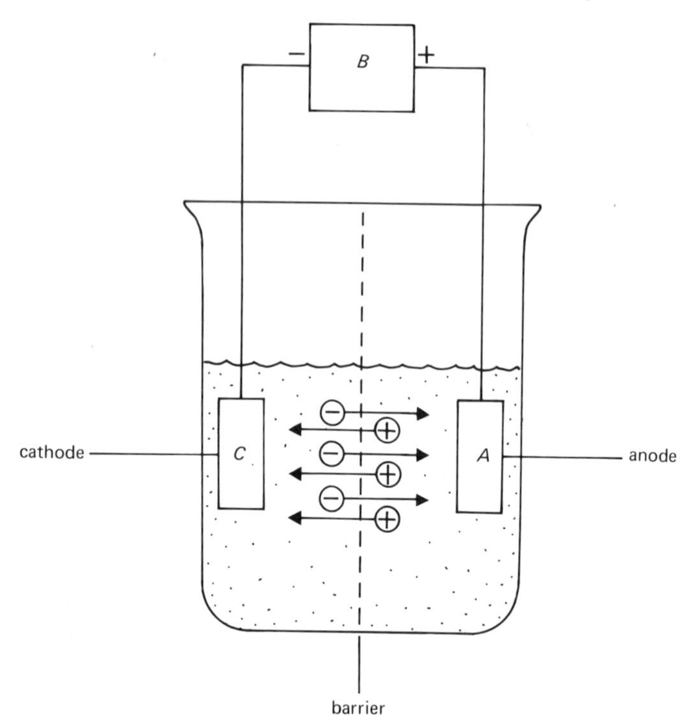

Figure 1.2 Divided (two-compartment) electrochemical cell. *B* is a source of direct current.

It is equal to 9.65×10^4 C/equivalent, but since the faraday is also a proportionality factor between chemical and electrochemical free energy ($\Delta G = -n\mathscr{F}E$), it is sometimes useful to express it in energy terms (23.06 kcal/volt-equivalent). Depending on experimental conditions, the same substrate may react by a variety of pathways involving different numbers of faradays:

$$R_2C{=}O + 1e^- + H^+ \longrightarrow \tfrac{1}{2}\underset{\text{HO}}{R_2C}{-}\underset{\text{OH}}{CR_2}$$

$$R_2C{=}O + 2e^- + 2H^+ \longrightarrow R_2CHOH$$

$$R_2C{=}O + 4e^- + 4H^+ \longrightarrow R_2CH_2 + H_2O$$

The number of electrons involved in a reaction provides information about the oxidation state of the product and hence can be of decisive importance in product identification. Experimental determination of the quantity of electricity (in coulombs) involved in an electrochemical reaction is known as *coulometry* (Chapter 3). Incidentally, in view of the fact that experimental conditions can substantially affect the products of electrolysis, an important component of subsequent discussions must be an analysis of the effects of changes in experimental parameters (pH, solvent, potential, etc.) on reactions (Chapter 4). An understanding of this area is critical to the design and interpretation of new electrochemistry.

One often finds the terms *working electrode* and *counter electrode* in the electrochemical literature. The working electrode is the one at which the electrochemical reaction of interest is carried out. When reductions are being studied, the cathode is the working electrode; in oxidations the anode is the working electrode. The counter electrode is then the other current-carrying electrode, whose presence is necessary to complete the circuit: the anode is the counter electrode when reductions are being studied, and the cathode is the counter electrode in oxidative studies. These terms obviously have no fundamental significance, but are often useful in descriptions of the detailed construction of electrolysis cells (Chapter 9).

Two other terms are often found in the older literature. The *current efficiency* is the ratio (expressed as a percent) of the theoretical amount of electricity required to effect a given reaction to the actual amount consumed. Thus, if reduction of 1 mole of a ketone to the alcohol is found experimentally to consume 2.5 $\mathscr{F}$, the current efficiency is 80% (2/2.5 × 100). Current efficiencies less than 100% can arise if other electroactive materials are present in solution; current efficiencies greater than 100% are rare, but are occasionally observed, for example, when chain processes are initiated at the electrode. The *current density* is equal to the electrolysis current divided by the area of the working electrode. With a few exceptions, current densities have no major significance for synthetic experimentation and will not normally be used here.

1.1.2 Potential

Consider the following simple redox couple:

$$\mathrm{O} + ne^- \rightleftharpoons \mathrm{R}$$

where O and R represent the oxidized and reduced forms of a given species. From thermodynamic considerations[6] we may write:

$$E = E^0 - \frac{RT}{n\mathscr{F}} \ln \frac{a_\mathrm{R}}{a_\mathrm{O}} \tag{1.1}$$

from which we see that E^0, the *thermodynamic standard potential*, is the potential where the activities of R and O are equal (some definitions of E^0 state that it is the potential where the activities of both species are equal to unity. Lingane[6] has commented further on this point, which will not be of importance in our discussions). Equation 1.1 is doubtless already known to the reader as the *Nernst equation.*

It is a common electrochemical practice to use a variation of the Nernst equation in which concentrations are used instead of activities. The potential where the concentrations of R and O are equal is known as the *apparent standard potential*, or *formal potential*, E'^0:

$$E = E'^0 - \frac{RT}{n\mathscr{F}} \ln \frac{[\mathrm{R}]}{[\mathrm{O}]} \tag{1.2}$$

There are a number of justifications for the use of apparent standard potentials. These can differ from thermodynamic standard potentials as a result of, for example, complexation or solvation of O and R, or other experimental variables, and hence one always measures E'^0, not E^0. The data necessary for correcting E'^0 to E^0 may not be available. Furthermore, potential data are generally used electrochemically only for *relative* comparisons, for example, to predict the position of an equilibrium such as the following:

$$\mathrm{O_1 + R_2 \rightleftharpoons R_1 + O_2}$$

The use of formal potentials is acceptable in relative comparisons such as these, as long as one uses them for comparisons of systems for which all values of E'^0 were measured under identical conditions and if one does not attempt to use them directly to compute absolute thermodynamic data. Because formal potentials are dependent upon the experimental conditions under which measurements are taken, these must always be specified as completely as possible.

Potentials can only be defined usefully with respect to an accepted reference point. The half-reaction that serves as the fundamental reference of potential is reduction of the proton to hydrogen,

$$\mathrm{H^+ + e^- \rightleftharpoons \tfrac{1}{2}H_2} \qquad E^0 = 0 \text{ V}$$

and the fundamental experimental reference point for potential measurements is the *normal hydrogen electrode* (N.H.E.), which consists of a platinized platinum electrode immersed in a solution with proton activity equal to unity and saturated with hydrogen gas at 760 mm pressure. By definition the potential of this electrode is 0 V. In agreement with IUPAC conventions, the sign of the standard potential for reduction of substances that are more difficult to reduce than the proton will be taken as negative in this book:

$$\mathrm{Na^+ + e^- \rightleftharpoons Na^0} \qquad E^0 = -2.71 \text{ V } (vs \text{ N.H.E})$$

The expression "cathodic of N.H.E." denotes an electrode whose potential is negative. As the potential of a cathode becomes increasingly negative, the cathode becomes a more powerful reducing agent.

The N.H.E. happens to be quite experimentally inconvenient for routine use[7]; hence, customary practice is to measure potentials with a *secondary reference electrode* whose potential relative to N.H.E. is accurately known. By far the most generally used of the secondary reference electrodes is the saturated calomel electrode (S.C.E.), which consists of a paste of mercurous chloride (calomel) in contact with both a mercury pool and a saturated potassium chloride solution[8a]:

$$Hg/Hg_2Cl_2/KCl_{sat.} \qquad E = +0.2412 \text{ V } (vs \text{ N.H.E.})$$

Besides being easy to prepare, the S.C.E. has the advantage that its potential is stable and rather insensitive to passage of small currents. Most of the potential data in the literature is reported relative to S.C.E., thus simplifying comparisons between compounds studied by different investigators. A number of reference electrodes other than the S.C.E. are used to some extent, however. These include, for example, a pool of mercury in the particular solvent system of interest, and electrodes based on silver and its salts, such as the silver/silver halide electrodes, which consist of a silver wire coated with the silver halide and immersed in a solution containing a soluble salt of the same halide, or the silver/silver(I) electrodes, in which the silver wire is immersed in a solution of a soluble silver salt. Extensive compilations of potentials of reference electrodes suitable for use in organic solvents are available.[8b,c] It is generally preferable to use the S.C.E. whenever possible.

One often encounters in the literature the term *overvoltage*, particularly *hydrogen overvoltage*. Certain aspects of this phenomenon are not well understood, but it is basically kinetic in origin. Reduction of protons to hydrogen is thermodynamically spontaneous at potentials negative of 0 V (*vs* N.H.E.). With a few metals, notably platinum electrodes of large surface area (platinized platinum), the expected hydrogen evolution is indeed observed at potentials negative of 0 V when $[H^+] = 1\ M$. At many other metallic electrodes, however, hydrogen evolution does not become detectable until potentials considerably more negative than the theoretical value are reached. It is said that there is an overvoltage for hydrogen evolution at these metals. Typical values[9] of hydrogen overvoltage in 1-*N* HCl are platinum, 0 V; copper, −0.19 V; lead, −0.40 V; cadmium, −0.39 V; and mercury, −0.80 V.* It is generally believed that this overvoltage phenomenon arises because on some metals hydrogen evolution is so negligibly slow at the theoretical potential that it is simply not experimentally detectable. Rates of reduction increase rapidly as cathode potentials become more negative (Chapter 2), however, so if the cathode potential is made increasingly negative, a potential will eventually be reached where the rate of reduction of protons is high enough for hydrogen evolution to be apparent. The difference between the theoretical and the actual potentials at which detectable hydrogen evolution takes place is the hydrogen overvoltage. For reasons that are not well understood, the hydrogen overvoltage is different for each metal. [It is undoubtedly relevant to this problem that metals that are good hydrogenation catalysts (Pt, Pd, Ni)

* These potentials must be regarded only as approximate, since they depend on a number of parameters such as pH and the history and physical composition of the electrode.

generally also have low hydrogen overvoltages.] Since hydrogen overvoltage arises because of an apparent activation barrier to proton reduction, it is often termed the *activation overpotential* for hydrogen evolution. We shall not normally be concerned with hydrogen overvoltage, except insofar as it leads to a choice of a convenient electrode substance for electrode reactions. Let us examine this point in the following manner. Suppose it is desired to reduce a substance at -0.3 V in 1-*N* HCl. At a copper electrode, the necessary reduction potential could never be reached, since hydrogen evolution would begin first, at -0.2 V. On the other hand, the reduction would be feasible at lead, cadmium, or mercury electrodes, since their hydrogen overvoltages are considerably cathodic of the desired reduction potential. Obviously, it is generally desirable to use an electrode material of high hydrogen overvoltage, since a wider range of electrochemical reactions is thereby accessible. Of the high overvoltage cathode materials, the one used most often is mercury, since because it is a liquid a clean electrode surface can be maintained with relative ease and it can be used in the important electrochemical technique known as polarography (Section 1.2.1). This is not to imply that mercury can be used universally as an electrode material. Certain reactions, such as the electrochemical reductions of acetylenes to *cis*-alkenes at spongy nickel, and of carbonyl compounds to saturated hydrocarbons at lead, do not occur at a mercury cathode. This is not related to overvoltage, however, but rather apparently to specific chemical involvement of the electrode itself in these reductions (Chapter 6). Furthermore, mercury is rather easily oxidized (ca. $+0.6$ V *vs* N.H.E.) and is therefore not suited for use as an anode material. Adams has discussed at length the anodic potential limits of a number of substances.[10] It appears that platinum is the substance best suited to studies of the electrochemical oxidation of organic compounds, despite certain difficulties involved in achieving a reproducible surface.

1.1.3 Current–Potential Curves (Voltammograms)

The most common record of electrochemical behavior is the *voltammogram*. As its name implies, this is a plot of current *vs* potential (voltage). As voltammetry is usually employed, an electrochemical cell is furnished with the necessary circuitry such that the electrolysis current and the potential of the working electrode relative to a chosen reference electrode can both be continually measured and recorded. The potential of the working electrode is then varied (*scanned*) over a certain range. Current flow is observed when the electrode potential reaches a value such that electrolysis of the substrate takes place. The generally accepted convention for representation of the resulting voltammetric data is a graph on which potential is plotted on the abscissa and current on the ordinate. Potentials that are cathodic, that is, negative with respect to the reference electrode are plotted to the right, and potentials that are anodic of the reference are plotted to the left. Cathodic currents are *positive* by convention and hence are plotted upward (Figure 1.3). Voltammograms of reduction processes are recorded by scanning the potential from left to right, that is, toward increasingly powerful reducing potentials. Conversely, anodic voltammograms are scanned from right to left, and the resulting anodic currents are by convention negative and downward. The zero of potential is the potential of the reference electrode, but of course it has no fundamental significance; reductions can take place at positive potentials, and oxidations can occur at negative potentials. Examination of the voltammograms for reduction of substances A and B (Figure 1.3) reveals,

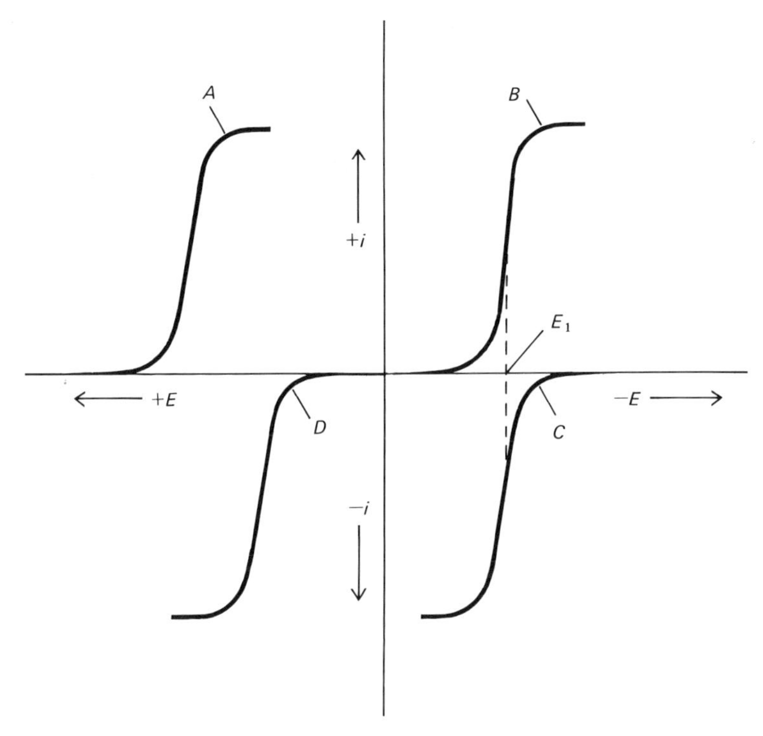

Figure 1.3 Representative voltammograms, with current and potential conventions indicated.

from their relative positions on the potential scale, that A is easier to reduce than B. The voltammograms of substances C and D were scanned from right to left and likewise demonstrate that C is oxidized more easily than D. It will also be noted that reduction of B and oxidation of C take place close to the same potential, E_1. When B and C represent a redox couple, that is, B is the oxidized form of C, and certain other conditions are also fulfilled, the potential E_1 may be shown to be equal to the standard potential (properly, the formal potential E'^0) for the redox couple (Chapter 2).

1.2 METHODOLOGY

Initial studies of the electrochemical behavior of previously uninvestigated systems tend to be quite similar. The basic features of the system are usually delineated by information from a few voltammetric techniques. Mechanistic hypotheses are then made based on the voltammetric data. These hypotheses are tested by coulometry, preparative-scale electrolysis, and identification of products (Chapter 3). At this point, the electrochemical behavior of the system may be known sufficiently well to permit immediate synthetic application. It is often useful, however, to investigate the effects of changes in any of a variety of variables, such as solvent, pH, or added reagents, on the course of the reaction, in order to gain further insights, particularly into possible ways of modifying the reaction or diverting it into new channels (Chapter 4). Even more detailed information than this, such as rates of electron transfer between the electrode and substrates or rates of fast chemical reaction of electrochemically generated intermediates, can be obtained by the use of a variety of other electrochemical instrumental techniques, but this area will not be treated here.

Two representative voltammetric techniques, *polarography* and *voltammetry in stirred solution*, are discussed below. These are introduced for several reasons. First, they represent typical voltammetric data and hence permit the introduction at this point of several terms used to describe various features of voltammograms. The differences and similarities between the two techniques will also suggest a number of areas that will require careful analysis in later chapters. To cite one example, each voltammogram consists of steps of current, called *waves*, with each wave followed by a current plateau. The plateaus arise because the rate of mass transport of electroactive material to the working electrode becomes the current-limiting process at certain potentials. It will be necessary to examine later the various modes of mass transport to develop some appreciation of the differences between electrochemical processes in quiet solution and those in stirred solution. A discussion of the relationship between potential and the rate of electron transfer is clearly also necessary, and will in fact be introduced as part of a general analysis of factors that affect the shapes of voltammetric waves (Chapter 2).

1.2.1 Polarography

Polarography, one of the most widely used electrochemical techniques, can provide a considerable amount of useful information with relatively little experimental effort. The technique is surprisingly simple. Its basic component is its unique working electrode, the *dropping mercury electrode* (D.M.E), which is simply a long thin (i.d. 0.05–0.10 mm) capillary attached to a mercury reservoir of adjustable height. Drops of mercury continually form at the end of the capillary, grow, and eventually break off when their weight exceeds the mercury surface tension supporting the drop at its neck. The lifetime of an individual drop depends on a number of factors, including the height of the mercury column and the composition of the solution, and is generally between 2 and 10 sec. The capillary is immersed in a solution containing the electroactive substance (10^{-3}–$10^{-4}M$). Circuitry is provided such that (1) the mercury drop serves as a working electrode, (2) its potential may be scanned gradually (ca. 5 mV/sec) to more negative potentials, and (3) the current flowing between the drop and the counter electrode is recorded as a function of the working electrode potential (relative to a nearby reference electrode). The resulting voltammogram is known as a *polarogram.** A typical polarogram is illustrated in Figure 1.4. Observe several features of this polarogram. The most obvious are the "oscillations," each one associated with the growth of an individual drop and the sharp decrease in the current as the drop falls. One also notes two voltammetric waves, *A* and *B*, which are rather symmetrical and of similar height. Several tenths of a volt past wave *B*, the current increases rapidly and goes off scale (*C*). One might wish to know at this point not only the cause of this behavior, but also why one would ever wish to use the D.M.E. when, for example, a solid cathode could be used to eliminate the current oscillations inherent to the D.M.E. To answer the first question, the waves represent discrete electrochemical processes and might arise from reduction either of two different components of the solution or of a single substance in successive stages. It happens that the latter is true in Figure 1.4, because it is the polarogram of benzophenone in dimethylformamide (DMF), and the waves correspond to the following reduction processes (*cf.* Chapters 4 and 6):

$$(C_6H_5)_2C{=}O + e^- \longrightarrow (C_6H_5)_2\dot{C}{-}\bar{O} \qquad \text{wave } A$$

$$(C_6H_5)_2\dot{C}{-}\bar{O} + e^- \longrightarrow (C_6H_5)_2\bar{C}{-}\bar{O} \qquad \text{wave } B$$

The final current rise at *C* is known variously as the *solvent discharge*, *solvent breakdown*, or *solvent decomposition* potential and represents the point where the cathode potential has become sufficiently negative to effect reduction of either the solvent itself or the supporting electrolyte (Chapter 2). The solvent discharge potential constitutes a natural limit on accessible potentials, and naturally there is great interest in the development of solvent–electrolyte systems in which very cathodic potentials can be reached. Generally speaking, aprotic solvents are superior to protic solvents and quaternary ammonium salts are superior to metal salts when very cathodic potentials are required.[11]

As for the reasons why the D.M.E. is quite widely used, there are several: (1) it is experimentally simple and inexpensive; (2) drop time and drop size are

* The literature is not completely consistent with respect to the terms "polarography" and "polarogram." Some authors use these terms to describe any experiment in which current is plotted as a function of potential. In this book, the latter type of experiment will be described by the general terms "voltammetry" and "voltammogram." Under this usage, polarography is the special case of voltammetry at the D.M.E.

quite reproducible with a given capillary; (3) the mercury surface is constantly renewed, so electrolysis products do not accumulate at the electrode surface; and (4) as a consequence of (2) and (3), successive drops grow in solution of

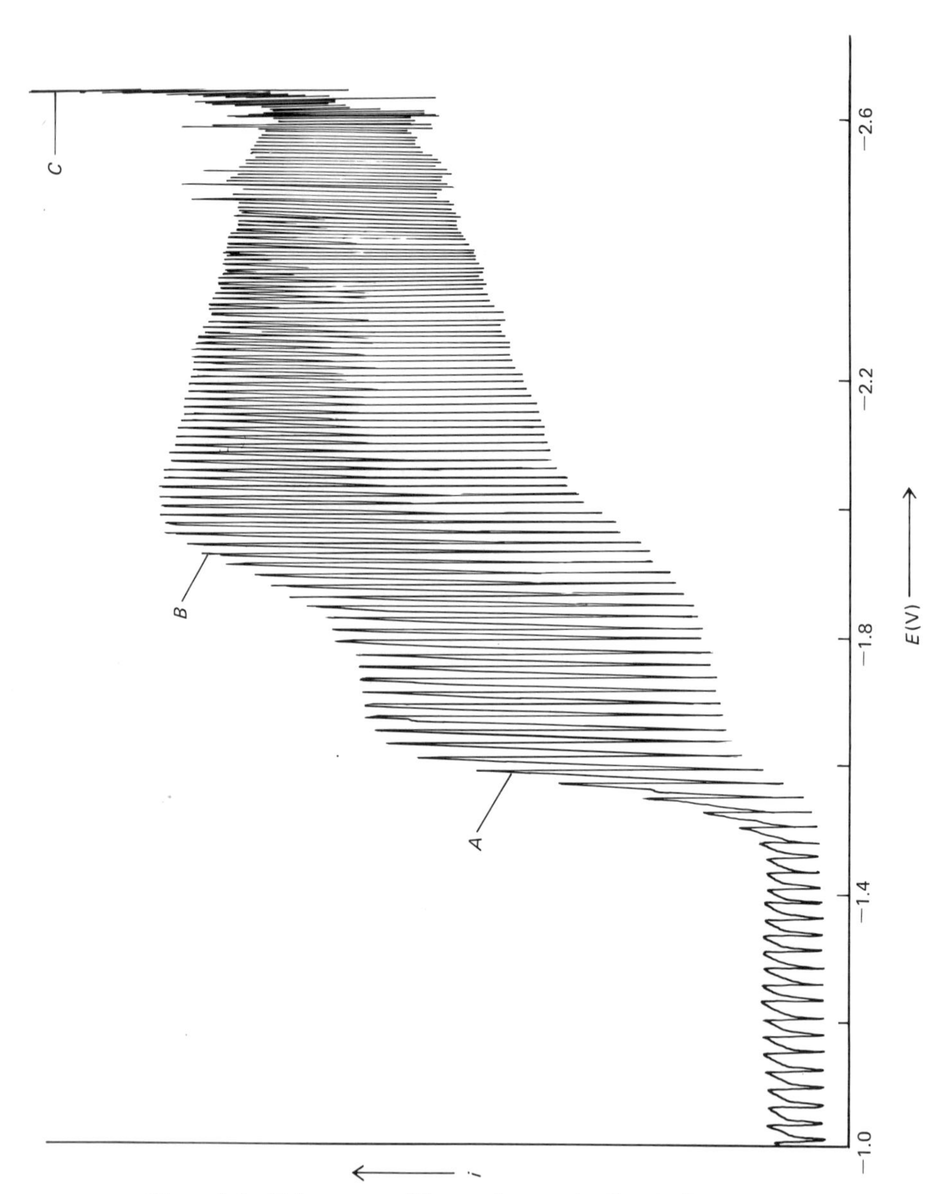

Figure 1.4 Polarogram of benzophenone in dimethylformamide.

identical composition at identical rates. Furthermore, the oscillations do not particularly complicate the analysis of polarograms, since current measurements can simply be made at the tops of individual oscillations. If desired, one may smooth the polarogram manually by simply drawing the envelope of the oscillations, that is, a line connecting the tops of the individual oscillations. It is also possible to damp the recorder, producing a smoothed curve of lower magnitude than the peaks of the undamped oscillations; a few instruments track and record the peak currents, giving a smoothed curve similar to that obtained manually. For the sake of simplicity, smoothed polarograms will be used throughout the remainder of this book. A few other features common to polarograms may be seen by reference to Figure 1.5 (in which the current oscillations have been smoothed). Notice that the current begins to increase slowly and regularly long before the reduction wave is reached. This phenomenon, which is exaggerated somewhat in the figure for clarity, is known as the *charging current*, and its origin is discussed in Chapter 2. The dashed line in Figure 1.5 is a polarogram measured under the same conditions as the solid line except without the electroactive substance added. It is seen that the charging current is still present, and hence it represents a background that must always be subtracted from the observed current. The difference between the total current (solid line)

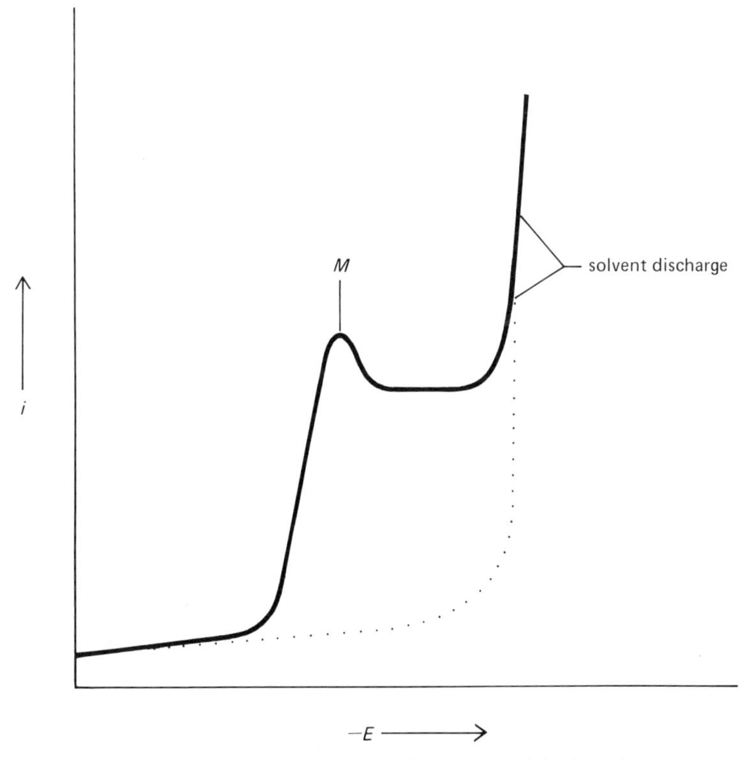

Figure 1.5 Representative damped polarogram, with charging current.

measured at any potential and the charging current at that potential is known as the *faradaic current*, that is, the current actually involved in electrolysis of the substrate.

The hump M atop the reduction wave is known as a *polarographic maximum*. The question of the origins of polarographic maxima is a complex and controversial subject. Several different phenomena, including adsorption, can be involved. One feature all have in common is stirring of the solution at the mercury–solution interface. Maxima can interfere with interpretation of polarographic behavior, but they may often be suppressed by addition of traces of surface-active agents to the polarographic solution or, better, by operating at low concentrations of electroactive material. Maxima are treated extensively elsewhere,[12] and discussion of the phenomenon is outside the scope of this book.

1.2.2 Voltammetry in Stirred Solution

If one stirs the solution used for the polarogram of Figure 1.4, using a stationary electrode, and scans the cathode potential toward increasingly negative values, the voltammogram shown in Figure 1.6 results. Although this outwardly

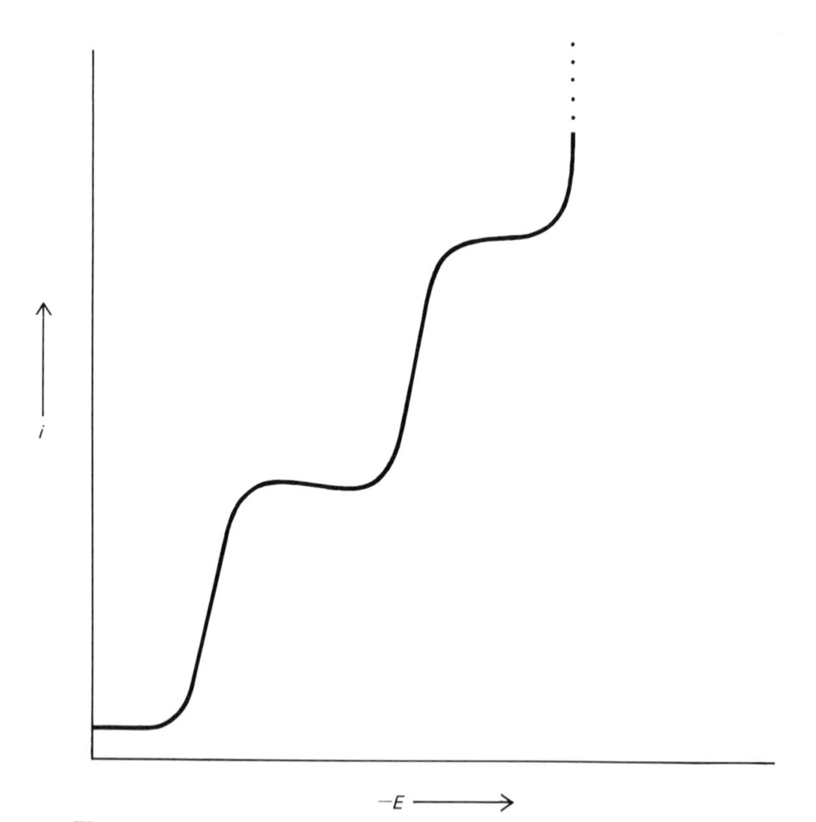

Figure 1.6 Voltammogram of benzophenone in stirred solution.

resembles a damped polarogram, analysis of such a voltammogram requires an entirely different set of assumptions about the nature of mass transport of the electroactive substance to the cathode. In particular, the mathematical models used to describe the shapes of waves in Figures 1.4 and 1.6 are quite different. It will be helpful to describe at length in Chapters 2 and 3 these respective models and the kinds of information that polarography and voltammetry in stirred solution can provide. Mathematical analysis of the latter technique will lead to principles that are especially useful in the design of efficient cells for preparative electrolysis (Chapters 3 and 9).

References

[1]A. J. Fry, M. A. Mitnick, and R. G. Reed, *J. Org. Chem.*, **35**, 1232 (1970).
[2]B. C. L. Weedon, *Advan. Org. Chem.*, **1**, 1 (1960).
[3]J. D. Anderson, J. O. Petrovich, and M. M. Baizer, *Advan. Org. Chem.*, **6**, 257 (1969).
[4]B. Belleau and N. L. Weinberg, *J. Amer. Chem. Soc.*, **85**, 2525 (1963).
[5a]L. Eberson, *J. Amer. Chem. Soc.*, **89**, 4669 (1967); [b]L. Eberson and B. Olofsson, *Acta Chem. Scand.*, **23**, 2355 (1969).
[6]J. J. Lingane, *Electroanalytical Chemistry*, 2nd ed., New York, Wiley, 1958, pp. 46–49.
[7]D. J. G. Ives and G. J. Janz, *Reference Electrodes*, New York, Academic, 1961.
[8a]Lingane, *op. cit.*, pp. 359–365; [b]J. N. Butler, in P. Delahay and C. A. Tobias, eds., *Advances in Electrochemistry and Electrochemical Engineering*, New York, Wiley, 1970, pp. 77–175; [c]C. K. Mann and K. K. Barnes, *Electrochemical Reactions in Nonaqueous Systems*, New York, Marcel Dekker, 1970, pp. 17–28.
[9]C. K. Mann and K. K. Barnes, *Electrochemical Reactions in Nonaqueous Systems*, New York, Marcel Dekker, 1970, p. 209.
[10]R. N. Adams, *Electrochemistry at Solid Electrodes*, New York, Marcel Dekker, 1969, chap. 2.
[11]C. K. Mann, in A. J. Bard, ed., *Electroanalytical Chemistry*, New York, Marcel Dekker, 1969, vol. 3, pp. 57–134.
[12]L. Meites, *Polarographic Techniques*, 2nd ed., New York, Wiley, 1965, chap. 6.

2 ELECTROCHEMICAL PRINCIPLES

The chemist is often confronted with the situation in which a reaction of interest does not proceed at a convenient rate under an initially selected set of conditions. Apart from the rate changes possible by manipulation of concentrations, solvent, and so on, the principal recourse in such a situation must be to slow or speed the reaction by a change in reaction temperature, since rate and temperature are interrelated by the familiar activation parameters. Routine chemical practice such as this differs very fundamentally from electrochemical experimentation because of the existence of an additional experimental parameter, the *electrode potential*, for manipulation of electrochemical reaction rates. Electron-transfer rates may be varied easily over many orders of magnitude in a single experiment by proper control of the electrode potential, using apparatus of simple design. In fact, electrode potential is so much more convenient and powerful a reaction parameter than temperature that most electrochemical data in the literature have been measured at or near room temperature, a feature of obvious advantage. A recognition of the nature of the potential dependence of electrode reaction rates is clearly central to the understanding of electrode processes and constitutes the central theme of this chapter. It will not be sufficient, however, merely to relate electron-transfer rates to potential, because, as a result of the dramatic changes of these rates with potential, it is generally found that at certain potentials electron transfer is so fast that the overall process is actually limited by the rate of mass transport of the substrate to the electrode. Since there exist different modes of mass transport of varying efficiency, it will be necessary to examine each of these so that it will be clear which is operating in a given experimental situation. Finally, as a consequence of the fact that electron transfer at an electrode surface is necessarily a heterogeneous process, it will be necessary to examine briefly the structure of the electrode–solution interface and its effects on the course of electrochemical reaction. With the preceding considerations in hand, it will be possible in the next chapter to examine specific methods for deduction of mechanistic information. Although electrode potential is an extremely important experimental parameter for exerting control over electrochemical reactions (and temperature is not), many other variables can affect the course of an electrode reaction. The effects of such variables (pH, solvent, added proton donors, surface-active agents, etc.) are discussed in Chapter 4. This area, it has become clear in recent years, offers many possibilities for creative synthetic design.

2.1 ELECTRODE KINETICS

It was indicated previously that, depending on the electrode potential, the rate-determining step in an electrochemical reaction might be either the electron-transfer rate or the mass-transport rate. For the sake of the present discussion it will be assumed that mass-transport rates are high enough that electron transfer is rate determining over the range of potentials discussed. What is desired here is a formal expression relating the rate of electron transfer to electrode potential.

Consider a reaction-coordinate diagram (Figure 2.1) on which curves *A*

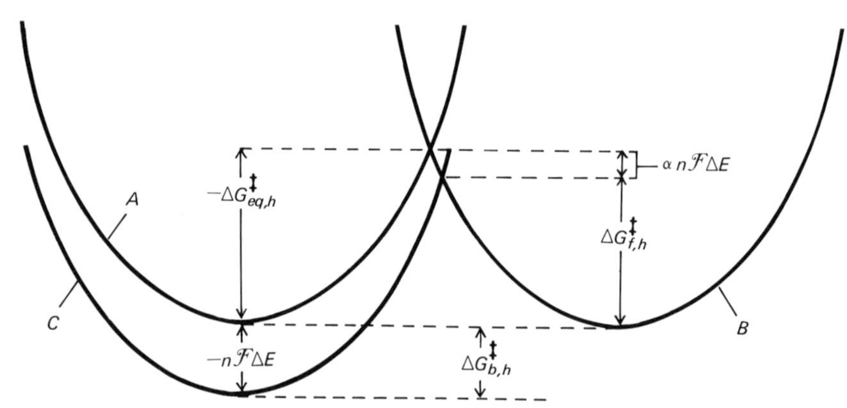

Figure 2.1 Reaction-coordinate diagram for an electron-transfer reaction at an electrode. (C. N. Reilley, in I. M. Kolthoff and P. J. Elving, eds., *Treatise on Analytical Chemistry*, New York, Wiley, 1963, part I, vol. 4, p. 2123.)

and *B* represent the potential-energy profiles of the reduced and oxidized forms, respectively, of the following redox couple:

$$\mathrm{O} + ne^- \underset{k_{b,h}}{\overset{k_{f,h}}{\rightleftharpoons}} \mathrm{R}$$

The potential-energy profiles are taken at the equilibrium potential of the system, where by definition the rate constants for reduction and oxidation, $k_{f,h}$ and $k_{b,h}$, are equal.* To effect electrochemical reduction or oxidation, an activation barrier, $\Delta G^{\ddagger}_{\mathrm{eq.},h}$, must be overcome. If this barrier is high, interconversion of the two species will be slow. Suppose, however, that the electrode potential is displaced from its equilibrium value by an amount ΔE in such a direction as to *favor* reduction. This may be represented as a new potential-energy profile, curve *C*, for the reduced species; the decrease in free energy for this species is, of course, $\Delta G = n\mathcal{F}\,\Delta E$. Notice that the new activation energy for reduction,

* The subscripts *f* and *b* stand, of course, for the forward and back half-reactions; the *h* indicates that these are *heterogeneous* rate constants.

$\Delta G^{\ddagger}_{f,h}$ is lower than $\Delta G^{\ddagger}_{\text{eq.},h}$ by only a fraction α of the total free-energy decrease imposed on the system. The remainder acts to raise the activation energy for oxidation. One may express the relationships shown in Figure 2.1 as

$$\Delta G^{\ddagger}_{f,h} = \Delta G^{\ddagger}_{\text{eq.},h} - \alpha n\mathscr{F}\,\Delta E \tag{2.1}$$

$$\begin{aligned}\Delta G^{\ddagger}_{b,h} &= \Delta G^{\ddagger}_{\text{eq.},h} + n\mathscr{F}\,\Delta E - \alpha n\mathscr{F}\,\Delta E \\ &= \Delta G^{\ddagger}_{\text{eq.},h} + (1-\alpha)n\mathscr{F}\,\Delta E\end{aligned} \tag{2.2}$$

where $0 < \alpha < 1$.

The sign of $n\mathscr{F}\,\Delta E$ in the second term of Eq. 2.2 is positive because the activation energy for oxidation is *increased* as a result of the imposed potential. The fraction α is termed the *transfer coefficient* and is related to the shapes of the potential energy surfaces of O and R at different potentials. In 1939 Eyring, Glasstone, and Laidler[1] showed, using absolute reaction-rate theory, that the rate constants for the cathodic and anodic reactions, $k_{f,h}$ and $k_{b,h}$, respectively, should depend exponentially on potential:

$$k_{f,h} = k^{\text{ref}}_{f,h} \exp \frac{-\alpha n\mathscr{F}(E - E_{\text{ref}})}{RT} \tag{2.3}$$

$$k_{b,h} = k^{\text{ref}}_{b,h} \exp \frac{(1-\alpha) n\mathscr{F}(E - E_{\text{ref}})}{RT} \tag{2.4}$$

where $k^{\text{ref}}_{f,h}$ and $k^{\text{ref}}_{b,h}$ are the rates of reduction and oxidation, respectively, at a potential E_{ref} selected as a reference point, and potential E is measured relative to E_{ref}. This *exponential* potential dependence of the electron-transfer rate is the basis for the previous statement that electrochemical reaction rates may be changed readily over many orders of magnitude by adjustment of the electrode potential.

It will be convenient for certain points in the following discussion to have an alternative form of the Eyring electrochemical rate equation that relates electrolysis current to potential. This is readily done; from Faraday's law (Chapter 1) and the expressions for the reaction rates, one finds

$$i_c = n\mathscr{F}AC^0_{\text{ox}}\, k^{\text{ref}}_{f,h} \exp \frac{-\alpha n\mathscr{F}(E - E_{\text{ref}})}{RT} \tag{2.5}$$

$$i_a = -n\mathscr{F}AC^0_{\text{red}}\, k^{\text{ref}}_{b,h} \exp \frac{(1-\alpha) n\mathscr{F}(E - E_{\text{ref}})}{RT} \tag{2.6}$$

where i_c and i_a are the cathodic and anodic components of the current at potential $(E - E_{\text{ref}})$, C^0_{ox} and C^0_{red} are the concentrations of oxidized and reduced species at the electrode surface (since mass transport is assumed to be much faster than electron transfer over the potential range covered by the present discussion, C^0_{ox} and C^0_{red} will be equal to their bulk concentrations), and A is the area of the electrode. Equations 2.5 and 2.6 permit the evaluation of the electrolysis currents at any potential in terms of the corresponding currents at some reference potential E_{ref}. Several such reference potentials are in common use. For example, the potential of the normal hydrogen electrode (N.H.E.), which is 0 V by definition, is very frequently used as the reference point. In this case,

$$i_c = n\mathscr{F}AC_{\mathrm{ox}}^0 k_{f,h}^0 \exp \frac{-\alpha n\mathscr{F}E}{RT} \tag{2.7a}$$

$$i_a = -n\mathscr{F}AC_{\mathrm{red}}^0 k_{b,h}^0 \exp \frac{(1-\alpha)n\mathscr{F}E}{RT} \tag{2.8a}$$

Notice that the rate constant for electron transfer at the potential of the N.H.E. is symbolized by $k_{f,h}^0$ or $k_{b,h}^0$, depending on whether reduction or oxidation is meant. It is often convenient experimentally to *measure* potentials with the saturated calomel electrode (S.C.E.), whose potential is +0.2412 V relative to N.H.E., while still reporting these potentials relative to N.H.E. as the reference point. In this case,

$$i_c = n\mathscr{F}AC_{\mathrm{ox}}^0 k_{f,h}^0 \exp \frac{-\alpha n\mathscr{F}(E+0.2412)}{RT} \tag{2.7b}$$

$$i_a = -n\mathscr{F}AC_{\mathrm{red}}^0 k_{b,h}^0 \exp \frac{(1-\alpha)n\mathscr{F}(E+0.2412)}{RT} \tag{2.8b}$$

where E is measured relative to S.C.E.

Another commonly used and useful reference potential may be defined with the aid of Figure 2.2, which represents a plot of i_c and i_a as a function of potential, using Eqs. 2.7a and 2.8a and arbitrary values of $k_{f,h}^0$, $k_{b,h}^0$, and α, with $C_{\mathrm{ox}}^0 = C_{\mathrm{red}}^0$. It will be noticed that there can exist only one potential at which the rates of oxidation and reduction, and hence i_c and i_a, are equal. This is, of course, the standard potential E^0; its definition here as a kinetic parameter, rather than from thermodynamic considerations[2] (Eq. 1.2), will be shown later to be more useful experimentally. The advantage of using E^0 as the reference potential is that one rate constant, the electron-transfer rate at E^0, symbolized by $k_{s,h}$, characterizes both the cathodic and the anodic half-reactions:

$$i_c = n\mathscr{F}AC_{\mathrm{ox}}^0 k_{s,h} \exp \frac{-\alpha n\mathscr{F}(E-E^0)}{RT} \tag{2.7c}$$

$$i_a = -n\mathscr{F}AC_{\mathrm{red}}^0 k_{s,h} \exp \frac{(1-\alpha)n\mathscr{F}(E-E^0)}{RT} \tag{2.8c}$$

Using any reference potential other than E^0, one must define the system by separate rate constants (e.g., $k_{f,h}^0$ and $k_{b,h}^0$ when N.H.E. is the reference point) for each branch. On the other hand, it will be noted that one must either know E^0 from thermodynamic data or be able to obtain it from measurements on both branches of Figure 2.2 in order to evaluate $k_{s,h}$. It often happens that for one

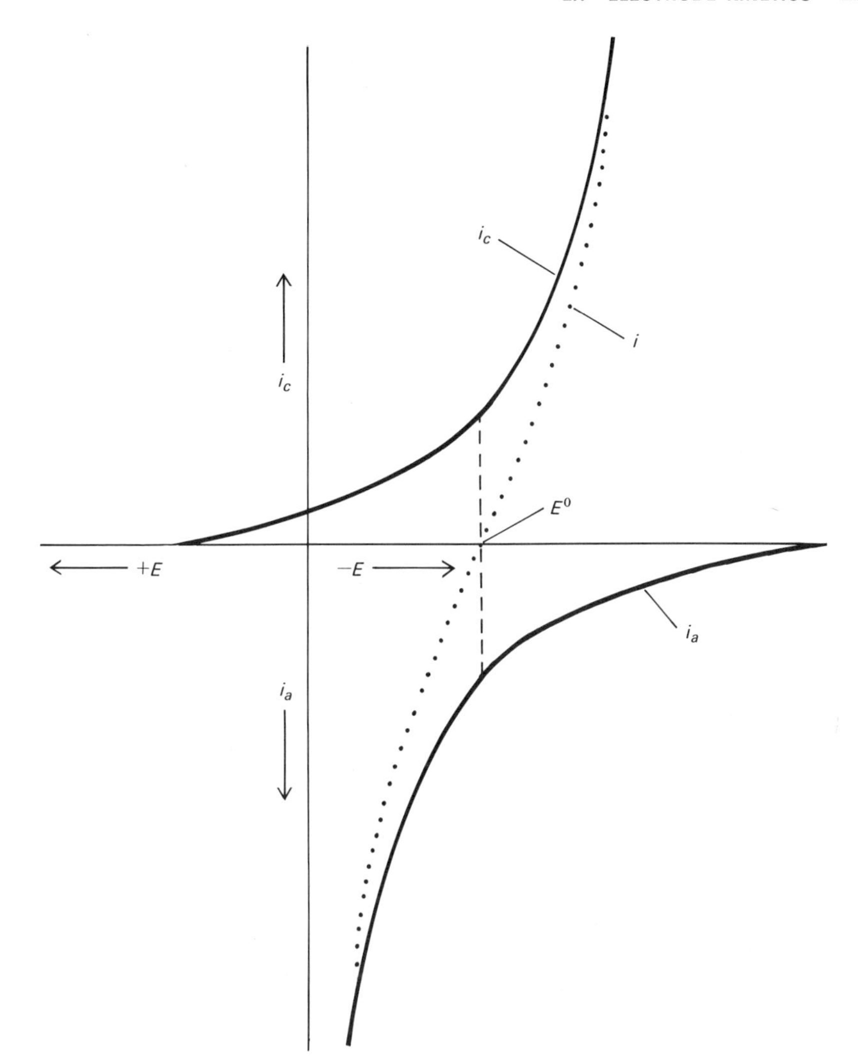

Figure 2.2 Potential dependence of anodic, cathodic, and net currents.

reason or another neither of these is possible. In that case, it is necessary to have recourse to some other reference point, usually N.H.E.

The dashed line in Figure 2.2 represents the *net* current, i, $(= i_a + i_c)$* as a function of potential. Notice that at potentials cathodic of E^0, a finite anodic current component always exists, and vice versa. It is clear that while the *net* current at E^0 is zero, there is a state of dynamic electrochemical equilibrium at this potential consisting of equal and opposite currents,** due to the equal rates

* Recall that by sign convention cathodic currents are positive and anodic currents are negative.

** The absolute value of the individual currents at E^0 is known as the exchange current, i_{exch}.

of oxidation and reduction. The existence of this equilibrium is readily demonstrated. For example, if an isotopically labeled metal electrode is immersed in a solution of the corresponding metal ion, and the electrode potential is maintained at E^0 for a short time, the solution will be found to have incorporated the label.[3]

At E^0, $i_a = -i_c$, and hence, from Eqs. 2.7a and 2.8a,

$$n\mathscr{F}Ak_{f,h}^0 C_{ox}^0 \exp\frac{-\alpha n\mathscr{F}E^0}{RT} = n\mathscr{F}Ak_{b,h}^0 C_{red}^0 \exp\frac{(1-\alpha)n\mathscr{F}E^0}{RT} \tag{2.9}$$

Rearrangement of Eq. 2.9 affords

$$\frac{C_{ox}^0 k_{f,h}^0}{C_{red}^0 k_{b,h}^0} = \exp\frac{n\mathscr{F}E^0}{RT} \tag{2.10}$$

Taking logarithms of both sides of Eq. 2.10 and solving for E^0,

$$E = \frac{RT}{n\mathscr{F}}\ln\frac{k_{f,h}^0}{k_{b,h}^0} - \frac{RT}{n\mathscr{F}}\ln\frac{C_{red}^0}{C_{ox}^0} \tag{2.11}$$

Since $C_{ox}^0 = C_{red}^0$ at $E = E^0$,

$$E^0 = \frac{RT}{n\mathscr{F}}\ln\frac{k_{f,h}^0}{k_{b,h}^0} \tag{2.12}$$

And, in general, combining Eqs. 2.11 and 2.12,

$$E = E^0 - \frac{RT}{n\mathscr{F}}\ln\frac{C_{red}^0}{C_{ox}^0} \tag{2.13}$$

Equation 2.13 is, of course, the Nernst equation; this derivation from electrode kinetics may be compared with the thermodynamic derivation in Chapter 1. Although Eqs. 2.11 and 2.13 both permit the calculation of the equilibrium concentrations of oxidized and reduced species at the electrode surface at any potential, the kinetic form of the Nernst equation, Eq. 2.11, is more useful operationally for the following reason. Suppose that the potential of an electrode in a solution containing the redox couple is suddenly changed to a new value, and that after a given time t the concentrations C_{ox}^0 and C_{red}^0 are measured (ignore for now the question of *how* these surface concentrations might be measured). If C_{ox}^0 and C_{red}^0 cannot attain their new equilibrium values in time t, the system will appear to deviate from the Nernst equation. That is, for electrochemical measurements that are made very rapidly, the *position* of equilibrium for the redox process is not as important as the *rate of approach* to equilibrium, which depends on the magnitudes of the electron-transfer rate constants $k_{f,h}^0$ and $k_{b,h}^0$ (or, equivalently, $k_{s,h}$). A system for which these rates are high enough that C_{ox}^0 and C_{red}^0 can reach their equilibrium (Nernstian) values at any potential *during the time scale of the experiment* is said to be *reversible*. Conversely, if electron transfer is slower than the time scale of the experiment, C_{ox}^0 and C_{red}^0

will be found to deviate from the values predicted by the Nernst equation, and the system is said to be irreversible. Notice the operational character of the definition of electrochemical reversibility: a system that appears reversible under one set of conditions might appear markedly irreversible under other conditions, for example, using faster measurement rates. All systems deviate from Nernstian behavior to some degree, but for many this deviation is less than the experimental error of the measurement. Under the preceding operational definition, such cases may be classified as reversible.

It is possible to illustrate graphically the distinction between the thermodynamic and kinetic definitions of reversibility (Figure 2.3). The dashed lines

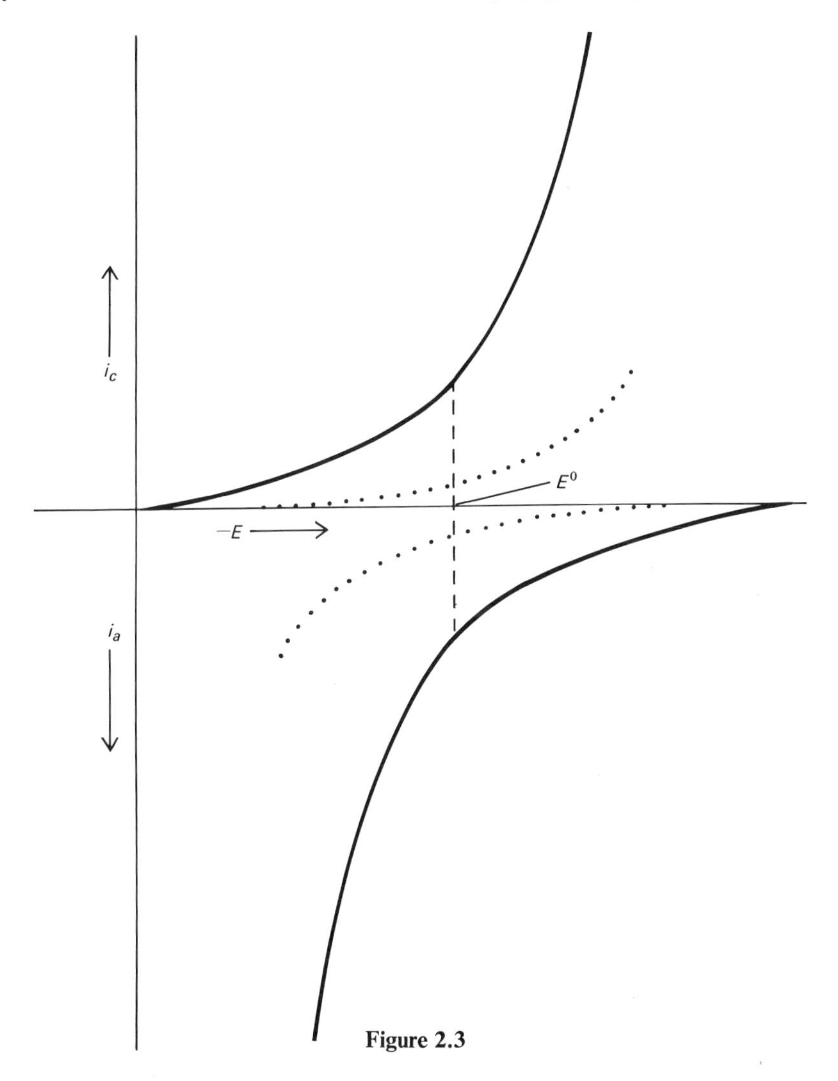

Figure 2.3

in Figure 2.3 represent a redox couple whose standard potential is E^0; the solid lines represent another redox couple whose standard potential is also E^0. (The concentrations of all species are assumed to be equal.) Despite the fact that both reactions have the same standard potential, they are clearly very different. The electron-transfer rate at E^0 is much higher for the redox couple indicated by a solid line, and in fact at all potentials $k_{f,h}$ and $k_{b,h}$ are higher for this couple. On a time scale intermediate between the electron-transfer rates of the two couples, one would appear reversible and the other (dashed lines) would be irreversible. Furthermore, in a slow measurement technique, such as potentiometric titration, both couples would appear identically reversible, while a very fast technique might not allow even the faster redox couple time to attain Nernstian equilibrium. In later sections it will be shown that experimental indications of reversibility or irreversibility can provide valuable clues to the mechanism of an unknown electrode mechanism (Chapter 3).

2.2 MASS TRANSPORT

It has been stated previously that the concentrations of reduced and oxidized forms used in the Nernst equation are those *at the electrode surface*. These may be very much different from concentrations in the bulk of solution. Furthermore, the rate-determining step in the overall electrode process at potentials where electron-transfer rates calculated from Eq. 2.3 or 2.4 are very high will actually be the rate of arrival of electroactive material at the electrode. For these reasons it will be necessary to examine the various modes of mass transport, *migration*, *diffusion*, and *convection*, by which the substrate is carried to the electrode surface from bulk solution. Depending on the experimental conditions, any or all of these might be operating in a given experiment. It will be seen shortly that the mass-transport properties of a given experiment have a considerable effect on the nature of the mechanistic information available from that experiment.

2.2.1 Migration

In order that a current may flow in an electrochemical experiment, there must exist a potential difference between the anode and cathode. Ions, since they are charged particles, will move in the electrical field associated with this potential gradient—positive ions moving to the cathode and negative ions moving to the anode.* This movement of ions in the electrical field between the anode and cathode is termed *migration*. It is, of course, the migration of charged species that constitutes a mechanism for conduction of current through the solution. The fraction of the current carried by a given cation or anion is known as its

* Hence the terms *cation* for a positive ion and *anion* for a negative ion.

transference number, t. The latter is a function of the concentrations and equivalent ionic conductances of all of the individual ionic species present in the solution (Eqs. 2.14 and 2.15).[4]

$$t_j = \frac{C_j \lambda_j}{\sum_i C_i \lambda_i} \tag{2.14}$$

and

$$\sum_j t_j = 1 \tag{2.15}$$

where C_j is the concentration of the jth ion and λ_j is its equivalent ionic conductance, as defined[4] by the relations

$$\Lambda = \sum_j \lambda_j \tag{2.16}$$

and

$$\Lambda = \frac{1000L}{C} \tag{2.17}$$

where Λ is the molar conductance of the solution, L is its specific conductance (the quantity actually measured), and C is the concentration of the electrolyte.

Although certain electroanalytical techniques, such as conductometric titrations,[5] make use of migration, the techniques generally used for elucidation of electrode mechanisms are based on the assumption that mass transport of the electroactive substance is not occurring by migration. Migration of a charged electroactive substance may be effectively eliminated by addition to the solution of an electrochemically inert electrolyte. If the electrolyte is present in sufficient excess, generally a concentration 100 or more times that of the electroactive ion, the transference number of the latter will be negligibly small. This *supporting electrolyte*, from Eq. 2.14, carries essentially all of the current passed through solution. Of course, if one is electrolyzing a *neutral* organic compound, migration of the electroactive substance cannot occur. In this event, however, a supporting electrolyte is still necessary, not only to carry the electrolysis current but to decrease the rather large resistances encountered in nonaqueous solvents.

2.2.2 Diffusion

Movement of a species through solution by random thermal motion is known as *diffusion*. Whenever a concentration gradient exists in solution, there also exists a driving force for diffusion of solute from regions of high concentration to regions of lower concentration.[6] In any experiment in which the electrode potential is such that the electron-transfer rate is very high, the region immediately surrounding the electrode will be depleted of the electroactive species, setting up a concentration gradient and thus the situation in which this species will constantly be arriving at the electrode surface by diffusion. Calculation of the cathodic electrolysis current at a given potential with the aid of Eq. 2.7a or the anodic current as given by Eq. 2.8a, when the electron-transfer rate reaches a sufficient magnitude that depletion by electrolysis causes the surface concentration to differ appreciably from bulk concentration, requires a knowledge of the surface concentration. When diffusion is the only means of mass transport, surface concentrations may be calculated given a few assumptions concerning the mathematical form of the concentration gradient. It will be shown later that certain of these assumptions require a knowledge of the geometry of the electrode toward which diffusion is taking place.

2.2.3 Convection

Convection, or *hydrodynamic transport*, involves movement of a substance by stirring of the solution, which in turn may be caused by, for example, density gradients (due to concentration or temperature gradients) or other mechanical disturbance. Convection is a considerably more effective mode of mass transport than is diffusion, and hence mass-transport limited electrolysis currents are considerably higher, at equal concentrations of electroactive substance, for stirred solutions than for quiescent solutions, in which diffusion is the only means of mass transport. Although convection is more efficient than diffusion, it is also far more difficult to treat mathematically.[7] Useful approximate methods for examining convective mass transport are, however, available and will be introduced later.

2.3 VOLTAMMETRY

It is now necessary to examine the quantitative and semiquantitative interrelationships between potential, electrochemical reaction rate, and mass transport. These relationships find expression in a wide variety of voltammetric experiments commonly used to characterize the electrochemical behavior of an electroactive substance. There will be examined in succession the factors affecting the relative heights and shapes of voltammetric waves such as those in Figures 1.4 and 1.6.

2.3.1 Limiting Currents

One of the striking features of Figures 1.4 and 1.6 is the current plateau, or limiting current, which is atttained in each case after a rather sharp initial increase of current with potential. A limiting current is always attained at potentials where the rate of mass transport has reached its maximum value under the conditions of the experiment. It will be necessary to distinguish limiting currents according to the mass-transport phenomena that are operating in a given experiment: either by diffusion only (quiet solution) or a combination of convection and diffusion (stirred solution). Throughout both discussions it will be assumed that electrode potentials are such that mass transport is so much slower than electron transfer that the latter may be ignored.

Quiet Solution (Mass Transport by Diffusion)

Stationary Electrode In the technique known as *chronoamperometry at fixed potential*, or the *potentiostatic* or *potential-step method*, one quickly changes (steps) the potential of an electrode from a potential where electron transfer is so slow as to be negligible to a potential where electron transfer is so fast that the surface concentration of the electroactive species is zero, and the maximum rate of mass transport by diffusion is obtained, and then measures the diffusion-limited electrolysis current as a function of time at fixed potential. A typical chronoamperometric curve is shown in Figure 2.4. Although chronoamperometry is not a voltammetric technique, it is introduced here as a prelude to the discussion of polarographic limiting currents in the next section.

It was indicated above that treatment of diffusion requires assumptions regarding the mathematical form of the expression relating the diffusion rate to concentration, and that certain of these assumptions in particular are related to electrode geometry. The fundamental relationships between diffusion rate and

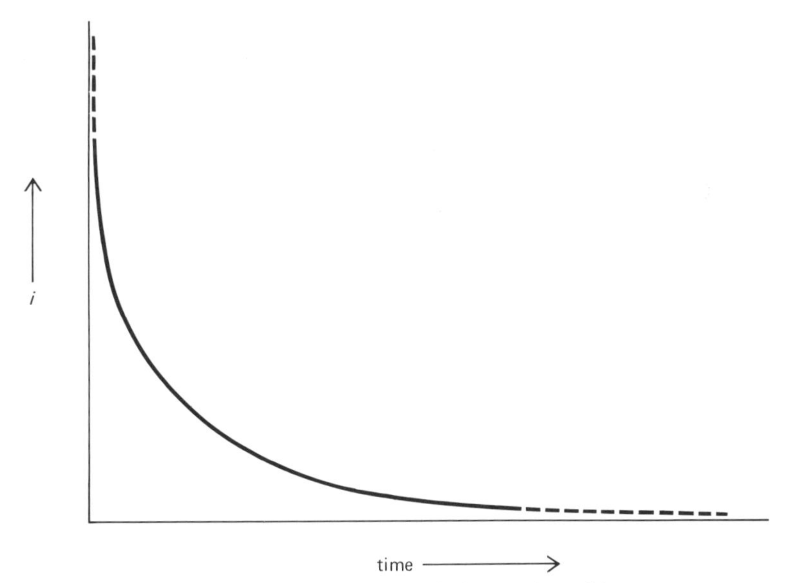

Figure 2.4 Chronoamperometric (potentiostatic) curve.

concentrations, known as *Fick's laws of diffusion*, have long been known.[8] They will merely be summarized here. Fick's first law is completely general and independent of electrode geometry; it states that the diffusion rate, or flux, defined as the number of molecules dN of substance diffusing across a surface of cross-sectional area A at distance X from the electrode in a given interval of time dt, is directly proportional to the magnitude of the concentration gradient responsible for diffusion:

$$\text{flux} = \frac{1}{A}\frac{dN}{dt} = D\left(\frac{\partial C_{x,t}}{\partial x}\right) \tag{2.18}$$

The surface of area A is assumed to be parallel to the electrode surface (Figure 2.5). Notice that the partial derivative is used because the concentration C of the

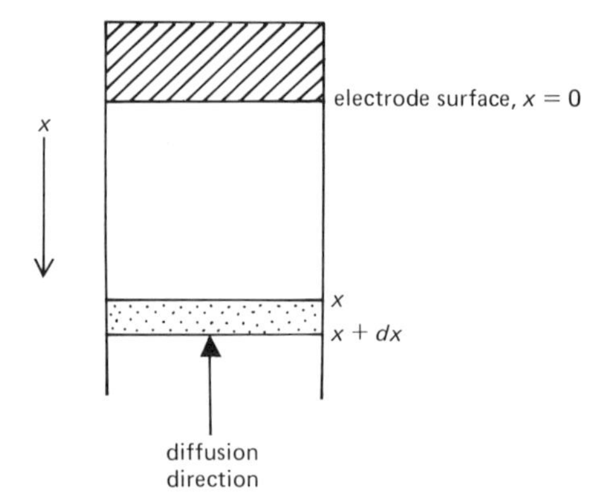

Figure 2.5 Graphical illustration of Fick's first law. (C. N. Reilley, in I. M. Kolthoff and P. J. Elving, eds., *Treatise on Analytical Chemistry*, New York, Wiley, 1963, part I, vol. 4, p. 2130.)

diffusing substance is a function of both x and t, and that diffusion is *toward* the electrode since the material is being removed at the electrode surface by electrolysis. The proportionality constant D is called the *diffusion coefficient* of the substance in the given medium and is expressed in units of square centimeters per second. Fick's first law may be put into a form appropriate for electrochemical use by noting that, since diffusion is the slow step, the overall electrolysis rate is equal to the diffusion rate and hence, from Faraday's law,

$$i = n\mathcal{F}\left(\frac{dN}{dt}\right)_{0,t} \tag{2.19}$$

and

$$i = n\mathcal{F}AD\left(\frac{\partial C}{\partial x}\right)_{0,t} \tag{2.20}$$

Equation 2.20 states the relationship between the electrolysis current and the concentration gradient at the electrode surface. However, one is really interested in the variation of current with *time* (as in Figure 2.4), since the concentration gradient and hence the current will change as electrolysis proceeds. It is at this point that a specific geometry of the electrode surface must be assumed. Suppose the restriction is imposed that the electrode be planar, and that net mass transport can occur only in a direction perpendicular to the electrode. For this situation, known as *linear diffusion*, it may be shown[9] that the time variation of concentration, *Fick's second law*, can be expressed as

$$\frac{\partial C_{x,t}}{\partial t} = D \frac{\partial^2 C_{x,t}}{\partial x^2} \tag{2.21}$$

which indicates that for linear diffusion there exists a simple relationship between the dependence of substrate concentration on time and distance from the electrode, respectively. If the boundary conditions are imposed that (1) at distances well removed from the electrode surface, $C = C^*$, the bulk concentration of electroactive species; (2) the electron-transfer rate at the new potential is sufficiently higher than the mass-transport rate that the concentration of electroactive substance at the electrode surface, C^0, is *zero* at any time after electrolysis is initiated; and (3) C^0 is equal to C^* immediately *before* electrolysis is initiated, one may solve Eq. 2.21 by standard mathematical techniques involving the application of Laplace integral transforms.[6,9] The solution, describing the time dependence of current in chronoamperometry at fixed potential, is the well-known *Cottrell equation*[10]:

$$i_t = \frac{n\mathscr{F}AC^*D^{1/2}}{\pi^{1/2}t^{1/2}} \tag{2.22}$$

The Cottrell equation predicts that the chronoamperometric (potentiostatic) current is not only directly proportional to the electrode area and bulk concentration, but is inversely proportional to the square root of the time elapsed since the beginning of the experiment. A convenient check on the Cottrell equation lies in the fact that it may be rearranged as

$$i_t t^{1/2} = \frac{n\mathscr{F}AD^{1/2}C^*}{\pi^{1/2}} \tag{2.23}$$

All the terms on the right-hand side of Eq. 2.23 will be constant in a given potentiostatic experiment, and hence the product $i_t t^{1/2}$ should be a constant. Adams[10b] has demonstrated that the Cottrell equation is satisfied nicely by, for example, a commercial planar platinum electrode for times up to 35 sec, beyond which time convection begins to assume increasing importance. Likewise, Figure 2.4, representing the potentiostatic reduction of anthracene in DMF at a mercury-pool electrode, shows the square-root-of-time decrease of current predicted by the Cottrell equation. Careful work, however, requires that the electrode be surrounded by a cylindrical shield to minimize both convection and nonlinear (nonperpendicular) diffusion to the electrode surface (Figure 2.6).[11]

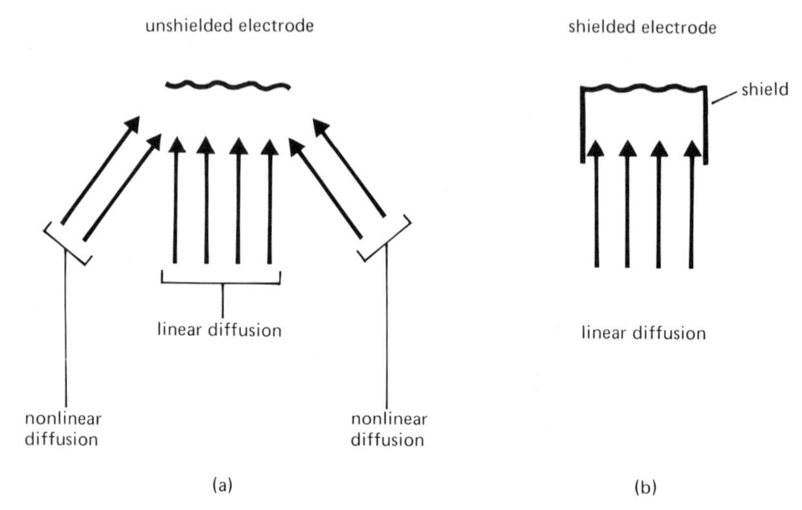

Figure 2.6 Diffusion to (a) an unshielded planar electrode; (b) a planar electrode with convection shield.

Polarography Many electrochemical experiments involve holding one of two variables (potential or current) constant and observing the other as a function of time. In chronoamperometry, for example, the potential is fixed and one measures the current as a function of time. Polarography (Section 1.2) is a somewhat more complex technique: in one sense it is a voltammetric technique, since current is plotted as a function of potential (Figure 1.4). At the same time, however, it will be noted that *during the lifetime of a single drop the electrode potential may be considered constant*, because of the moderate potential-sweep rates employed; hence, on the current plateau *each drop consists of an individual chronoamperometric experiment*. The current–time curve for a single drop does not have the simple shape of Figure 2.4, however, since the mercury-drop area is continually increasing with time. To convert the Cottrell equation into a form compatible with the time dependence of the drop size, one assumes both that the drop is spherical and that the mercury flow rate is uniform, that is, that the drop volume increases linearly with time. The drop area and volume are given by

$$A = 4\pi r^2 \tag{2.24}$$

and

$$V = \tfrac{4}{3}\pi r^3 \tag{2.25}$$

respectively, and furthermore,

$$V = \frac{M}{d} = \frac{10^{-3}mt}{d} \tag{2.26}$$

where V is the volume of the sphere at time t in cubic centimeters, M is the weight of the drop in grams, m is the rate of mercury flow into the drop in milligrams per second, and d is the density of mercury in grams per cubic centimeter. Combining Eqs. 2.25 and 2.26:

$$r = \left(\frac{3 \times 10^{-3}mt}{4\pi d}\right)^{1/3} \tag{2.27}$$

Introducing the latter value of r in Eq. 2.24 and combining terms:

$$A = \left(\frac{6 \times 10^{-3}\pi^{1/2}mt}{d}\right)^{2/3} \tag{2.28}$$

Substituting this value for the time-dependent drop area into the Cottrell equation:

$$i_t = \left(\frac{n\mathscr{F}C^*D^{1/2}t^{1/6}}{\pi^{1/6}}\right)\left(\frac{6 \times 10^{-3}m}{d}\right)^{2/3}\left(\frac{7}{3}\right)^{1/2} \tag{2.29}$$

The final term in Eq. 2.29, $(\frac{7}{3})^{1/2}$, which is introduced to account for the fact that the drop is continually expanding into fresh solution, has been deduced by a number of investigators.[12] Substituting values for the constants in Eq. 2.29 and defining the *diffusion current*, i_d, as the current at the end of the drop life t (in seconds), one finds

$$i_d = 708nD^{1/2}t^{1/6}m^{2/3}C^* \tag{2.30a}$$

where i_d is in microamperes and C^* is the concentration of electroactive material in millimoles per liter. Equation 2.30a is the well-known *Ilkovic equation.*[13] It is often found written in a slightly different form:

$$i_d = 607nD^{1/2}t^{1/6}m^{2/3}C^* \tag{2.30b}$$

The difference between the numerical constants arise because the diffusion current described by Eq. 2.30b is the average current over the drop life,* i_{ave}, while that of Eq. 2.30a is, as indicated above, the maximum current, i_{max} (Figure 2.7). Either i_{max} or i_{ave} may be defined as the diffusion current as long as the same usage is used consistently and one states specifically which form of the

* Equation 2.30b was generally used in the older literature because current-measuring devices then available (galvanometers) were too slow to track the drop oscillations; a properly damped galvanometer gives i_{ave} as the midpoint of its oscillations. Modern recorders are fast enough, however, to follow the current oscillations during polarography, and to measure i_{ave} one must actually damp the recorder intentionally.

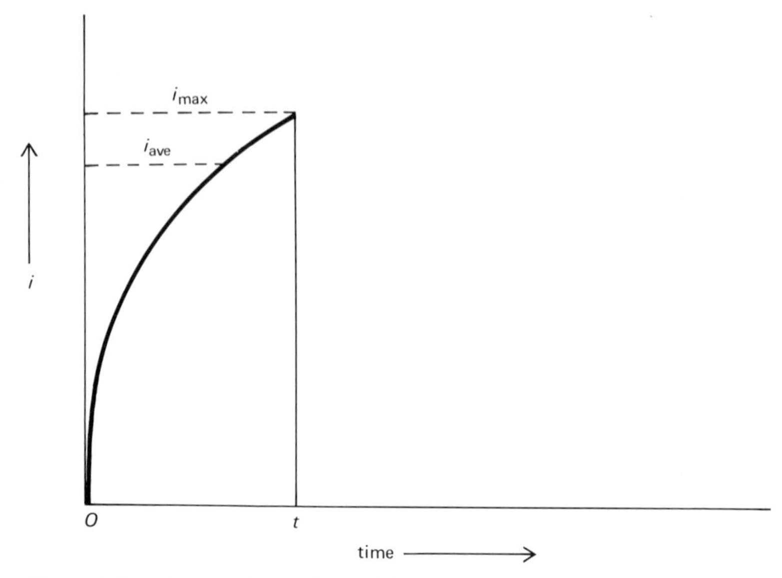

Figure 2.7 Current–time relationship at a single drop, in polarography.

Ilkovic equation is being used. There are, though, some situations in which i_{max} is preferred over i_{ave} on theoretical grounds (see p. 45); for this reason i_d will be defined throughout this book as i_{max}, and the Ilkovic equation will be used in the form given in Eq. 2.30a. In precise work, diffusion currents calculated from the Ilkovic equation are found to be in error by as much as 5%, principally because the Cottrell equation on which it is based assumes linear diffusion, that is, a planar electrode; Ilkovic justified use of the Cottrell equation by assuming that the thickness of the layer of material around the drop actually being electrolyzed was far smaller than the drop radius. A more exact rederivation, the *Koutecky equation*, in which Fick's second law is written in a form appropriate for spherical diffusion is available,[14] but is more complex than the Ilkovic equation and is unnecessary for the uses to which polarography will be put in the present volume.

The reader will observe that, although the Ilkovic equation presents at first sight an imposing set of variables, almost all will be known or easily measured in a single experiment. The value of polarography in quantitative chemical analysis lies in the direct proportionality between diffusion current and concentration. The utility of the Ilkovic equation in mechanistic studies rests, on the other hand, on the fact that through proper analysis of polarographic limiting currents one may obtain the number of electrons, n, involved in the overall electrode process represented by the polarographic wave. The next chapter will discuss specific techniques for extracting this important clue to electrode mechanism. Suffice it for the present to point out that in general only n and D will be unknown for a given polarographic experiment, and that D can usually be estimated with sufficient accuracy to allow for the evaluation of n via substitution in the Ilkovic equation.

Stirred Solution (Mass Transport by Convection and Diffusion)

As has been indicated previously, an exact mathematical treatment of hydrodynamic mass transport (convection) is impossible, except for a few special cases.[7] A very useful, albeit approximate, treatment involves the *Nernst diffusion-layer hypothesis.*[15] Nernst postulated that there exists a stagnant layer of solution at the electrode surface, of thickness δ. The hypothesis assumes that the electroactive material moves through bulk solution to the boundary of the diffusion layer by convection, but is carried to the electrode surface by diffusion alone (Figure 2.8a). It is assumed further that there exists a linear concentration gradient across the layer and that the concentration everywhere outside the diffusion layer is equal to the bulk concentration (Figure 2.8b). These assumptions are tantamount to the statement that Fick's first law of diffusion may be applied to the supposed stagnant layer. The

$$\frac{dC}{dx} = \frac{C^* - C^0}{\delta}$$

solid line in Figure 2.8b illustrates the linear concentration gradient predicted by Fick's first law; the dashed line, on the other hand, represents a much closer approximation to the actual situation. In particular, notice that the diffusion layer is not of fixed thickness δ. Given the approximate nature of the Nernst diffusion-layer theory, however, Faraday's law (Eq. 2.20) states that

$$i = n\mathscr{F}AD\left(\frac{C^* - C^0}{\delta}\right) \tag{2.31}$$

At potentials where electron transfer is so much faster than mass transport that C^0 is equal to zero, and the maximum mass-transport rate has been achieved,

$$i_{\text{lim}} = \frac{n\mathscr{F}ADC^*}{\delta} \tag{2.32}$$

It will be seen from Eq. 2.32 that limiting currents in stirred solution are proportional to both bulk concentration and electrode area, but inversely proportional to the diffusion-layer thickness. Since δ is decreased by more efficient stirring, limiting currents in stirred solution increase with stirring rate.[16] Limiting currents in stirred solution are, of course, much higher than those in quiet solution, as a direct consequence of the superiority of convection over diffusion as a means of mass transport. Even though the Nernst concept of a

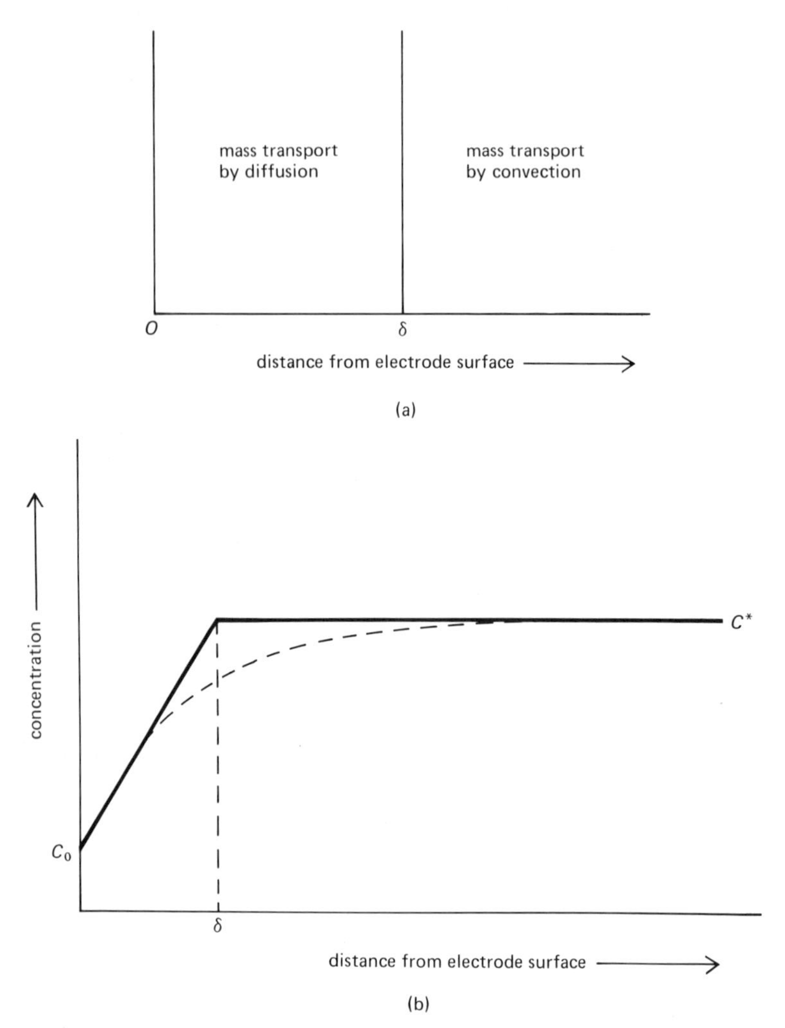

Figure 2.8 Nernst diffusion-layer hypothesis. (a) Mass-transfer processes in the vicinity of the electrode. (b) Actual (dashed line) and assumed (solid line) concentration gradients.

diffusion layer of fixed thickness is not correct, it is nevertheless possible to make use of Eq. 2.32 in the design of cells for electrochemical experimentation, as will be seen in the next chapter.

2.3.2 Shapes of Voltammetric Waves

As the preceding discussion has emphasized, limiting currents contain information about the total numbers of electrons involved in the overall electrochemical process. From measurements taken on the rising portion of the voltammetric wave, however, it is possible to obtain information about the *rate-determining* electrochemical step. The following discussion will be devoted to the interrelation between the wave shape and the nature of the rate-determining electrochemical step. Since it will be shown that the shapes of voltammetric waves are intimately related to rates of electron transfer, it will be necessary to recognize two extreme classes of electrode process, based on electron-transfer rates at E^0: reversible (fast electron transfer) and irreversible (slow electron transfer). The distinction between stirred and unstirred solutions will also be retained, so that four separate categories of behavior will be discerned in all.

Reversible Processes

Consider again the electrode process illustrated by the solid lines in Figure 2.3. When the rates of oxidation and reduction in the vicinity of E^0 are fast relative to the time scale of the measurement, the Nernst equation will be satisfied over the entire length of the voltammetric wave. Since

$$E = E^0 - \frac{RT}{n\mathscr{F}} \ln \frac{C^0_{\text{red}}}{C^0_{\text{ox}}} \tag{2.13}$$

it remains simply to write C^0_{red} and C^0_{ox}, the concentrations of the reduced and oxidized forms of the electroactive species at the electrode surface, in terms of parameters that can be measured experimentally.

Polarography In polarography as it is usually performed, one examines a solution containing only the oxidized member of a redox couple, and potentials are scanned to increasingly cathodic values. These conditions are assumed in the following discussion, although neither is, strictly speaking, necessary. From Faraday's law, the current will be proportional to the amount of the oxidized form reduced at the electrode surface:

$$i = k_{\text{ox}}(C^*_{\text{ox}} - C^0_{\text{ox}}) \tag{2.33}$$

where k_{ox} is a proportionality constant, and C^*_{ox} and C^0_{ox} represent the bulk and surface concentrations, respectively, of the oxidized species. At $i = i_d$, that is, at potentials corresponding to the polarographic limiting current plateau, C^0_{ox} is zero and, from Eqs. 2.33 and 2.30a,

$$i_d = k_{\text{ox}} C^*_{\text{ox}} = 708nD^{1/2}_{\text{ox}} C^*_{\text{ox}} m^{2/3} t^{1/6} \tag{2.34}$$

so that

$$k_{\text{ox}} = 708nD^{1/2}_{\text{ox}} m^{2/3} t^{1/6} \tag{2.35}$$

From Eq. 2.33, the current at any point on the rising portion of the wave is

$$i = k_{\text{ox}} C^*_{\text{ox}} - k_{\text{ox}} C^0_{\text{ox}}$$

and hence

$$C^0_{ox} = \frac{k_{ox} C^*_{ox} - i}{k_{ox}} = \frac{i_d - i}{k_{ox}} \tag{2.36}$$

For the anodic process occurring at the electrode surface,

$$i = -k_{red}(C^*_{red} - C^0_{red})$$

where the negative sign is used because of the customary sign convention for anodic processes. Since the bulk concentration of reduced species is zero,

$$i = k_{red} C^0_{red} \tag{2.37}$$

where

$$k_{red} = 708 n D^{1/2}_{red} m^{2/3} t^{1/6}$$

Given the definitions of C^0_{ox} and C^0_{red} in Eqs. 2.36 and 2.37, one may rewrite Eq. 2.13 as

$$E = E^0 - \frac{RT}{n\mathcal{F}} \ln \frac{i/k_{red}}{(i_d - i)/k_{ox}}$$

$$= E^0 - \frac{RT}{n\mathcal{F}} \ln\left(\frac{i}{i_d - i} \cdot \frac{k_{ox}}{k_{red}}\right)$$

Rearranging and substituting for k_{ox} and k_{red},

$$E = E^0 - \frac{RT}{n\mathcal{F}} \ln \frac{D^{1/2}_{ox}}{D^{1/2}_{red}} - \frac{RT}{n\mathcal{F}} \ln \frac{i}{i_d - i} \tag{2.38}$$

Notice that* when $i = 0.5 i_d$,

$$E = E_{1/2} = E^0 - \frac{RT}{n\mathcal{F}} \ln \frac{D^{1/2}_{ox}}{D^{1/2}_{red}} \tag{2.39}$$

so that, from Eqs. 2.38 and 2.39,

$$E = E_{1/2} - \frac{RT}{n\mathcal{F}} \ln \frac{i}{i_d - i} \tag{2.40a}$$

At 25°C, the temperature at which polarographic measurements are normally made, Eq. 2.40a becomes

$$E = E_{1/2} - \frac{0.0592}{n} \log \frac{i}{i_d - i} \tag{2.40b}$$

The half-wave potential is significant for a number of reasons. It is independent of the bulk concentration of the electroactive substance (Figure 2.9b),

* The potential, $E_{1/2}$, at which $i = 0.5 i_d$, is known as the *half-wave* potential (Figure 2.9a).

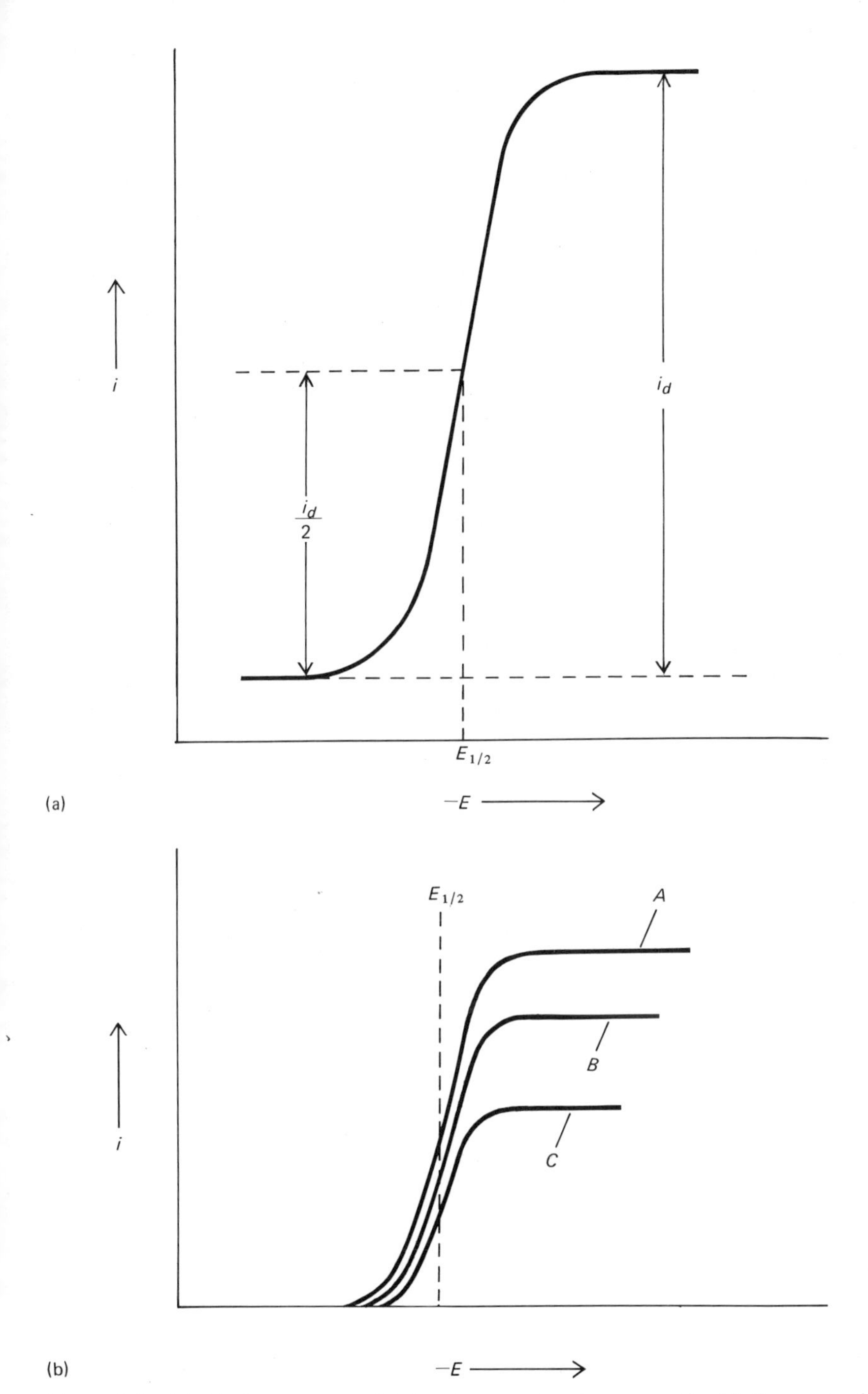

Figure 2.9 (a) Half-wave potential (neglecting charging current). (b) Concentration independence of the half-wave potential. Curves A, B, and C represent solutions of increasing concentration in the electroactive substance.

and hence is characteristic for a given substance under a specific set of conditions (solvent, temperature, electrolyte) and may be used in qualitative identification. Furthermore, $E_{1/2}$ is very close to E^0 (see Eq. 2.39), enabling one to obtain thermodynamic data from polarographic measurements.* Finally, since Eq. 2.40b is based on the assumption of reversible electrode behavior, it may be made the basis for an experimental test of reversibility, as well as a guide to the number of electrons, n_a, involved in the reversible electrochemical reaction.** A plot of log $i/(i_d - i)$ *vs* potential, called a "log plot," *for a reversible process* should have a slope of $59.2/n_a$ mV (Eq. 2.40b). Since n_a must be an integer, log plots for reversible processes can only have slopes of 59.2, 29.6, 19.7, etc. mV, and slopes deviating from these values by more than a few millivolts are therefore indicative of non-Nernstian behavior.

Actually, this test of reversibility is even simpler, because n is almost always equal to 1 or 2 for organic electrode reactions. One strict caution should, however, be raised at this point concerning the proper interpretation of log plots: while a value for the log plot slope other than 59.2, 29.6, 19.7, etc. mV clearly is diagnostic of irreversible behavior, *the converse is not true. That is, the satisfaction of Eq. 2.40b is a necessary, but not sufficient, criterion of reversibility.* The reasons for this will be clear after the discussion of irreversible processes on pp. 40–45. Other supplemental evidence is always necessary to establish reversibility when a reversible log plot is observed.

Another experimental test of reversibility, also based on Eq. 2.40b but much more rapid and convenient than construction of a log plot, was suggested by Tomes in 1937.[18a] Suppose that one defines $E_{1/4}$ and $E_{3/4}$, by analogy to $E_{1/2}$, as the potentials where, after correction for the residual current, $i = i_d/4$ and $i = 3i_d/4$, respectively. Then, from Eq. 2.40b,

$$E_{1/4} = E_{1/2} - \frac{0.0592}{n} \log \frac{1}{3}$$

and

$$E_{3/4} = E_{1/2} - \frac{0.0592}{n} \log \frac{3}{1}$$

Therefore, at 25°C,

$$E_{1/4} - E_{3/4} = \frac{0.0592}{n} \log 9 = \frac{0.0564}{n} \tag{2.41}$$

* Even in the (unlikely) extreme case where the diffusion coefficients of the oxidized and reduced species differ by as much as a factor of 2, E° and $E_{1/2}$ will differ by only 8 mV at 25°C.

** For an irreversible process, the value of n_a determined by the log plot corresponds to the number of electrons involved in the rate-determining electrochemical step. This can differ from n_{tot}, the number of electrons involved in the *overall* electrochemical process (the Ilkovic equation provides n_{tot}). Meites[17] has cited a case where $n_a = 1$, but n_{tot} is 18. For a reversible process, of course, $n_a = n_{tot}$.

Tomes's criterion, Eq. 2.41, then, involves merely evaluation of the difference $E_{1/4} - E_{3/4}$; as with the log plot, values of 56.4, 28.2, 18.8, etc. mV are a necessary but not sufficient criterion of polarographic reversibility.

Stirred Solution The shape of the current–potential curve in stirred solution for a reversible process can readily be shown to resemble that of the corresponding polarographic wave. It is first necessary to write the relationships governing C^0_{ox} and C^0_{red} in order to cast Eq. 2.13 in a form appropriate to stirred solution. For a solution containing initially only the oxidized form of a redox couple, one may write, from Eq. 2.31,

$$i = n\mathscr{F}D_{ox}A\left(\frac{C^*_{ox} - C^0_{ox}}{\delta}\right) \tag{2.42}$$

Expanding Eq. 2.42 and combining with Eq. 2.32,

$$i = i_{lim} - \frac{n\mathscr{F}D_{ox}AC^0_{ox}}{\delta}$$

and

$$C^0_{ox} = \frac{(i_{lim} - i)\delta}{n\mathscr{F}D_{ox}A} \tag{2.43}$$

The corresponding anodic current, applying the usual sign convention, is given by

$$i = -n\mathscr{F}D_{red}A\frac{C^*_{red} - C^0_{red}}{\delta}$$

and, since $C^*_{red} = 0$,

$$C^0_{red} = \frac{i\delta}{n\mathscr{F}D_{red}A} \tag{2.44}$$

Substituting the values of C^0_{ox} and C^0_{red} from Eqs. 2.43 and 2.44 into Eq. 2.13 and rearranging:

$$E = E^0 - \frac{RT}{n\mathscr{F}}\ln\left(\frac{i}{i_{lim} - i}\cdot\frac{D_{ox}}{D_{red}}\right)$$

$$= E^0 - \frac{RT}{n\mathscr{F}}\ln\frac{D_{ox}}{D_{red}} - \frac{RT}{n\mathscr{F}}\ln\frac{i}{i_{lim} - i} \tag{2.45}$$

At 25°C and again defining $E_{1/2}$ as the potential where $i = i_{lim}/2$,

$$E = E_{1/2} - \frac{0.0592}{n}\log\frac{i}{i_{lim} - i} \tag{2.46}$$

where

$$E_{1/2} = E^0 - \frac{RT}{n\mathscr{F}}\ln\frac{D_{ox}}{D_{red}} \tag{2.47}$$

and hence for a reversible process the half-wave potential in stirred solution will be a characteristic physical property of the species. It will be noted from Eq. 2.47 that $E_{1/2}$ will be independent of both concentration and stirring rate. Because the diffusion coefficients of the reduced and oxidized species, D_{red} and D_{ox}, respectively, appear to the one-half power in Eq. 2.39 and to the first power in Eq. 2.47, an $E_{1/2}$ as measured in stirred solution will differ slightly from a polarographic $E_{1/2}$. For the case mentioned on p. 38, where the diffusion coefficients of oxidized and reduced species differ by a factor of 2, the half-wave potential in stirred solution will differ from the polarographic half-wave potential by 10 mV (and from E^0 by 18 mV), an amount small enough to be ignored in most work. (Incidentally, this discussion has been based on the premise that the bulk concentrations C_{ox}^* and C_{red}^* remain constant. This requires that a voltammogram in stirred solution must be measured sufficiently rapidly, and/or that the ratio of solution volume to electrode area must be large enough that the electroactive species is not appreciably depleted during the time necessary to carry out the measurement.)

Irreversible Processes

Referring once more to Figure 2.3, consider the redox couple indicated by dashed lines, for which electron-transfer rates are low at potentials near E^0. If these rates are so low that approach to Nernstian equilibrium after a change in electrode potential is very slow relative to the voltage scan rate, the redox couple is called an *irreversible system.* If, in addition, the rate of the back reaction may be neglected, it is called a *totally irreversible* system. In this section the nature of the voltammetric wave shapes for totally irreversible processes in stirred and unstirred solution will be discussed.

Another type of process must also be distinguished as irreversible. This arises when the Nernstian equilibrium cannot be maintained either because electron transfer is slow, for example, if bond breakage occurs in the electron-transfer step, as in

$$\text{A—B} + \text{e}^- \longrightarrow \text{A} + \text{B}^-$$

or because one of the two components of the redox couple is shortlived on the time scale of the experiment, for example,

$$\begin{aligned} \text{O} + \text{e}^- &\rightleftharpoons \text{R} \\ \text{R} &\xrightarrow{\text{rapid}} \text{X} \end{aligned} \tag{2.48}$$

The former situation may be treated by the methods of this section; the latter requires a more detailed treatment than will be given here, but certain properties of the wave shape, particularly its broadening, are as predicted for the other irreversible processes considered here.

Polarography Since, by definition, irreversible processes do not exhibit Nernstian behavior, wave-shape analysis must proceed from a different tack from that involved in the derivation of Eq. 2.40b for reversible processes. First, recall that because of the existence of a residual (background) current in all polarograms (*cf.* Figure 1.5), the faradaic current is not normally observable in a polarogram until it has reached a value of a few tenths of a microampere, that is, sufficient to be distinguished from the residual current. Given the relationship between current and electrochemical reaction rate (Eq. 2.7a), one may estimate the value of $k_{f,h}$ necessary to produce a cathodic current of this magnitude. Suppose that this rate is depicted by $k'_{f,h}$ in Figure 2.10, which consists of plots of electrochemical rate constants *vs* potential for two reactions

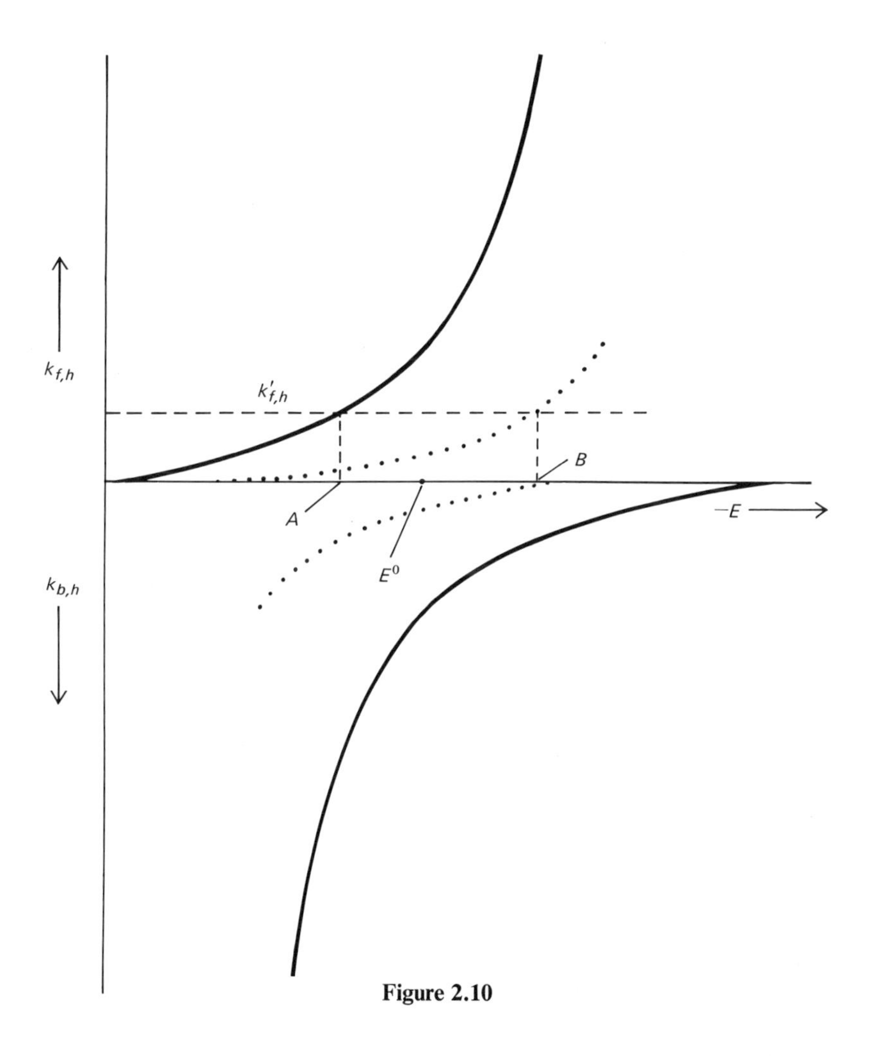

Figure 2.10

with the same E^0 but substantially differing values of $k_{s,h}$. Notice that for the redox couple represented by solid lines, the anodic rate constant is very high at potential A, where the cathodic rate constant has reached $k'_{f,h}$. At potential A, therefore, the rates of both the cathodic and anodic processes are fast, and the Nernstian equilibrium can be maintained. Likewise, at potential B, cathodic of E^0, the reduction rate is very high, while the anodic rate is still sufficiently high to allow maintenance of electrochemical equilibrium. Over the range of potentials A to B, this system will therefore behave reversibly, according to Eq. 2.40b. The situation is very much different for the redox couple represented by the dashed lines in Figure 2.10. Notice that when the cathodic rate finally becomes large enough, at potential B, to allow passage of a measurable cathodic current, the *anodic rate is negligibly low*, and as $k_{f,h}$ increases at potentials more cathodic than B, the disparity between cathodic and anodic rates becomes increasingly greater. Consequently, as the cathodic current in a polarogram of this substance is rising (at potentials cathodic of B) toward its eventual diffusion-controlled limit given by Eq. 2.30a, the Nernst equilibrium cannot be maintained because of the low oxidation rates at these potentials. Likewise, at potentials anodic of E^0, such as A, the reduction rate is too low for proper maintenance of electrochemical equilibrium. This redox couple, therefore, is by the preceding definition totally irreversible. (Given typical polarographic drop times, the value of $k_{s,h}$ below which Nernstian behavior cannot be maintained is ca. 2×10^{-2} cm^2/sec.)

Since for a totally irreversible couple anodic rates are negligibly small at potentials where measurable cathodic currents can be observed, the polarographic wave will have the simple exponential shape predicted by Eq. 2.7a at potentials near the foot of the polarographic wave, where C^0_{ox} may be assumed equal to C^*_{ox}. Of course, as the reduction rate becomes increasingly larger at potentials cathodic of B, mass transport also becomes increasingly prominent in controlling the overall reduction rate, until, just as with reversible processes, diffusion becomes exclusively rate determining, and a current plateau is reached.

It follows from the preceding discussion, incidentally, that, unlike the situation for reversible processes, where $E_{1/2}$ is almost exactly equal to E^0, the half-wave potential for reduction of the oxidized half of an irreversible redox couple will be *considerably cathodic of* E^0 (the difference depending on the magnitude of $k_{s,h}$). Likewise, for polarographic oxidation of the reduced form $E_{1/2}$ will be considerably anodic of E^0 (Figure 2.11). The form of the current–potential curve at potentials where the cathodic rate has increased sufficiently that the solution in the vicinity of the electrode surface begins to be appreciably depleted by electrolysis (this is roughly at potentials where $i = 0.1i_d$) is related to both diffusion and electron-transfer rates. A useful treatment of shapes of totally irreversible polarographic waves is due to Meites and Israel,[19] who cast into a convenient form certain calculations of Koutecky.[20] Observe that, owing to the low electron-transfer rate at any potential on the rising portion of the wave, the current i at the end of the life of each drop will be less than the corresponding

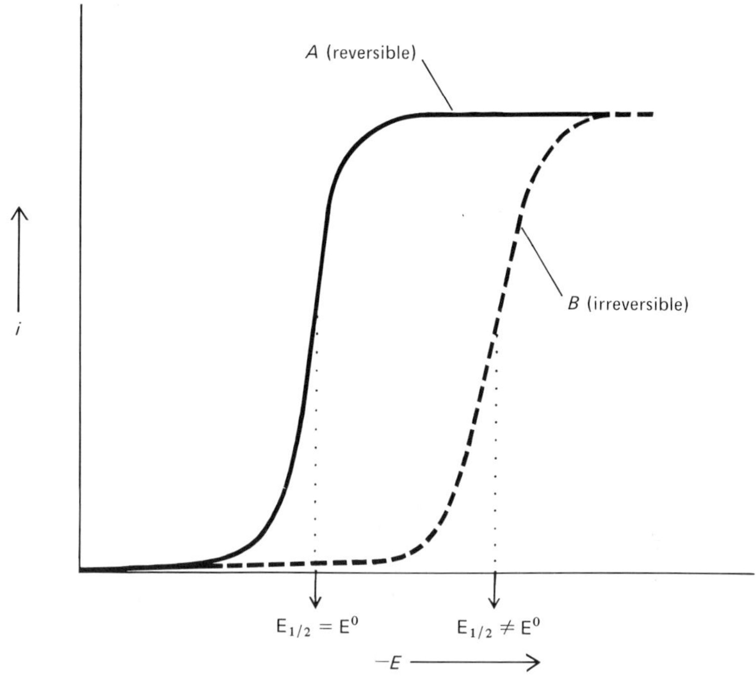

Figure 2.11 Polarographic reduction of two substances of identical E^0. Curve A: reversible process, $E_{1/2} = E^0$. Curve B: irreversible process, $E_{1/2}$ is cathodic of E^0.

current i_∞ to be expected for a very fast (reversible) cathodic process at the same potential. Koutecky tabulated values of the parameters $F(\chi)$ and χ. $F(\chi)$ is the fraction by which i is less than i_∞ because of this slow electron transfer; χ is a measure of the instantaneous faradaic current at any point during the drop life, and is a function of $k^0_{f,h}$, α, time, and potential:

$$i = F(\chi) i_\infty \tag{2.49}$$

$$\chi = \left(\frac{12}{7}\right)^{1/2} \frac{t^{1/2}}{D_{\mathrm{ox}}^{1/2}} k^0_{f,h} \exp\left(\frac{-\alpha n \mathscr{F} E}{RT}\right) \tag{2.50}$$

Meites and Israel[19] pointed out that a plot of log χ *vs* $\log(F(\chi)/[1 - F(\chi)])$ is linear over the range $0.1 < F(\chi) < 0.9$; least-squares analysis showed the line to fit closely the equation

$$\log \chi = -0.0130 + 0.9163 \log \frac{F(\chi)}{1 - F(\chi)} \tag{2.51}$$

For a totally irreversible process, the initial polarographic current rise will begin sufficiently cathodic of E^0 that i_∞, the current for a fast reversible process, will be equal to i_d, the diffusion-controlled limiting current. In other words, for the two processes of Figure 2.10, the reversible wave will have already reached its current plateau before a measurable polarographic current is observed for the totally irreversible process (Figure 2.11). Therefore, substituting i_d for i_∞ and combining Eqs. 2.49 and 2.51,

$$\log \chi = -0.0130 + 0.9163 \log \frac{i}{i_d - i} \tag{2.52}$$

Combining Eqs. 2.50 and 2.52,

$$\log\left(\frac{12}{7}\right)^{1/2} \frac{t^{1/2}}{D_{\text{ox}}^{1/2}} k_{f,h}^0 \exp \frac{-\alpha n \mathscr{F} E}{RT} = -0.0130 + 0.9163 \log \frac{i}{i_d - i}$$

Setting $T = 25°\text{C}$, this may be rearranged to

$$E = \frac{0.0592}{\alpha n} \log \frac{1.349 k_{f,h}^0 t^{1/2}}{D_{\text{ox}}^{1/2}} - \frac{0.0542}{\alpha n} \log \frac{i}{i_d - i} \tag{2.53}$$

Equation 2.53 may be written in the simple and familiar form

$$E = E_{1/2} - \frac{0.0542}{\alpha n} \log \frac{i}{i_d - i} \tag{2.54}$$

where*

$$E_{1/2} = \frac{0.0592}{\alpha n} \log \frac{1.349 k_{f,h}^0 t^{1/2}}{D_{\text{ox}}^{1/2}} \tag{2.55}$$

Observe that although the derivation of the current–potential relationship for a totally irreversible process (Eq. 2.54) involves an entirely different set of assumptions than that for a reversible process (Eq. 2.40b), the form of the two is quite similar, the only difference being the coefficient of the log $i/(i_d - i)$ term. Since the slope of a plot of E *vs* log $i/(i_d - i)$ is equal to $54.2/\alpha n$ mV, one may obtain α, the transfer coefficient, if n can be measured from auxiliary data. The similarity between Eqs. 2.40b and 2.54 is the basis for the caution on p. 38 regarding deduction of reversibility *solely* from a log-plot slope of 59 mV, for one would also obtain such a slope for an irreversible couple with $\alpha n = 0.9$ (e.g., $\alpha = 0.9$, $n = 1$, or $\alpha = 0.45$, $n = 2$). It should also be kept in mind that although Eq. 2.40b and 2.41 for a reversible process are obeyed in principle

* Meites and Israel wrote this equation for potentials measured relative to S.C.E. and referred to N.H.E.:

$$E_{1/2} = -0.2412 + \frac{0.059}{\alpha n} \log \frac{1.349 k_{f,h}^0 t^{1/2}}{D_{\text{ox}}^{1/2}}$$

over the entire polarographic wave,* the calculations of Koutecky on which Eq. 2.54 is based were shown by Meites and Israel[19] to produce a linear correlation only over a range of values of i of ca. 10–90% of i_d. Therefore, Eq. 2.54 does not hold for the bottom and top 10% of the polarographic wave for an irreversible process. Equation 2.7a nicely complements Eq. 2.54, however, since, as indicated previously, it describes the shape of the polarographic wave when *i is* less than i_d. It should also be noted that Eq. 2.54 was derived for currents measured at the end of the drop life, that is, at $i_d = i_{max}$, not i_{ave}.[19,20]

Tomes's criterion, $E_{1/4} - E_{3/4}$, may be evaluated for irreversible systems from Eq. 2 54 in a manner exactly analogous to the derivation of Eq. 2.41 for reversible systems. The computation gives

$$E_{1/4} - E_{3/4} = \frac{0.0517}{\alpha n} \tag{2.56}$$

Equation 2.56 provides a rapid means of evaluating αn; notice also that if, for example, $\alpha n = 0.91$, $E_{1/4} - E_{3/4} = 56.4$ mV and the system would appear reversible. Once again, it is clear that satisfaction of Eq. 2.40b or 2.41 indicates only that a system *may* be reversible, and further evidence supportive of reversibility must then be sought.

A caution with respect to the measurement and use of $E_{1/2}$ to characterize irreversible processes may be made by reference to Eq. 2.55, which states that *the measured polarographic half-wave potential for an irreversible process is dependent on the mercury drop time.* Since drop times are themselves dependent on both the *electrode potential* (see Figure 2.12 and Section 2.4.2) and on the characteristics of the particular capillary being used, comparison of waves for two different substances measured at different potentials or the same wave measured in different laboratories for a single substance really ought to be made only after adequate correction for the potential dependence of t; of course, this correction may not be necessary for polarographic measurements carried out at potentials where t is changing only slowly with potential, such as in the region near the maximum** in Figure 2.12. Meites[21] has shown that the drop-time correction may be made in the following manner. Equation 2.53 may be rewritten in the form†

$$E = E^0_{1/2} - \frac{0.0542}{\alpha n}\left(\log \frac{i}{i_d - i} - 0.546 \log t\right) \tag{2.57}$$

* Experimental errors in measurement of $i/(i_d - i)$, however, become large at the very top and bottom of the wave. Meites[18b] has recommended that the log plot be constructed for values of i ranging roughly from 3 to 97% of i_d. The point where $\log i/(i_d - i) = 0$ is, of course, $E_{1/2}$; this constitutes a graphical procedure for measurement of $E_{1/2}$ with greater accuracy than direct measurement from the polarogram, as in Figure 2.9a.

** The significance of this point, the so-called *electrocapillary maximum*, will be discussed in Section 2.4.2.

† Meites wrote Eq. 2.57 with an additional term, −0.2412, on the right side, since potentials were measured relative to S.C.E. but referred to N.H.E.

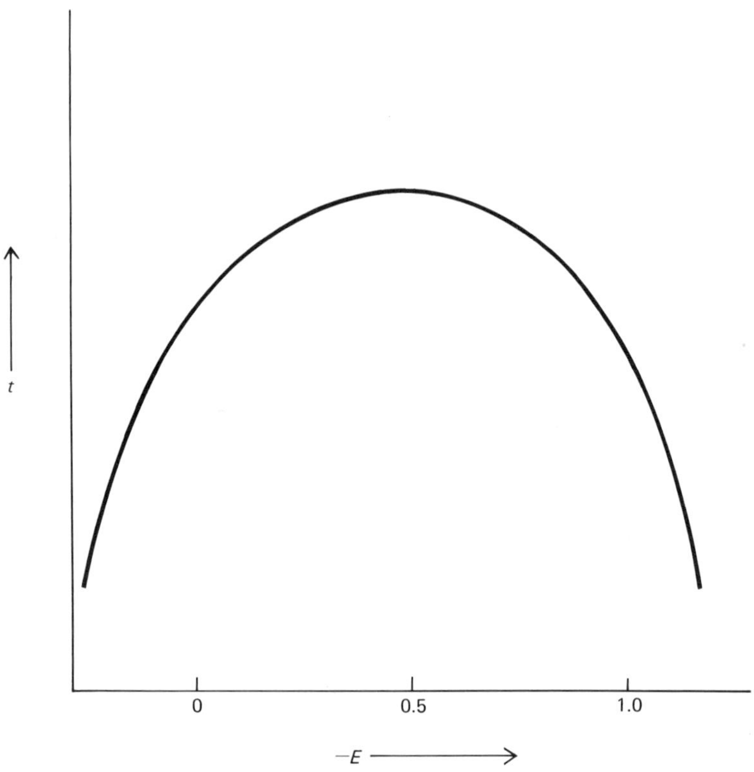

Figure 2.12 Dependence of polarographic drop time on electrode potential for a solution of 10^{-3}-M $HgCl_2$ in saturated KCl. (J. Heyrovsky and J. Kuta, *Principles of Polarography*, Prague, Academia, 1966, p. 452.)

where

$$E^0_{1/2} = \frac{0.0592}{\alpha n} \log \frac{1.349 k^0_{f,h}}{D^{1/2}_{\mathrm{ox}}} \tag{2.58}$$

Notice from Eq. 2.58 that $E^0_{1/2}$ is drop-time independent, as required. Meites has outlined two routes to $E^0_{1/2}$. The first,[21a] which is considerably faster, consists of measuring the drop time $t_{1/2}$ at the experimental half-wave potential. Inspection of Eq. 2.57 will reveal that

$$E^0_{1/2} = E_{1/2} - \frac{0.02957}{\alpha n} \log t_{1/2} \tag{2.59}$$

and substitution of αn (as found by application of Eq. 2.56) into Eq. 2.59 then affords $E^0_{1/2}$ with an accuracy satisfactory for most work.[21b] In very precise work one would use the alternative procedure, which involves measurement of the drop time t at a number of potentials on the rising portion of the wave and substitution of these values into Eq. 2.57. A plot of E *vs* $\log i/(i_d - i) - 0.546 \log t$ is constructed, and $E^0_{1/2}$ is by definition the potential where the latter term is equal to zero (Figure 2.13).[21c]

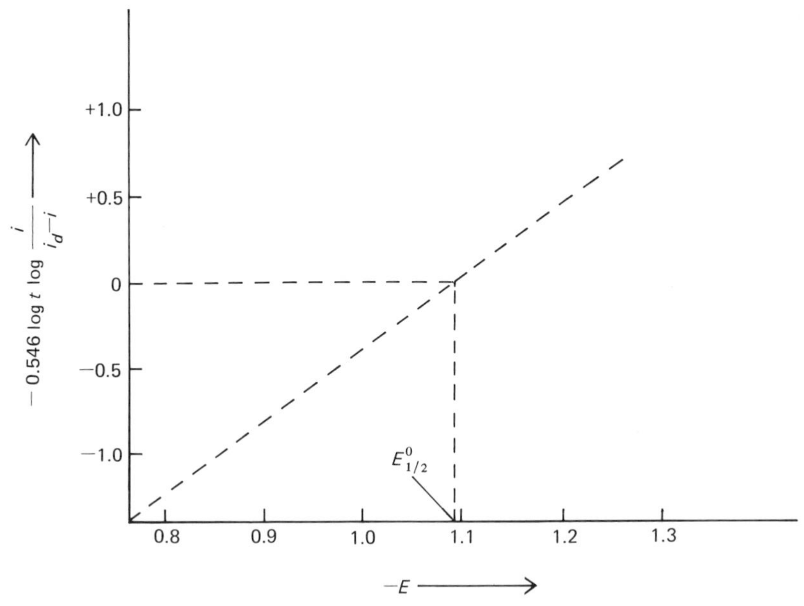

Figure 2.13 Graphical determination of $E^0_{1/2}$ for the reduction of 1.0-mF chromate ion in 0.10-F sodium hydroxide. (L. Meites and Y. Israel, *J. Amer. Chem. Soc.*, **83**, 4905 [1961]).

Whether or not one chooses to make the drop-time correction depends ultimately on the use to which the polarographic data is to be put. If one merely needs the half-wave potential in order to select the proper electrode potential for a preparative-scale electrolysis (Section 3.3), then there is no real reason to compute $E^0_{1/2}$, since half-wave potentials in stirred solution are in any event only approximately equal to those measured polarographically (see below). On the other hand, computation of $E^0_{1/2}$ is advisable when polarographic data are to be used in any sort of quantitative correlations, such as substituent effects on half-wave potential for a series of homologous compounds.[22]

Stirred Solution The current–potential relationship for electrolysis of a stirred solution containing only the oxidized form of a totally irreversible redox couple may be derived by assuming first that, as in the polarographic case, anodic rates will be negligible at potentials where measurable cathodic currents are observed. Thence, combining Eqs. 2.7a and 2.43,

$$i = \frac{k^0_{f,h}\delta(i_{\text{lim}} - i)}{D_{\text{ox}}} \exp\left(\frac{-\alpha n \mathscr{F} E}{RT}\right)$$

Taking the logarithm of both sides and rearranging,

$$E = \frac{RT}{\alpha n \mathscr{F}} \ln \frac{k^0_{f,h}\delta}{D_{\text{ox}}} - \frac{RT}{\alpha n \mathscr{F}} \ln \frac{i}{i_{\text{lim}} - i} \tag{2.60a}$$

At 25°C,

$$E = E_{1/2} - \frac{0.059}{\alpha n} \log \frac{i}{i_{\text{lim}} - i} \tag{2.60b}$$

where

$$E_{1/2} = \frac{0.059}{\alpha n} \log \frac{k^0_{f,h}\delta}{D_{\text{ox}}} \tag{2.61}$$

and potentials are, of course, relative to N.H.E. As in polarography, the shape of the voltammetric wave for an irreversible process in stirred solution (Eq. 60b) is similar in shape, but stretched out relative to the corresponding wave for a reversible process (since $\alpha < 1$). Most noteworthily, and in striking contrast to the reversible case, the *half-wave potential for an irreversible process in stirred solution is dependent on δ and hence on the stirring rate.* Furthermore, comparison of Eqs. 2.55 and 2.61 will reveal that, unlike the reversible case, there is no simple relationship between the polarographic half-wave potential for an irreversible process and the corresponding half-wave potential measured for the same process in stirred solution. The magnitude of the difference between these quantities will vary from system to system, but it might in certain cases amount to several tenths of a volt. It is evident that, for the sake of reproducibility, voltammetric data in stirred solution must be measured under carefully controlled conditions, particularly with regard to stirring rate and cell geometry. This point is of occasional importance in preparative-scale electrolysis.

2.4 THE ELECTRICAL DOUBLE LAYER AND ADSORPTION

Figure 2.14 consists of the polarogram of a solution of potassium chloride in water, measured relative to an S.C.E. reference electrode. Notice that the current appears to increase linearly over the range +0.2 to −0.5 V, and again linearly over the range −0.5 to −1.5 V, but that the slopes of these two linear segments are not equal. Clearly then, the inflection potential (−0.5 V) has some special significance. This is corroborated by comparison of this polarogram with Figure 2.12, which indicates that the inflection point in Figure 2.14 coincides with the

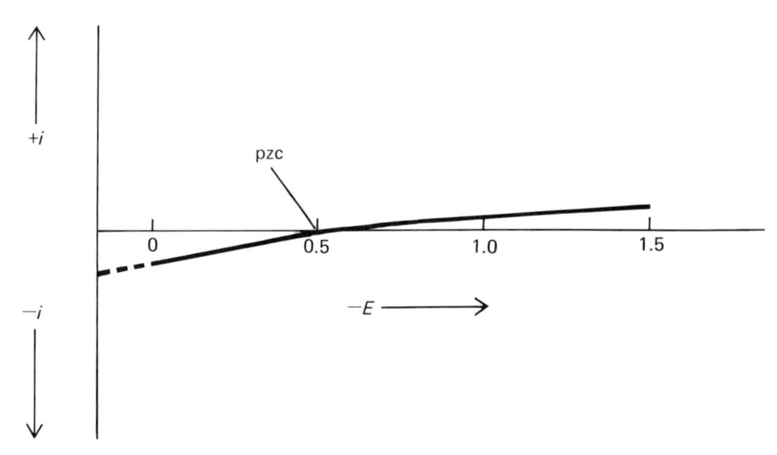

Figure 2.14 Potential dependence of the polarographic charging current (pzc is the potential of zero charge).

electrocapillary maximum of Figure 2.12. The phenomena observed in Figures 2.12 and 2.14 are a direct consequence of potential-dependent changes in the structure of the mercury-solution interface. Because electron transfer is a heterogeneous process, it can be affected substantially, sometimes even profoundly, by changes in the composition of the interfacial region. It will now be necessary to examine the potential-dependent electrostatic and adsorptive processes that, operating in concert, determine this composition. This treatment will, incidentally, account for the phenomena of Figures 2.12 and 2.14, but, more relevant to the synthetic aims of this book, it will also become evident that the mercury–solution interfacial structure may be considered as one more variable that may occasionally be manipulated to affect the course of an electrode process.

2.4.1 The Double Layer

The potential corresponding to the inflection point in Figure 2.14 and the maximum in Figure 2.12 is of considerable theoretical interest. As well as being called the electrocapillary maximum, it is often referred to as *the potential of zero charge*, or pzc. The latter is more descriptive of its true significance, as can be seen in the following way. Consider a mercury electrode immersed in a solution containing anion X^-, which has a tendency to be more or less strongly adsorbed on mercury. At equilibrium,

$$\mathrm{Hg} + n\mathrm{X}^- \rightleftharpoons \mathrm{Hg(X)}^{n-}_{\mathrm{ads.}}$$

$$\Delta G = -n\mathscr{F}E$$

As a consequence of this adsorption, the mercury will accumulate both an excess negative charge and an excess of X^- over its bulk concentration. The potential of the electrode measured relative to a reference electrode (N.H.E. or S.C.E.) will be characteristic for a given specific combination of metal, anion, solvent, temperature, and so on. If one now applies to the electrode a potential E to balance the potential developed by adsorption, X^- will be desorbed, that is, will be no longer in excess at the electrode surface. The electrode no longer has an excess negative charge under these conditions, which is why E is termed the potential of zero charge. At this potential the distribution of cations and anions at the electrode surface is identical to that in the bulk of solution. Suppose, however, that one applies to the mercury electrode a potential B, which is more negative than E (Figure 2.15). At this potential the electrode will possess an

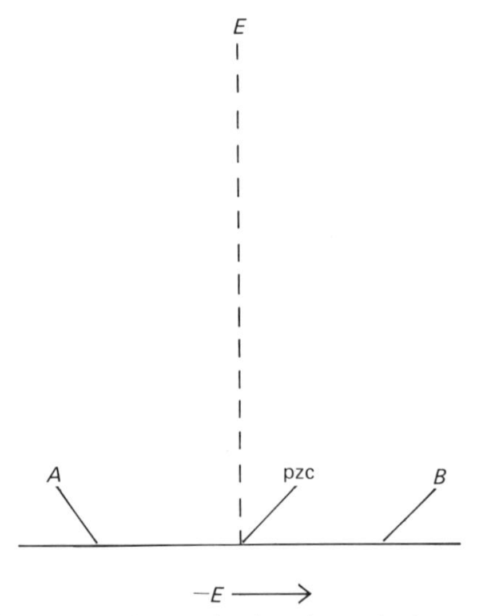

Figure 2.15 Potential of zero charge (pzc). Electrode bears a positive charge at potentials anodic of pzc, and a negative charge at potentials cathodic of pzc.

excess negative charge. As a consequence of this charge, the electrode will be able electrostatically to orient charged constituents of the solution (e.g., ions of the supporting electrolyte) and neutral dipolar components of the solution, including the solvent itself. Conversely, at potential A, anodic of E, the electrode will possess an excess positive charge and will be equally able to exert an electrostatic structure-making effect on the solution immediately nearest it. Of course, it will be apparent that the excess charge of the electrode, and hence its ability to structure the solution in which it is immersed, will increase as its potential increases either cathodically or anodically from the electrocapillary maximum.

Consider now the detailed structure of the oriented region of solution near the surface of the electrode, the so-called *electrical double layer*. At present, the most generally accepted model for the structure of the double layer is that of

Grahame,[23] and is essentially a refinement of earlier models suggested by Stern, Gouy and Chapman, Helmholtz, and Quincke (Figure 2.16). (This model, it must be said, was suggested for aqueous solution, where ions are generally well solvated and relatively free. It can serve here only as a rough guide to double-layer structure in nonaqueous solvents, which have been far less well investigated, but for which ion pairing and ion solvation must clearly play different roles than in water.)[24] Grahame suggested that, although at potentials other than pzc ions will be attracted to the electrode (cations or anions, depending on whether the electrode potential is cathodic or anodic, respectively, of pzc), the distance of closest approach of solvated ions is limited both by a layer of

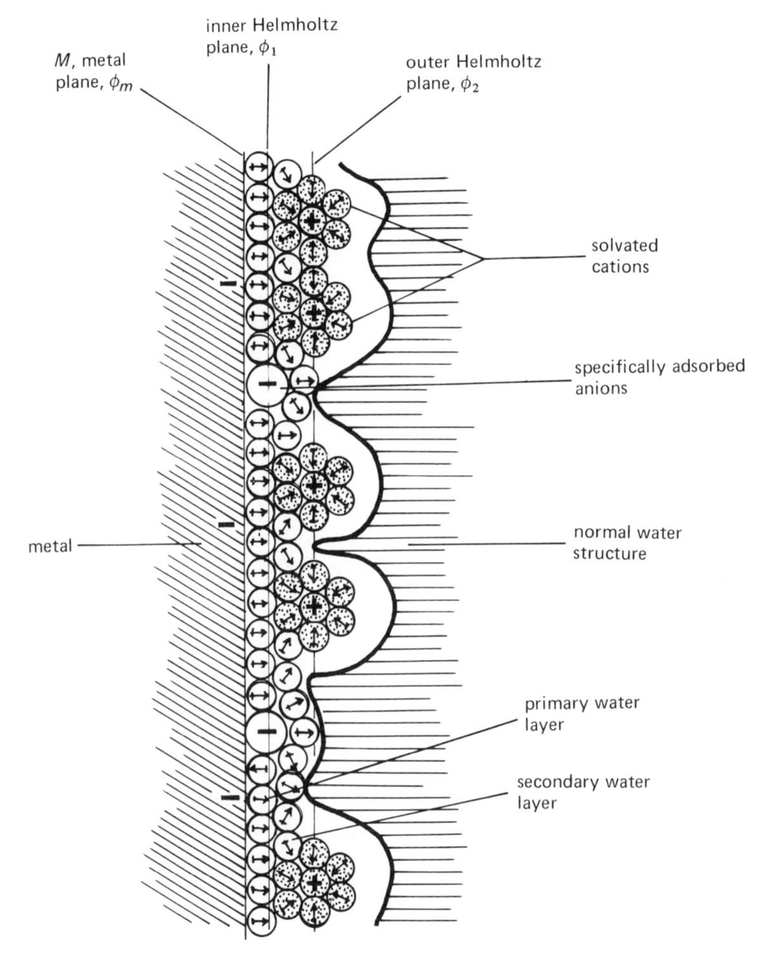

Figure 2.16 Structural model for the electrical double layer. (J. O'M. Bockris, M. A. V. Devanathan, and K. Muller, *Proc. Roy. Soc.*, (A) **274**, 68 [1963]. *Reprinted by permission of the Royal Society.*)

dipole-oriented solvent molecules at the electrode surface and by the solvation sheath of the ions themselves, which increases their effective radius considerably. The plane passing through the centers of the solvated ions is known as the *outer Helmholtz plane* (OHP). To maintain overall charge neutrality, the layer of ions at the OHP will then attract to itself a layer of oppositely charged ions, which will in turn attract another layer of ions, and so on. The extent of electrostatic structuring will, in general, diminish rather quickly with increasing distance from the electrode, because of thermal agitation and dipolar competition from the solvent itself. This region of gradually diminishing ionic organization extending outward from the OHP and merging eventually into bulk solution is known as the *diffuse layer*. The thickness of this region is a function of the charge on the electrolyte in question, the dielectric properties of the solvent, temperature, and the electrolyte concentration. The border between the diffuse layer and the bulk solution is not well defined, but for a 0.1-*M* aqueous solution of a 1:1 electrolyte, it has been estimated that, at 25°C, 99.99% of the diffuse double layer is contained in the region within 88 Å of the electrode surface.[25] This figure increases to 280, 890, and 2800 Å for 10^{-2}-, 10^{-3}-, and 10^{-4}-*M* solutions, respectively.

It was indicated previously that the region immediately nearest the electrode surface consists of a layer of dipole-oriented solvent molecules. Other substances may also penetrate closer than the OHP. Neutral molecules, for example, lacking a solvent sheath, may approach and be adsorbed at the electrode surface. It has been known for some time, furthermore, that a variety of anions (Cl^-, Br^-, I^-, SCN^-, etc.) are more or less strongly adsorbed on mercury *even at potentials cathodic of the pzc.*[26] Since the electrode is negatively charged at these potentials, this adsorption is obviously not of electrostatic origin. This phenomenon, known as *specific adsorption*, is not well understood, although a number of theories have been advanced.[27] Specific adsorption is widely believed to involve short-range chemical forces and may be related to polarizibility, since the degree to which an anion exhibits "soft" base character tends to parallel its tendency to be specifically adsorbed. [Specific adsorption of cations is uncommon; some organic cations, e.g., tetraalkylammonium ions, and a few inorganic ions, e.g., thallium(I) and lead(II), are specifically adsorbed.][28] It is also believed that the specifically adsorbed anion must lose its solvent sheath, at least on the side facing the electrode, in order for the ion to approach closer than the OHP. The region immediately nearest the electrode, consisting both of oriented solvent and other neutral molecules, and specifically adsorbed unsolvated ions, is known as the *inner layer*. The plane through the centers of the specifically adsorbed ions is known as the *inner Helmholtz plane* (IHP). Of course, at potentials farther and farther cathodic of pzc, the electrostatic repulsion between the electrode and anion increasingly opposes the forces responsible for specific adsorption, and indeed most specifically adsorbed anions are completely desorbed at potentials not very cathodic of the pzc. A few anions, however, such as iodide ion, require quite negative potentials for complete desorption.[29]

2.4.2 Electrocapillary Curves

The size that a drop growing from a dropping mercury electrode can attain depends ultimately on the surface tension of the mercury–solution interface, since the drop falls only when the downward pull of gravity exceeds the surface tension tending to keep the drop attached to the mercury in the column. If the rate of flow of mercury through the column is uniform, as it is under polarographic conditions, the lifetime of the drop will be directly proportional to its weight and can serve as a measure of surface tension. If the drop is charged, either positively or negatively, the repulsion between the like charges in the double layer, particularly at the nearby IHP, decreases the amount of energy necessary to stretch the mercury–solution interface, that is, by definition, the mercury–solution interfacial tension. The drop time, therefore, is longest at the pzc, where the electrode surface is uncharged; this is the phenomenon responsible for the potential dependence of the drop time (Figure 2.12). One important use of such electrocapillary data is in studies of adsorption. The surface tension of mercury is often decreased substantially when a constituent of the solution is adsorbed on mercury (Figure 2.17).[30] Qualitatively, the decrease in drop times for *n*-butanol over the potential range of -0.1 to -1.2 V provide evidence that it is adsorbed on mercury, and the fact that the two curves coincide at potentials more cathodic than -1.2 V suggests strongly that *n*-butanol is

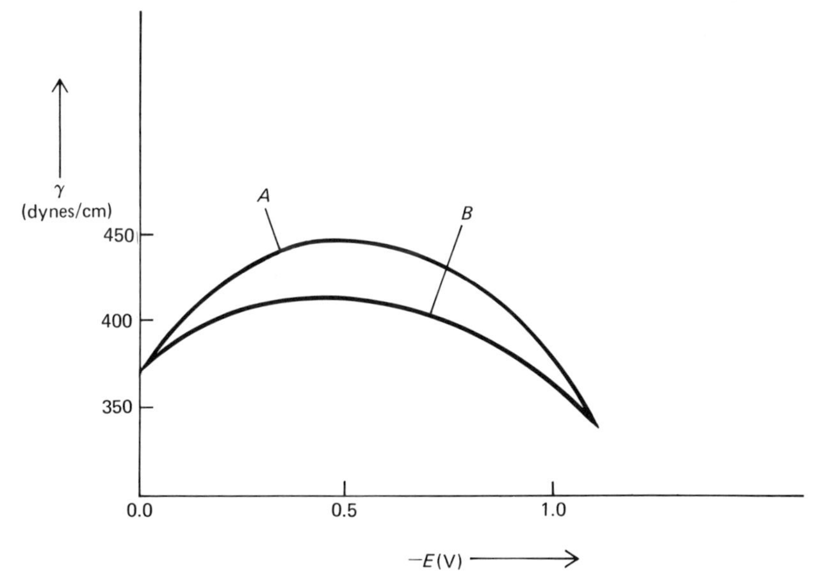

Figure 2.17 Plots of mercury-solution surface tension as a function of potential. Curve *A*: 0.001-*F* $HgCl_2$ in saturated aqueous potassium chloride. Curve *B*: same solution, containing 0.1-*M* *n*-butanol. (J. Heyrovsky and J. Kuta, *Principles of Polarography*, Prague, Academia, 1966, p. 452).

desorbed from the mercury surface at these potentials. Accurate measurements of either drop times[31] or surface tension[32] as a function of concentration can, in fact, be made to yield data concerning the nature and extent of adsorption at the mercury surface.

2.4.3 Charging Current

Recall, from Figure 2.14, that the background (residual) current in a polarogram is neither zero nor constant over most of its length. It consists, rather, of two linear segments, with the slope changing at a potential at which the current is zero, and which can now be recognized as the pzc by reference to Figure 2.12. This behavior can be understood in the following way. At potentials on either side of the pzc, a more or less highly structured double layer is formed in response to the excess charge on the electrode. The electrical work involved in the charge separation necessary to construct the double layer is of the same form as that involved in charging a capacitor.* The amount of electricity required to charge the double layer of an electrode of area A is given by

$$q = \kappa A(E_{\text{max}} - E)$$

where q is the quantity of electricity in microcoulombs, κ is the differential capacity of the double layer in microfarads per square centimeter, A is in square centimeters, and E_{max} is the potential of the electrocapillary maximum of pzc. If κ_+ and κ_- are the differential capacities on the anodic and cathodic sides of E_{max}, it is approximately true that $\kappa_+ \neq \kappa_-$, but that both are independent of potential. (This accounts for the linearity in the slope of the two segments.) The average charging current required to change the potential of an electrode of area A from E_{max} to a new potential E, in t sec, is given by

$$i_t = \frac{A\kappa}{t}(E_{\text{max}} - E)$$

The residual current is, of course, zero at pzc, since construction of a double layer is unnecessary. The sign of the current is negative on the anodic side of pzc and positive on the cathodic side as a consequence of the fact that the double layer must be charged in an opposite sense on each side of the pzc. Notice, finally, that for a solid electrode held at potential E, one need only charge the double layer at the beginning of the experiment; no further electrical work is then necessary to maintain the double layer at constant potential. In a polarographic experiment, on the other hand, the double layer is destroyed when the drop falls off, so the full charging current at each potential is required for every drop across the entire width of the polarogram.

* Indeed, changes in the double-layer capacity as a function of experimental conditions (potential, etc.) are readily followed by an ac impedance bridge and constitute another method used to probe double-layer structure.[33]

2.4.4 Kinetics of Electron Transfer

If there were no electrical double layer surrounding the electrode, one would expect that a species approaching an electrode polarized at potential E (relative to bulk solution) would be at bulk solution potential until actually in contact with the electrode, at which point it would abruptly experience potential $E - E_{max}$, where E_{max} is the potential of the electrocapillary maximum or pzc. In fact, however, as a consequence of the relatively highly structured ionic atmosphere in the vicinity of the electrode, there is no such discontinuity of potential at the electrode surface. Rather, there exists a potential gradient extending from the electrode surface outward into bulk solution, except when $E = E_{max}$, when the electrode potential is equal to bulk solution potential. The nature of this potential gradient is quite dependent on experimental conditions. It is illustrated in Figure 2.18a for the case in which specific adsorption is absent. Recall from the preceding discussion of the double-layer structure that hydrated ions cannot approach closer to the electrode than the OHP. (It is also often assumed that nonadsorbed neutral molecules are also reduced at the OHP.) It will be evident that it is the potential experienced by a given species *at the OHP*, not the actual electrode potential, that affects the rate of electron transfer. If one calls the potential of the OHP relative to bulk solution the *Ψ-potential* or *Φ_2-potential*,* then the potential difference between the electrode surface and the OHP is $(E - E_{max}) - \Psi$, and it is this potential that in fact ought to be used in expressions relating currents or electron-transfer rates to potential, for example, Eq. 2.3 or 2.7a. The fact that the potential imposed on the electrode and that actually experienced by a species undergoing electrochemical reaction at the OHP are not the same can normally be neglected in most synthetic and mechanistic experiments, since it is usually *relative* potentials that are of interest in experiments of this type. One important qualification must, however, be added to this statement. The E_{max} will differ, often substantially, for an electrode immersed in different solvent–electrolyte combinations. Hence, comparison of measured potentials for one compound in several different electrochemical techniques, or for several compounds in the same technique, such as polarography, must be made under conditions of identical double-layer structure. This is closely approximated in the absence of adsorption when the solvent, temperature, electrolyte, and concentration of the latter are identical for all experiments. The rates of electron transfer to a given substrate may otherwise vary over several orders of magnitude at the same applied potential.

The situation is considerably more complex in the presence of specific adsorption (Figure 2.18b). As a consequence of incorporation of anions into the inner layer, there is no longer a smooth potential gradient extending outward from the electrode surface. Rather, the excess of anions in the inner layer may

* Under this nomenclature, Φ_1 is the potential of the IHP relative to bulk solution.

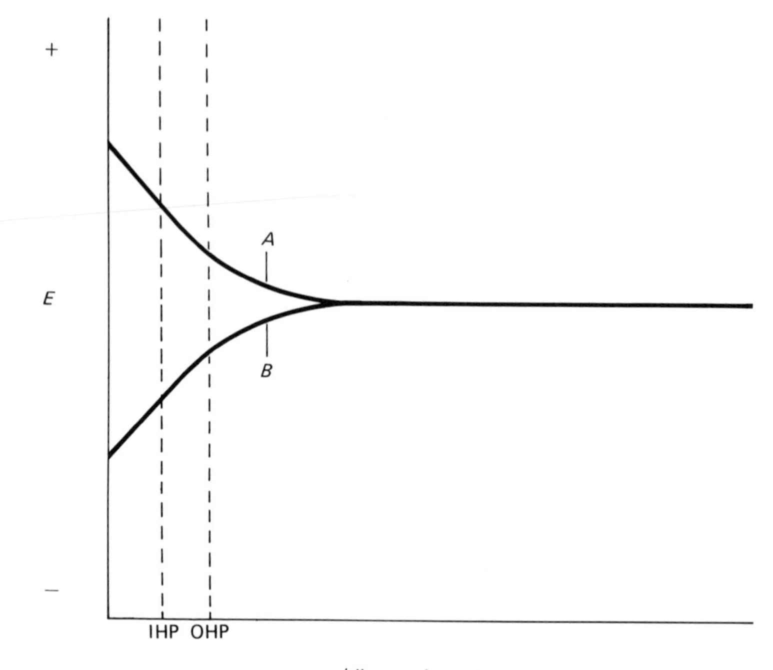
+
E
−
A
B
IHP
OHP
x (distance from electrode surface)

(a)

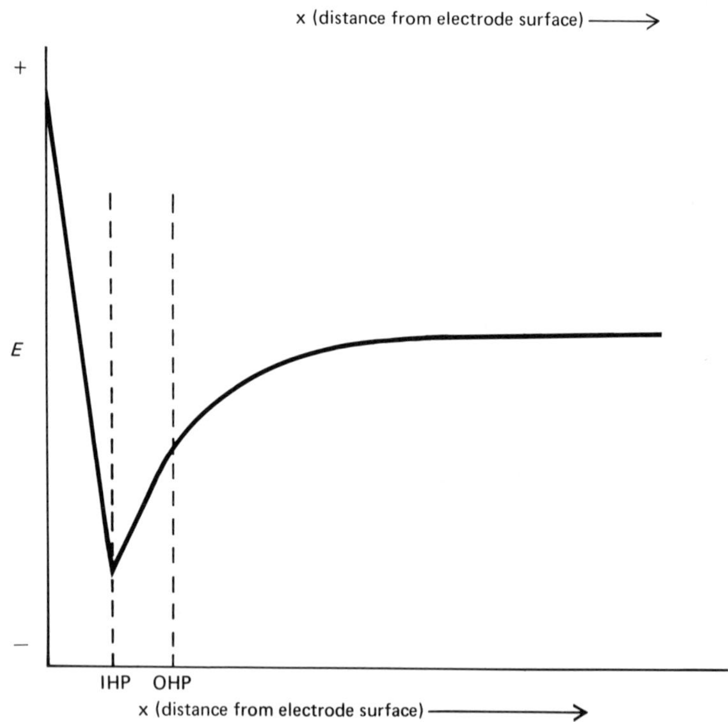
+
E
−
IHP
OHP
x (distance from electrode surface)

(b)

in this extreme case affect even the sign of the solution potential a short distance from the electrode. Notice that a species approaching the electrode of Fig. 2.18b would feel increasingly negative potentials, even though the electrode itself is positive. The situation is further complicated by doubts at present concerning whether species are reduced at the OHP in the presence of specific adsorption,[34] and, of course, if the electroactive species itself is adsorbed the double-layer structure will not even be the same as that for the same solvent–electrolyte combination in the absence of electroactive species. These considerations underline again the fact that potentials in electrochemical experiments must be measured under conditions of similar double-layer structure.

A corollary to the last statement is that one may, on occasion, change the course of an electrochemical reaction through changes in the double-layer structure. Given present knowledge of the double layer, particularly in nonaqueous solvents, however, such effects, when discovered, are largely empirical. Qualitatively, it is sufficient at this point to remind the reader that any change in experimental conditions that can affect double-layer structure may affect the relative and/or absolute rates of heterogeneous electron transfer to components of the solution. Such changes might include variations in temperature, solvent, nature and concentration of supporting electrolyte, addition of adsorbable (*surface-active*) species, and so on.[35] Surface-active agents can affect the nature of the electron-transfer process in a variety of ways, as will be seen in Chapter 4.

2.4.5 Field Effects

The potential difference between the electrode surface and bulk solution illustrated in Figure 2.18 may amount to a volt or more, over a rather short distance of ca. 50–5000 Å (the thickness of the double layer). This is an extremely steep gradient, amounting to as much as 10^6 V/cm, which is an electrical field of considerable intensity. It would be unsurprising to find the chemistry of highly polarizable species affected by the field gradient at such high field strengths.[36] The course of at least two organic electrode reactions is indeed thought to depend

---

Figure 2.18 Potential distribution in the double layer. Ordinate: potential experienced by an electroactive substance approaching the electrode. Abscissa: distance from the electrode surface. (a) Absence of specific adsorption. Curve *A*: potential distribution near a positively charged electrode. Curve *B*: potential distribution near a negatively charged electrode. (b) Potential distribution near a positively charged electrode in the presence of a specifically adsorbed anion. (D. C. Grahame, *Chem. Rev.*, **41**, 490, 491 [1947]).

on field strength.[37] Field effects are probably more widespread than present data would indicate, but systematic searches for such effects have not been made, and it is difficult to sort these out from other effects when the double-layer structure is changed in an experiment.

2.5 RELATIVE EFFICIENCY OF ELECTRODE PROCESSES

It is often of interest to compare the relative efficiency of two or more electrode processes. One may, for example, wish to rank several functional groups in a single molecule according to ease of electrochemical reduction. For this purpose, it is usually sufficient to carry out polarographic measurements on the substance of interest plus a number of model compounds. A far different situation arises when one wishes to make quantitative or semiquantitative conclusions about the relative efficiency of electrochemical reaction for a series of related compounds. This is often the case when attempts are to be made to deduce the nature of the transition state for electron transfer from substituent effects on electrochemical parameters.[22] Ideally, one ought to measure the electron-transfer rate at a single potential for each member of the series of compounds (e.g., for irreversible processes, by measurement of $k^0_{f,h}$ through appropriate analysis of the polarographic data, using either Eq. 2.7a or 2.54), followed by a Hammett or Taft treatment of the substituent effects on rate.[38] In practice, an alternative procedure has arisen, that of ordering the series of compounds according to polarographic half-wave potentials. For reversible processes uncomplicated by coupled chemical reactions, this procedure is quite satisfactory, since these potentials are related to relative E^0's. For example, it has been shown that for reversible processes one may with reasonable accuracy predict the approximate position of equilibrium, and even equilibrium constants, for the process

$$\mathrm{Ox_1 + Red_2 \rightleftharpoons Red_1 + Ox_2}$$

from the polarographic half-wave potentials of Ox_1 and Ox_2.[39] With certain qualifications, polarographic half-wave potentials for irreversible processes may also be used to rank such processes according to ease of electrochemical reaction. The nature of these qualifications will be clear from inspection of Eq. 2.55. The assumption that polarographic half-wave potentials are a measure of relative ease of electron transfer amounts ultimately to the assumption that $E_{1/2}$ is proportional to $\log k^0_{f,h}$. This will only be true if αn, D_{ox}, and t are constant over the series of compounds investigated. In such an investigation, one can and should always convert $E_{1/2}$ to $E^0_{1/2}$ (e.g., by the use of Eq. 2.59) to eliminate the drop-time dependence of $E_{1/2}$. Diffusion coefficients are usually not appreciably different for a series of homologous compounds (unless a member of the series is abnormally solvated). The remaining variable in Eq. 2.55, αn, must also be the same for all members of the series, and must be checked for each compound by application of Tomes's criterion or, in very precise work, log-plot slopes.

Since, as it will be recalled from the beginning of this chapter, α is related to the barrier to electron transfer, a value of αn for a compound deviating from αn for a series of similar compounds strongly suggests a different mode of electron transfer for the deviant compound. This difference might arise from the molecule undergoing, for some reason, a different electrochemical reduction pathway, or, possibly more likely, a different degree of adsorption from the other members of the set of compounds. The preceding discussion of electrode kinetics indicated that adsorption, changing the double-layer structure, may also change the nature of the electron-transfer process, and this often results in a measurably different αn value. The latter will then, from Eq. 2.55, shift the polarographic $E_{1/2}$ from that expected in the absence of adsorption. Likewise, any such study should include a test of diffusion control of the limiting current for each member of the series (Chapter 3). It is well known, for example, that organic sulfur compounds are strongly adsorbed on mercury[32b,40]; this alone, rather than any special electronic property of sulfur-containing substituents, such as the thiomethyl group ($-SCH_3$), could account for deviation of such substituents from a σ–$E_{1/2}$ line.

A plot of $E_{1/2}$ *vs* Hammett substituent constants is found for many series of compounds to have a slope in the range 0.1–0.6 V^{-1}. This should not be taken to mean that the transition state for such electrode reactions is relatively insensitive to electronic effects, because the slope of such a plot is equal to $0.059\rho/\alpha n$.

$$\frac{dE_{1/2}}{d\sigma} = \frac{0.059}{\alpha n}\rho \tag{2.62}$$

If for a series of compounds $\alpha n = 0.6$ and the slope of the $E_{1/2}$ *vs* σ plot is $+0.3$, the reaction constant ρ is $+3.1$, corresponding to a reduction involving in fact considerable electron demand in the transition state.[38]

References

[1a]H. Eyring, S. Glasstone, and K. J. Laidler, *J. Chem. Phys.*, **7**, 1053 (1939); [b]S. Glasstone, K. J. Laidler, and H. Eyring, *The Theory of Rate Processes*, New York, McGraw-Hill, 1961, chap. 10.

[2]J. J. Lingane, *Electroanalytical Chemistry*, 2nd ed., New York, Wiley, 1958, pp. 46–49.

[3]P. Delahay, *Double Layer and Electrode Kinetics*, New York, Wiley, 1966, p. 159.

[4]D. A. MacInnes, *Principles of Electrochemistry*, New York, Dover, 1961.

[5]Lingane, *op. cit.*, chap. 9.

[6]W. Jost, *Diffusion in Solids, Liquids, Gases*, rev. ed., New York, Academic, 1960.

[7]V. G. Levich, *Physicochemical Hydrodynamics* Englewood Cliffs, N.J., Prentice-Hall, 1962.

[8]A. Fick, *Pogg. Ann.*, **94**, 59 (1855).

[9]R. W. Murray and C. N. Reilley, *Electroanalytical Principles*, New York, Wiley, 1966.

[10a]F. G. Cottrell, *Z. Physik. Chem.*, **42**, 385 (1902); [b]R. N. Adams, *Electrochemistry at Solid Electrodes*, New York, Marcel Dekker, 1969.

[11]L. Meites, *Polarographic Techniques*, 2nd ed., New York, Wiley, 1965, p. 417.
[12]D. MacGillavry and E. K. Rideal, *Rec. Trav. Chim.*, **56**, 1013 (1937).
[13a]D. Ilkovic, *J. Chim. Phys.*, **35**, 129 (1938); [b]D. Ilkovic, *Coll. Czech. Chem. Commun.*, **6**, 498 (1934).
[14a]J. Koutecky, *Czech. Cas. Fys.*, **2**, 50 (1953); [b]J. Koutecky and M. von Stackelberg, in P. Zuman and I. Kolthoff, eds., *Progress in Polarography*, New York, Wiley, 1962, vol. 1, p. 21.
[15]Meites, *op. cit.*, pp. 422–423.
[16]Murray and Reilley, *op. cit.*, pp. 2145–2147.
[17]L. Meites and T. Meites, *Anal. Chem.*, **28**, 103 (1956).
[18a]J. Tomes, *Coll. Czech. Chem. Commun.*, **9**, 12 (1937); [b]Meites, *op. cit.*, pp. 217–225.
[19]L. Meites and Y. Israel, *J. Amer. Chem. Soc.*, **83**, 4903 (1961).
[20]J. Koutecky, *Coll. Czech. Chem. Commun.*, **18**, 597 (1953).
[21a]Meites, *op. cit.*, pp. 241–245; [b]L. Meites, private communication; [c]Meites and Israel, *op. cit.*, p. 4905.
[22]P. Zuman, *Substituent Effects in Organic Polarography*, New York, Plenum, 1967.
[23a]D. C. Grahame, *Chem. Rev.*, **41**, 441 (1947); [b]R. B. Whitney and D. C. Grahame, *J. Chem. Phys.*, **9**, 827 (1941).
[24]E. M. Kosower, *Introduction to Physical Organic Chemistry*, New York, Wiley, 1967.
[25a]D. M. Mohilner, in A. J. Bard, ed., *Electroanalytical Chemistry*, New York, Marcel Dekker, 1966, vol. 1, pp. 241–409; [b]R. Payne, *Advan. Electrochem. Electrochem. Eng.*, **7**, 1 (1970).
[26]Mohilner, *op. cit.*, pp. 244–245 and 331–352.
[27]D. J. Barclay, *J. Electroanal. Chem.*, **19**, 318 (1968).
[28a]E. B. Weronski, *Trans. Faraday Soc.*, **58**, 2217 (1962); [b]Mohilner, *op. cit.*, p. 334.
[29a]V. J. Puglisi, G. L. Clapper, and D. H. Evans, *Anal. Chem.*, **41**, 279 (1969); [b]E. B. Weronski, *Electrochim. Acta*, **14**, 259 (1968).
[30]Mohilner, *op. cit.*, pp. 251–254.
[31a]P. Corbusier and L. Gierst, *Anal. Chim. Acta*, **15**, 254 (1956); [b]A. J. Bard and H. B. Herman, *Anal. Chem.*, **37**, 317 (1965).
[32a]B. E. Conway and R. G. Barradas, *Electrochim. Acta*, **5**, 319 (1961); [b]E. Blomgren, J. O'M. Bockris, and C. Jesch, *J. Phys. Chem.*, **65**, 2000 (1961).
[33]R. Parsons and P. C. Symons, *Trans. Faraday Soc.*, **64**, 1077 (1968).
[35]Meites, *op. cit.*, pp. 254–256.
[35a]L. Holleck, B. Kastening, and R. D. Williams, *Z. Elektrochem.*, **66**, 396 (1962); [b]S. R. Missan, E. I. Becker, and L. Meites, *J. Amer. Chem. Soc.*, **83**, 58 (1961).
[36a]G. Kortun and J. O'M. Bockris, *Textbook on Electrochemistry*, New York, Elsevier, 1951, vol. 1, pp. 186–189; [b]H. S. Harned and B. B. Owen, *Physical Chemistry of Electrolytic Solutions*, 3rd ed., New York, Van Nostrand Reinhold, 1958, p. 138ff.
[37a]A. Vincenz-Chodkowska and Z. R. Grabowski, *Electrochim. Acta*, **9**, 789 (1964); [b]G. J. Hoijtink, *Rec. Trav. Chim.*, **76**, 869, 885 (1957).
[38]J. E. Leffler and E. Grunwald, *Rates and Equilibria of Organic Reactions*, New York, Wiley, 1963.
[39]J. M. Fritsch, T. P. Layloff, and R. N. Adams, *J. Amer. Chem. Soc.*, **87**, 1724 (1965).
[40a]I. M. Kolthoff, W. Stricks, and N. Tanaka, *J. Amer. Chem. Soc.*, **77**, 4739 (1955); [b]P. S. McKinney and S. Rosenthal, *J. Electroanal. Chem.*, **16**, 261 (1968).

3 TECHNIQUES FOR INVESTIGATION OF ELECTRODE PROCESSES

It is now appropriate to consider a number of specific techniques by which one may apply the principles of Chapter 2 to gain the mechanistic information necessary for proper design of a successful electrochemical synthesis. Chapter 4 will continue this discussion, with emphasis on manipulation of experimental parameters as a means of controlling the course of an electrode reaction. Throughout the following discussion, the kinds of information that can be obtained by using each instrumental technique will be illustrated by application to diphenylfulvene (**1**).

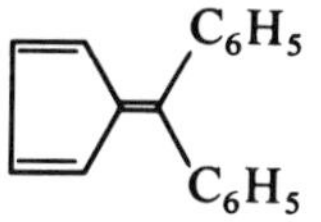

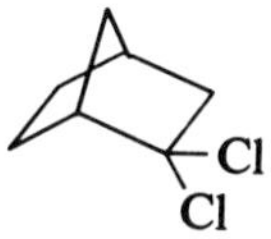

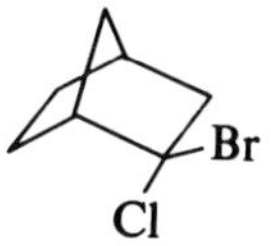

C_6H_5 C_6H_5 Cl Cl Br Cl

1 **2** **3**

It will be shown in this chapter that four complementary techniques (polarography, cyclic voltammetry, controlled-potential electrolysis, and coulometry) normally provide ample evidence of the mechanism of a given electrode process to allow one both to use that process intelligently in synthesis and to devise new chemistry based on it.*

3.1 POLAROGRAPHY

Measurement of the polarogram of a substance is normally the first step in characterization of its behavior upon electrochemical reduction, since polarography can provide a considerable amount of mechanistic information with relatively little experimental effort.** The number of polarographic waves indicates immediately the number of discrete electrode processes undergone by the substance, and the wave heights and shapes indicate the numbers of electrons involved in the overall and rate-determining steps, respectively, for each wave.

* A considerable degree of experimental simplification, incidentally, arises from the fact that polarography and cyclic voltammetry may, if one designs the experiment properly (Chapter 9), be carried out on the same sample of solution, while controlled-potential electrolysis and coulometry must in fact be carried out simultaneously.

** Conversely, initial characterization of the oxidative behavior of a substance is usually performed via cyclic voltammetry (Section 3.2) at a solid electrode because of the relative ease with which mercury undergoes electrochemical oxidation.

3.1.1 Wave-Height Analysis

Tests for Diffusion Control

The methods of this section are based on application of the Ilkovic equation (Eq. 2.30a) to polarographic limiting currents. Before this can be done, however, one must verify that the Ilkovic equation is indeed valid under the experimental conditions. Specifically, it is always necessary to test the underlying assumption that limiting currents are controlled by the rate of diffusion of the electroactive substance to the electrode surface. This is true in the large majority of cases, but, as will be seen, situations often arise where the rate of a process preceding the electron-transfer step actually controls the overall rate of reduction. Although the Ilkovic equation may not be applied to such cases, failure to exhibit diffusion control itself constitutes an important clue to the nature of the electrode process.

The most commonly employed test for diffusion control of the limiting current, because of its ease and simplicity, is correlation of the limiting polarographic current with the height of the mercury column above the tip of the D.M.E. (Figure 3.1). What is actually to be correlated is the variation in the

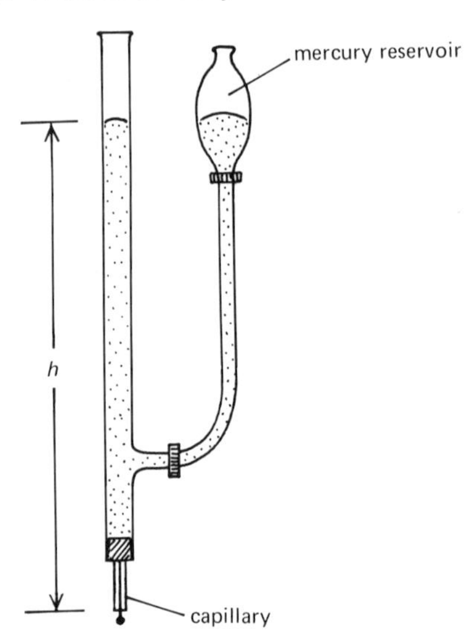

Figure 3.1 Dropping mercury electrode, showing measurement of h.

mercury flow rate m (in milligrams per second) and the drop time t (in seconds) with the height of the column of mercury above the drop. It may be shown from hydrostatics that m is proportional to P_{corr}, defined as

$$P_{corr} = P_h - P_b \tag{3.1}$$

where P_h is the pressure exerted by the column of mercury of height h, and P_b is the back pressure due to the surface tension of mercury, which tends to prevent the growth of the drop.[1] P_h is directly proportional to h, and P_b may be converted into an equivalent height, h_b, so that

$$h_{corr} = h - h_b \tag{3.2}$$

where h_{corr} is then the height of the mercury column corrected for the back pressure due to the mercury–solution interfacial tension.* The Ilkovic equation (Eq. 2.30a) predicts that limiting currents should be proportional to $m^{2/3}t^{1/6}$ when diffusion is the current-limiting process. It now remains to inquire into the nature of the changes of m and t with h_{corr}. From Eqs. 3.1 and 3.2, m is directly proportional to h_{corr}. The drop weight, mt, however, while dependent on the mercury surface tension and hence the potential, is independent of h_{corr}. Since mt is constant at a given potential and m is directly proportional to h_{corr}, t must be inversely proportional to h_{corr}. From these considerations and the $m^{2/3}t^{1/6}$ dependence of i_d predicted by Eq. 2.30a, the variation of i_d with h_{corr} is given by

$$i_d \propto h_{corr}^{2/3}\left(\frac{1}{h_{corr}}\right)^{1/6}$$

Hence,

$$i_d \propto h_{corr}^{1/2} \tag{3.3}$$

Equation 3.3 constitutes the basis for a rapid, convenient test of diffusion control. After measurement of the polarographic limiting current at a series of mercury column heights such that t is between 2 and 8 sec (Figure 3.1) and estimation of the back-pressure term, one constructs a plot of i_{lim} *vs* $h_{corr}^{1/2}$; a straight line with zero intercept is indicative of diffusion control. Typical plots are shown in Figure 3.2. The data are also tabulated in Table 3.1; $i/h_{corr}^{1/2}$ should be constant for a diffusion-controlled process. Notice that **1** and 2,2-dichloronorbornane (**2**) obey the strictures of the Ilkovic equation, but that the $i_{lim} - h_{corr}^{1/2}$ plot is clearly nonlinear for 2-*exo*-bromo-2-*endo*-chloronorbornane (**3**). Given the data of Figure 3.2, one may proceed** to a detailed analysis of the polarographic behavior of **1** or **2** by the methods to be described below. The nonlinear plot for **3** indicates, however, that the rate of some process other than diffusion determines the overall rate of its polarographic reduction. Since limiting currents are therefore not diffusion currents for this compound, interpretations based on the Ilkovic equation cannot be made. It is, however, still worth

* The back-pressure term is on the order of 0.5–1.5 cm in typical organic solvents, and, since it is directly proportional to the mercury–solution interfacial tension, may be substantially smaller at potentials far removed from the electrocapillary maximum (Section 2.4.2). Typical values of h are from 40 to 80 cm, so the correction term can be significant.

** A rigorous test of diffusion control would include a number of other experiments, such as the concentration and temperature dependences of the limiting current.[2] The mercury-column height dependence is normally adequate, however.

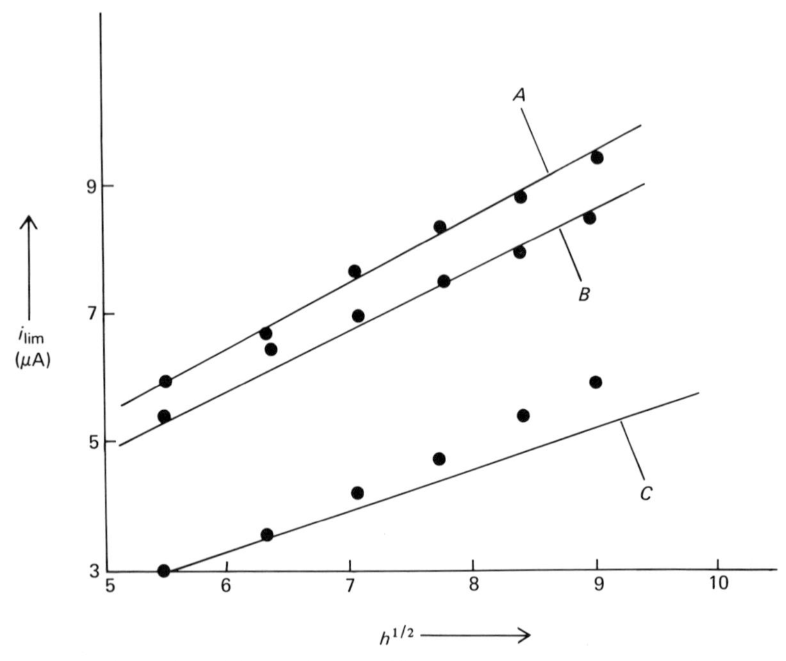

Figure 3.2 Plots of limiting polarographic current *vs* square root of the mercury column height. Plot *A*, 6,6-diphenylfulvene (**1**); plot *B*, 2,2-dichloronorbornane (**2**); plot *C* (nonlinear), 2-*exo*-bromo-2-*endo*-chloronorbornane (**3**). Solvent: dimethylformamide containing 0.1-*F* tetraethylammonium bromide.

Table 3.1 Limiting polarographic current as a function of the mercury column height

	6,6-Diphenylfulvene		2,2-Dichloronorbornane		2-*exo*-bromo-2-*endo*-chloronorbornane	
h(cm)	i(μA)	$i/h^{1/2}$	i(μA)	$i/h^{1/2}$	i(μA)	$i/h^{1/2}$
30	6.0	1.10	5.5	1.00	3	0.55
40	6.8	1.07	6.6	1.04	3.6	0.57
50	7.8	1.11	7.1	1.00	4.2	0.59
60	8.4	1.08	7.6	0.98	4.8	0.62
70	8.9	1.06	8.0	0.96	5.4	0.64
80	9.6	1.07	8.6	0.96	6.0	0.67

Source: A. J. Fry, W. E. Britton, and M. A. Mitnick, unpublished results.

inquiring into the possible causes of such behavior, that is, the kinds of processes other than diffusion that can limit polarographic currents, and to develop criteria for recognizing such situations. Deviations from diffusion control are generally due to *kinetic*, *catalytic*, or *adsorption* processes.

Kinetically controlled limiting currents arise when the rate of electrochemical reaction is governed not by the rate of diffusion of the electroactive material to the electrode but by the rate of *its formation near the electrode surface* by a preceding chemical reaction. The electrochemical reduction of many weak acids, for example, actually involves the electrochemical reduction of the proton formed by dissociation:

$$\mathrm{HA} \underset{k_b}{\overset{k_f}{\rightleftharpoons}} \mathrm{H^+ + A^-} \qquad K = \frac{k_f}{k_b}$$

$$\mathrm{H^+ + e^-} \longrightarrow \tfrac{1}{2}\mathrm{H_2}$$

If K is small, the bulk concentration of H^+ will be very low. Its concentration may be so low, in fact, that diffusion of H^+ from bulk solution to the electrode cannot give rise to measurable reduction currents. In this case, the only current observed will be due to the reduction of H^+ formed by dissociation near the electrode surface. If under these circumstances k_f is too low to permit H^+ to be supplied by dissociation in the quantities demanded by the expanding drop, then the rate of dissociation itself represents the limit to the currents that can be passed. A polarographic wave whose height is thus limited by the rate of a prior chemical reaction is called a *kinetic* wave. Kinetic waves were first observed by K. Wiesner in the polarographic reduction of sugars and quinones, and are also observed for many other compounds that are extensively hydrated in aqueous solution, such as chloral and formaldehyde.[3,4] In all such cases the carbonyl group is polarographically reducible; the hydrate (or for sugars the cyclic hemiacetal) is, however, not reducible, and hence the magnitude of the polarographic wave (indeed, whether or not a wave is observed at all) depends on the values of k_f and K.

The very data used in construction of the i *vs* $h_{\mathrm{corr}}^{1/2}$ plot as a test for diffusion control may then, it so happens, be applied to a test of kinetic control when a nonlinear plot is observed. This may be seen in the following way. First, note that the reduction current for a kinetic process will vary directly with the area of the electrode surface. From Eq. 2.28, $A \propto m^{2/3}t^{2/3}$, and, given the direct and inverse dependence of m and t, respectively, on h_{corr}, one finds that

$$i_{\mathrm{lim}} \propto h_{\mathrm{corr}}^{2/3}\left(\frac{1}{h_{\mathrm{corr}}}\right)^{-2/3} \tag{3.4}$$

That is, the limiting current for a kinetically controlled process is *independent* of h_{corr}; such processes are therefore often easily recognized (Figure 3.3).

Kinetic waves are sensitive to experimental conditions, since they involve the equilibrium constants for the prior reaction. For example, imagine the case

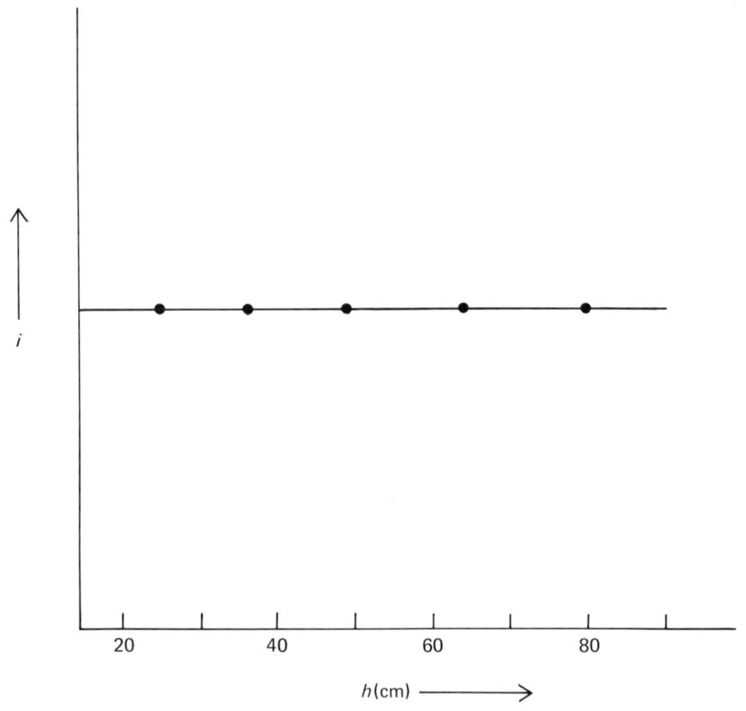

Figure 3.3 Plot of limiting polarographic current *vs* mercury column height for formaldehyde in aqueous buffer, pH 9.2. (J. Heyrovsky and J. Kuta, *Principles of Polarography*, Prague. Academia, 1966, p. 343.)

where a polarographic wave is observed for the conjugate acid of a weak base:

$$B + H^+ \underset{k_b}{\overset{k_f}{\rightleftharpoons}} BH^+$$

$$BH^+ + e^- \longrightarrow C$$

In neutral solution, the bulk concentration of BH^+ may be low enough so that the reduction wave is kinetically controlled, depending, of course, on the magnitudes of k_f and K. In alkaline solutions, a reduction wave for BH^+ may not even be observed; in strongly acidic solutions, the bulk concentration of BH^+ will be so high that the reduction is diffusion controlled. At intermediate acidities, reduction will be limited by a combination of kinetic and diffusion processes. In such cases, the limiting current dependence on mercury column height will also be intermediate between that expected for either kinetic or diffusion control. Such intermediate behavior will usually be obvious in a given case. Study of the pH dependence of the limiting current behavior often provides a convenient test of the nature of the electroactive species. Many such investigations have been reported.[5]

Incidentally, the situation is often encountered where two polarographic waves are observed: one due to a given species and the other due to its conjugate acid (or base). The relative heights of the two waves depend on both the equilibrium concentrations of both species and their *rates of interconversion.* Measurement of the relative heights over a range of pH values offers a convenient tool for obtaining kinetic and equilibrium data (e.g., k_f, k_b, and K in the example cited above) and has been used extensively for this purpose. Rates of dissociation and recombination as fast as 10^4–10^8 sec^{-1} are accessible via the polarographic method. The reader is referred to the literature for details.[4,6,7]

Catalytic currents arise when the product of an electrochemical reaction is *reconverted to the starting material* at the electrode surface through reaction with a second species that is itself electroinactive at the potentials used. This behavior may be symbolized (for initial reduction) as

$$O + e^- \longrightarrow R$$

$$R + Z \xrightarrow{k_1} O + X$$

(with an arrow returning O to the first step)

Observe that, in effect, O merely serves to mediate electron transfer to Z; hence the term "catalytic process." The reader will no doubt notice the similarity between the function of O in this net conversion of Z to X and the corresponding role of photosensitizers often used to effect indirect photochemical reaction by electronic energy transfer. Since the starting material, O, is regenerated at the electrode surface, it may then be reduced again, and since a number of such cycles may occur during the lifetime of a drop, polarographic limiting currents can be many times higher than for the case where $k_1 = 0$; abnormally high diffusion currents, then, constitute a method of recognizing catalytic processes.

Catalytic behavior may, at first, seem thermodynamically paradoxical, i.e., that Z, not O, should be reduced here. A clue to what is actually happening appears when one notes that, generally speaking, the reduction of Z is always considerably more irreversible than the reduction of O. Since the polarographic $E_{1/2}$ for an irreversible process is considerably cathodic of E^0 (p. 42), while $E_{1/2} = E^0$ for a reversible process, no violence need be done to thermodynamic intuition, as will be evident from inspection of Figure 3.4.

Catalytic processes are uncommon in organic electrochemistry, but are often met in the inorganic domain. A typical example was reported by Rechnitz and Laitinen,[8] who observed that the wave for polarographic reduction of molybdenum(VI) is many times higher in the presence of perchlorate ion than in the absence of perchlorate, presumably because of facile oxidation of molybdenum(V) by perchlorate. A clearly catalytic organic process was observed during the polarographic reduction of naphthalene (**4**) in the presence of *n*-butyl chloride (**5**).[9] Although **5** is electrochemically inert at potentials where naphthalene is reduced to the naphthalene radical anion (**6**), the limiting current for reduction of naphthalene in dimethylformamide (DMF) is considerably

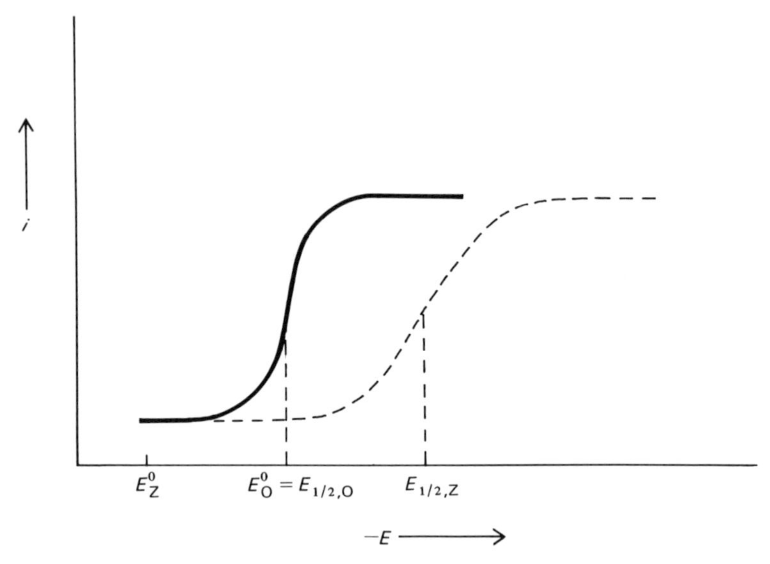

Figure 3.4 Half-wave and standard potentials for irreversible and reversible redox couples participating in an electrocatalytic cycle. Solid line, reduction of O (reversible); dashed line, reduction of Z (irreversible).

increased by the addition of butyl chloride. It has been established by a number of investigators that alkyl halides (RX) are easily reduced to hydrocarbons (RH) by **6**.[10] This chemical reaction regenerates naphthalene, which in the polarographic experiment is reduced again to continue the cycle:

$$\mathbf{4} + e^- \longrightarrow \mathbf{6}$$

$$\mathbf{6} + n\text{-}C_4H_9Cl \longrightarrow \mathbf{4} + n\text{-}C_4H_9\cdot + Cl^-$$

$$\mathbf{6} + n\text{-}C_4H_9\cdot \longrightarrow \mathbf{4} + n\text{-}C_4H_9{}^-$$

$$n\text{-}C_4H_9{}^- + H^+ \longrightarrow n\text{-}C_4H_{10}$$

Many organic compounds are more or less strongly adsorbed at a mercury–solution interface. The current required to reduce an adsorbed species will be limited not by the diffusion rate but by the number of molecules that can be physically accommodated, usually in a unimolecular film, on the electrode surface. The polarographic current will, in such cases, be related to the areas of both the electrode and the adsorbed species. Elementary considerations based on

Faraday's law indicate that the quantity of electricity, Q, in coulombs, necessary to reduce the adsorbed material will be given by

$$Q = i_{\text{ads}}\, t$$
$$= \frac{n\mathcal{F}A}{aN}$$

where i_{ads} is the average adsorption-controlled limiting current (in amperes) at the end of the drop life, A is the area (in square centimeters) of the drop at time t, a is the area (in square centimeters) taken up by the adsorbed molecule on the electrode surface, and N is Avogadro's number. Since the area of the drop is proportional to $m^{2/3}t^{2/3}$ (Eq. 2.28),

$$i_{\text{ads}} = \frac{n\mathcal{F}m^{2/3}t^{2/3}}{aNt}$$

$$= \frac{n\mathcal{F}}{aN}\frac{m^{2/3}}{t^{1/3}} \tag{3.5}$$

that is,

$$i_{\text{ads}} \propto m^{2/3}t^{-1/3}$$

Given the previously discussed relationship between m and t and h_{corr},

$$i_{\text{ads}} \propto h_{\text{corr}}^{2/3}\left(\frac{1}{h_{\text{corr}}}\right)^{-1/3}$$

that is,

$$i_{\text{ads}} \propto h_{\text{corr}} \tag{3.6}$$

According to Eq. 3.6, the fundamental criterion for adsorption control of the polarographic limiting current is the direct proportionality between i_{lim} and h_{corr}. When limiting currents are plotted against h_{corr} for 2-*exo*-bromo-2-*endo*-chloronorbornane (**3**), for example, a good linear correlation is observed (Figure 3.5). This accounts, then, for the failure of this material to exhibit diffusion-controlled behavior (Figure 3.2).* If a wave has been shown to be adsorption controlled, Eq. 3.5 may then be applied to estimate the area taken up by a molecule of the adsorbed species; hence, by such measurements one may deduce the orientation of the adsorbed species on the electrode surface. Many aromatic hydrocarbons, for example, have been found by these and related measurements to lie flat on the electrode when adsorbed (Chapter 4).

* It is unsurprising to find that reductions of **2** and **3** differ in being diffusion and adsorption controlled, respectively, despite their structural similarity. It has been known for some time that the order of decreasing ease of adsorption of alkyl halides on mercury is RI > RBr > RCl.[11]

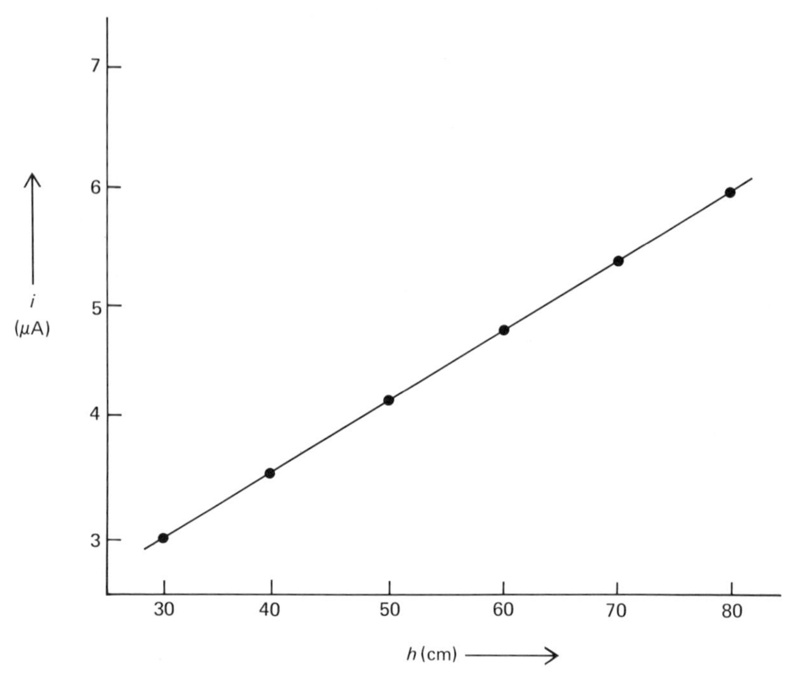

Figure 3.5 Plot of limiting polarographic current *vs* the mercury column height for 2-*exo*-bromo-2-*endo*-chloronorbornane (**3**). Solvent: dimethylformamide containing 0.1-*F* tetraethylammonium bromide.

The reader ought not to be intimidated by the preceding discussion of kinetic, catalytic, and adsorption phenomena. Most polarographic waves are diffusion controlled, and measurements made on them may be used in conjunction with the Ilkovic equation to obtain mechanistic information. It is necessary to be able to recognize these modes of anomalous limiting-current behavior when they do arise, and to realize that failure to obey diffusion control can itself provide information into the course of the electrode reaction. It should be stressed here that each new compound to be investigated polarographically should be tested for adherence to the Ilkovic equation, since even compounds that are closely related structurally may exhibit different limiting current behavior (*cf.* **2** and **3**).

Estimation of Polarographic n Values

Suppose that it has been established, from a linear plot of i_{lim} *vs* $h_{\text{corr}}^{1/2}$ (and also perhaps from the concentration or temperature dependence of i_{lim})[1] that a given wave is indeed diffusion controlled. It is desired to estimate the number of electrons involved in the overall electrode process represented by the polarographic wave. The Ilkovic equation may be rearranged to the form

$$n = \frac{i_d}{708 D^{1/2} m^{2/3} t^{1/6} C} \tag{3.7}$$

Although Eq. 3.7 contains six variables, four (i_d, C, m, and t) will be known or easily measured in a single polarographic experiment.* Given a value for the fifth variable, D, n could be calculated from Eq. 3.7. A number of procedures may be used to evaluate the diffusion coefficient. (Notice that since n must be an integer and since D appears to the one-half power in Eq. 3.7, high accuracy is not required in the value of D to be used in the calculation.) It should be stated at the outset that the *least* commonly employed procedure for obtaining D is direct measurement, both because of the labor involved in such measurements and the fact that it is not necessary to have highly accurate diffusion-coefficient data for the present purpose. Unambiguous assignment of n is usually feasible through estimation of D by one of the following three procedures. First, one may use, if the datum is available, the diffusion coefficient for a molecule of similar size and structure, measured in the same solvent–electrolyte system used for polarography of the unknown compound. When it can be applied, this is probably the best generally useful method of obtaining polarographic diffusion coefficients. Typical data, illustrating the kinds of variations to be expected, are illustrated in Table 3.2 (notice the differences that may be observed in D for a single compound in several solvents). Selection of a model compound is not critical, particularly when one is dealing with solutes, such as hydrocarbons, that do not associate strongly with the solvent. Notice, for example, that the diffusion coefficient of stilbene in DMF differs from those of azobenzene and diphenylacetylene by only 22 and 18%, respectively. These differences are small enough that stilbene could easily be used as a model compound for the evaluation of n for either of the other two. Second, one may use a method[12] suggested by both Lingane and Walkley, computation of the *diffusion current constant*.** If Eq. 3.7 is rearranged in the form

$$708 n D^{1/2} = \frac{i_d}{C m^{2/3} t^{1/6}} \tag{3.8}$$

it will be noticed that the numerical *constants* for a given system appear on the left side of the equation, while the right side consists of the four experimental *variables*. Due to irreproducible variations in construction, no two individual

* The solution is made up to known concentration; i_d is measured from the polarogram; by collecting and weighing a definite number of drops (25–100) over a measured time interval one may obtain m, the mercury flow rate in milligrams per second, and the drop time, t.

** It will be seen that this method is fundamentally equivalent to the preceding one; it is discussed separately here because of its widespread use in the literature.

Table 3.2 Diffusion characteristics of representative organic compounds in typical solvents

Compound	I_d[a]	$D, \times 10^5$ (cm^2/sec)	Electrolyte[b]	Reference
		Dimethylformamide		
Acetophenone	2.6	1.84	0.1-*F* TBAI	1
Azobenzene	2.76	1.52	0.1-*F* TBAP	2
Benzaldehyde	2.3	1.44	0.1-*F* TEAI	1
Naphthalene	–	1.27	0.1-*F* TEAP	3
Anthracene	–	1.19	0.1-*F* TEAP	3
trans-Stilbene	2.09	1.18	0.155-*F* TBAI	4
Phenylacetylene	8.32	1.18	0.155-*F* TBAI	4
Naphthalene	2.02	1.11	0.155-*F* TBAI	4
trans-Stilbene	1.92	1.00	0.2-*F* TBAI	5
Diphenylacetylene	1.92	1.00	0.155-*F* TBAI	4
2,2′-Azonaphthalene	2.03	0.83	0.1-*F* TBAP	2
Benzophenone	1.7	0.79	0.1-*F* TEAI	1
		Acetonitrile		
Aniline	–	2.62	0.1-*F* TEAP	6
Naphthalene	–	2.74	0.1-*F* TEAP	3
Ethyl iodide	6.15	2.56	0.1-*F* TBAB	5
Anthracene	–	2.55	0.1-*F* TEAP	3
1,1-Diphenylethylene	5.69	2.20	0.1-*F* TBAB	1
Triphenylethylene	4.84	1.58	0.1-*F* TBAB	1
Triphenylamine	–	1.3	0.1-*F* TEAP	7
		Dimethyl sulfoxide		
Allyl chloride	2.02	0.28	0.1-*F* TEAP	8
Allyl bromide	0.92	0.26	0.1-*F* TEAP	8
Allyl tosylate	1.14	0.09	0.1-*F* TEAP	8
		75% Dioxane–Water		
Naphthalene	–	0.56	0.1-*F* TEAP	3
Anthracene	–	0.52	0.1-*F* TEAP	3
Phenylacetylene	5.10	0.44	0.175-*F* TBAI	1
		50% Dioxane–Water		
Benzaldehyde	2.12	0.36	Phosphate buffer	1
		50% Ethanol		
Nitrobenzene	–	1.0	Buffer, pH 6.2	9
o-Nitrophenol	–	0.82	Buffer, pH 6.2	9
Dibenzoylmethane	1.47	0.43	HCl(pH 1.4)	10
		10% Ethanol		
Acetophenone	3.2	0.70	Buffer, pH 7.1	1
p-Bromoacetophenone	2.9	0.57	Buffer, pH 7.1	1
Nitrobenzene	5.35	0.49	Buffer, pH 3	1

Table 3.2 – *Continued*

		Water		
Benzoquinone	–	1.27	2-*F* KCl	7
Benzoquinone	–	0.86	0.1-*F* KNO_3	7
Bromoacetone	3.3	0.74	Acetate buffer, pH 4.6	1
Iodoacetone	3.3	0.74	Acetate buffer, pH 4.6	1
Chloroacetone	2.9	0.57	Acetate buffer, pH 4.6	1

[a] $I_d = i_d/C^*m \quad t^{1/6}$, where i_d is in microamperes, C^* is in millimoles per liter, m is in milligrams, and t is in seconds.

[b] TBAI = tetrabutylammonium iodide; TEAI = tetraethylammonium iodide; TEAP = tetraethylammonium perchlorate; TBAB = tetrabutylammonium bromide; TBAP = tetrabutylammonium perchlorate.

[c] References:

1. L. Meites, *Polarographic Techniques*, 2nd ed., New York, Wiley, 1965, pp. 671–711. (Reprinted by permission of John Wiley and Sons, Inc.)
2. J. L. Sadler and A. J. Bard, *J. Amer. Chem. Soc.*, **90**, 1979 (1968).
3. T. A. Miller, B. Prater, J. K. Lee, and R. N. Adams, *J. Amer. Chem. Soc.*, **87**, 121 (1965).
4. S. Wawzonek and D. Wearring, *J. Amer. Chem. Soc.*, **81**, 2067 (1959).
5. S. Wawzonek, E. W. Blaha, R. Berkey, and M. E. Runner, *J. Electrochem. Soc.*, **102**, 235 (1955).
6. J. Bacon and R. N. Adams, *Anal. Chem.*, **42**, 524 (1970).
7. R. N. Adams, *Electrochemistry at Solid Electrodes*, New York, Marcel Dekker, 1969, pp. 221–231. (Reprinted by permission of Marcell Dekker, Inc.)
8. J. P. Petrovich and M. M. Baizer, *Electrochim. Acta*, **12**, 1249 (1967).
9. A. C. Testa and W. H. Reinmuth, *J. Amer. Chem. Soc.*, **83**, 784 (1961).
10. D. H. Evans and E. C. Woodbury, *J. Org. Chem.*, **32**, 2158 (1967).

capillaries will give the same values of m and t even when all other conditions are the same, and of course i_d will vary accordingly. The left-hand side of Eq. 3.8, as computed from the values of the variables on the right side, will, however, always be constant for a given substrate (assuming that the solvent and electrolyte are the same). This constant, $708nD^{1/2}$, is called I_d, the *diffusion current constant*, and it permits measurements taken with different capillaries to be placed on a common basis.* Estimation of n from the diffusion current constant requires comparison of the value measured for the unknown substance (via Eq. 3.8) with that in the literature for a molecule of similar size for which n is known. For example, during polarographic reductions in DMF containing tetraethylammonium bromide (TEAB) as supporting electrolyte, it is usually found that I_d lies in the range 2–3 when $n = 1$ and the range 4–6 when $n = 2$ (Table 3.2).[13] As has been noted, this method is equivalent to the use of values of D for model compounds but is often more convenient, since I_d has been reported in the literature for many more compounds than has D. (Indeed, I_d is, and should be, reported routinely in any polarographic investigation of a

* I_d, therefore, plays a role similar to the extinction coefficient in spectrophotometry.

substance not previously studied.) Finally, if model compounds are not available and one is willing to make the approximation that molecules of the electroactive substance are both spherical and much larger than molecules of the solvent, the diffusion coefficient may be estimated from the *Stokes–Einstein equation*,[14]

$$D = \left(\frac{4\pi d}{3}\right)^{1/3} \frac{RT}{6\pi\eta K^{2/3}M^{1/3}} \tag{3.9a}$$

where d is the density of the electroactive substances in grams per cubic centimeter; M is its molecular weight; R and K are the gas constant (in ergs per degree-mole) and Avogadro's number, respectively; T is the absolute temperature; and η is the viscosity of the solvent in dyne-seconds per square centimeter. At 25°C, Eq. 3.9a reduces to

$$D = \frac{2.96 \times 10^{-7}}{\eta}\left(\frac{d}{M}\right)^{1/3} \tag{3.9b}$$

Santhanam, Wheeler, and Bard used the Stokes–Einstein equation successfully in calculating n for the polarographic reduction of tris-(p-nitrophenyl)-phosphate (**7**), which is presumably sufficiently large and symmetrical to justify the assumptions on which the Stokes–Einstein equation is based.[15] Of course, one needs for this calculation the density of the substrate and the viscosity of the solvent

$(p\text{-}NO_2\text{-}C_6H_4O)_3PO$
7

containing electrolyte. Viscosities of typical solvents used in organic electrochemistry are tabulated in Table 3.3. In view of the approximations involved, the Stokes–Einstein equation is not nearly as preferable for obtaining D as the preceding methods based on model compounds, but may occasionally be of use.

3.1.2 Wave-Shape Analysis

When a polarographic wave has been found to be diffusion controlled and the polarographic n value has been measured by the methods of the preceding section, one may then proceed to an analysis of the wave shape (preferably with the wave recorded on an expanded scale for more precise measurement).

Tomes's Criterion

The quantity $E_{1/4} - E_{3/4}$, it will be recalled (Section 2.3.2), is equal to $56.4/n$ mV for a reversible process or $51.7/\alpha n$ mV for a totally irreversible process. Tomes's criterion may be measured readily from the polarogram. Keep in mind from the previous discussion that this datum, while it can establish irreversibility, can only suggest reversibility; in the latter event, corroborative evidence must be sought. Cyclic voltammetry (Section 3.2) is a convenient source of supportive evidence. Application of Tomes's criterion to the polarographic

Table 3.3 Viscosities of solvents commonly employed in organic electrochemistry

Solvent	Viscosity (centipoises, 25°C)	Electrolyte[a]	References[b]
Acetonitrile	0.343	none	1
	0.390	0.1-*F* TEAP	2
Dimethylformamide	0.796	none	1
	0.855	0.1-*F* TEAP	3
	0.860	0.1-*F* TEAB	6
Dimethyl sulfoxide	1.95	none	1
	2.1	0.1-*F* LiCl	4
	5	1.0-*F* LiCl	4
75% Dioxane–Water	1.88	0.1-*F* TEAP	3
Ethanol	1.075	none	1
	1.38	0.25-*F* $LiNO_3$	1
Propylene carbonate	3.6	0.5-*F* TEAP	5
Pyridine	0.880	none	1
Water	0.890	none	1
	0.899	0.10-*m* TMAB	1
	0.897	0.1-*F* HCl	1

[a] TEAP = tetraethylammonium perchlorate; TMAB = tetramethylammonium bromide; TEAB = tetraethylammonium bromide.

[b] References:
1. R. H. Stokes and R. Mills, *Viscosity of Electrolytes and Related Properties*, Elmsford, New York, Pergamon, 1965, pp. 76, 77, and 126. (Reprinted by permission of Pergamon Press.)
2. J. Bacon and R. N. Adams, *Anal Chem.*, **42**, 524 (1970).
3. T. A. Miller, B. Prater, J. K. Lee, and R. N. Adams, *J. Amer. Chem. Soc.*, 87, 121 (1965).
4. J. N. Butler, *J. Electroanal. Chem.*, **14**, 89 (1967).
5. R. N. Adams, *Electrochemistry at Solid Electrodes*, New York, Marcel Dekker, 1969, p. 218. (Reprinted by permission of Marcel Dekker, Inc.)
6. A. J. Fry and W. E. Britton, unpublished result.

reduction of diphenylfulvene (**1**) (reversible) and 2,2-dichloronorbornane (**2**) (irreversible) is illustrated in Figure 3.6. Notice the decreased steepness of the rising portion of the irreversible wave, relative to the reversible wave.*

Log Plots

Although Tomes's criterion is normally a satisfactory test of reversibility and is considerably faster for this purpose than construction of a log plot, the latter may be of interest in very precise work, especially when it is desired to locate $E_{1/2}$ with high accuracy (Section 2.5). The principal caution to keep in mind in construction of the log plot is that one should use values of i and i_d measured at the end of the drop life, rather than average currents. This point arises because, while Eq. 2.40b is valid for reversible processes at either i_{max} or i_{ave}, Eq. 2.54, for irreversible processes, was derived for maximum currents, i_{max}, and

* Failure to compensate for the iR drop through organic solvents of high resistance has the same result (Chapter 9).

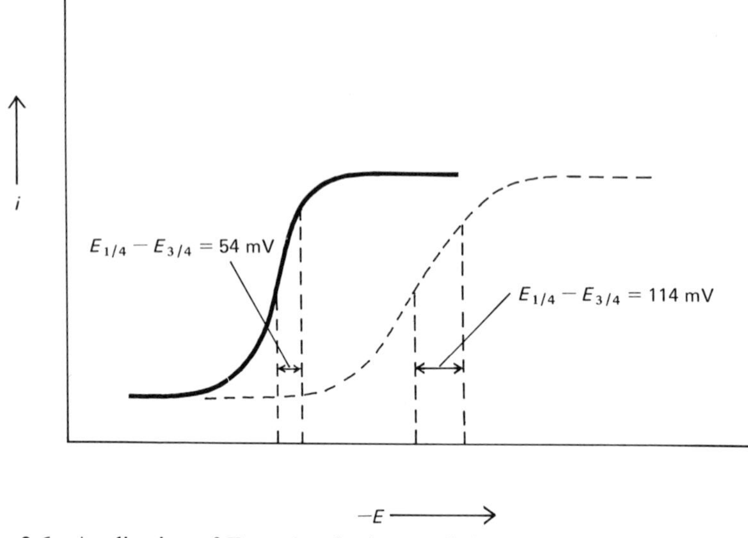

Figure 3.6 Application of Tomes's criterion to diphenylfulvene (**1**) (first wave) (solid line) and 2,2-dichloroborbornane (**2**) (dashed line). Solvent: dimethylformamide.

of course one does not know in advance whether an unknown process is reversible or irreversible.

3.1.3 Experimental Procedures

As will be evident from the discussion up to this point, the usual polarographic study includes a number of steps, consisting of both experiment and calculation. If these are properly planned, all can be carried out in a few hours. For this reason, and because several points of technique may need clarification, it is well at this point to summarize the usual steps in a polarographic investigation, in the approximate order in which they would normally be carried out. [As mentioned at the beginning of the chapter, reasons of experimental efficiency strongly suggest carrying out polarography and cyclic voltammetry (Section 3.2) on the same solution in a cell designed for this purpose (Chapter 9).] Meites's lucid text on polarographic techniques describes these steps in considerable detail.[1] Where more than one wave is observed, all experiments are carried out on each wave.

Experiments

1. *Measurement of the polarogram*, usually at substrate and supporting electrolyte concentrations of ca. 0.001 and 0.1 *M*, respectively. Dissolved oxygen should first be removed from the solution by purging with an inert gas, usually nitrogen, for 15 min. (It may be wise, if nondiffusion-controlled behavior is suspected or polarographic analysis of the electroactive species is contem-

plated, to verify the direct proportionality between diffusion current and substrate concentration implied in the Ilkovic equation by measuring polarograms at several substrate concentrations, in the range 0.0001–0.003 M approximately.)

2. *Measurement of m and t*, the mercury flow rate in milligrams per second and drop time in seconds with the D.M.E. maintained (polarized) at a potential on the limiting current plateau just past the wave. The importance of polarizing the electrode while measuring m and t should be emphasized, since the drop weight mt is related to the mercury–solution interfacial tension, which is in turn potential dependent (Section 2.4).

3. *Recording of each wave on an expanded scale* to facilitate later computations of $E_{1/2}$, $E_{1/4} - E_{3/4}$, and (sometimes) $\log i/(i_d - i)$.

4. *Measurement of the limiting current* over a range of values of the mercury column height such that t stays within the range 2–8 sec. For a polarogram consisting of more than one wave, it is simplest to record the complete polarogram at each height; for a polarogram consisting of one wave, it suffices to record the limiting current plateau at each height. Notice that a multiwave polarogram might consist of a mixture of adsorption, kinetic, catalytic, and diffusion waves, so that it does not suffice to measure the limiting current dependence on column height for just one wave.[15]

5. *Measurement of the residual current* on a solution of the supporting electrolyte in the polarographic solvent.

6. *Other supplementary experiments* will often suggest themselves, for example, the temperature dependence of i_{lim},[16] or the effects on polarographic behavior of additives such as proton donors, surface-active compounds, or trapping agents (Chapter 4).

Computations

1. *The half-wave potential*, $E_{1/2}$, may be determined graphically from the scale-expanded polarogram after suitable correction for the residual current.

2. *A plot of* i_{lim} *vs* $h_{\text{corr}}^{1/2}$ is constructed; if it is a straight line, the reduction is diffusion controlled, and the calculations below, which are all based on the Ilkovic equation, may be carried out. If this plot is not linear, one may plot i *vs* h_{corr} as a test for adsorption control.

3. *The number of electrons n* involved in the overall polarographic wave is computed, using either Eq. 3.7 (with an estimated value of the diffusion coefficient D) or the magnitude of the diffusion current constant, I_d.

4. *Tomes's criterion* is measured from the expanded polarographic wave; if desired, a log plot is also constructed to obtain a more accurate value for $E_{1/2}$. If the wave appears to be irreversible, conversion of $E_{1/2}$ to $E_{1/2}^0$ may be effected by first measuring the drop time at $E_{1/2}$ and then applying Eqs. 2.57 and 2.59.

Sample Calculation: Diphenylfulvene[17] The polarogram of a 0.001-M solution of diphenylfulvene (**1**) in DMF containing 0.1-M TEAB is shown in Figure 3.7a along with the residual current curve. After correction for the

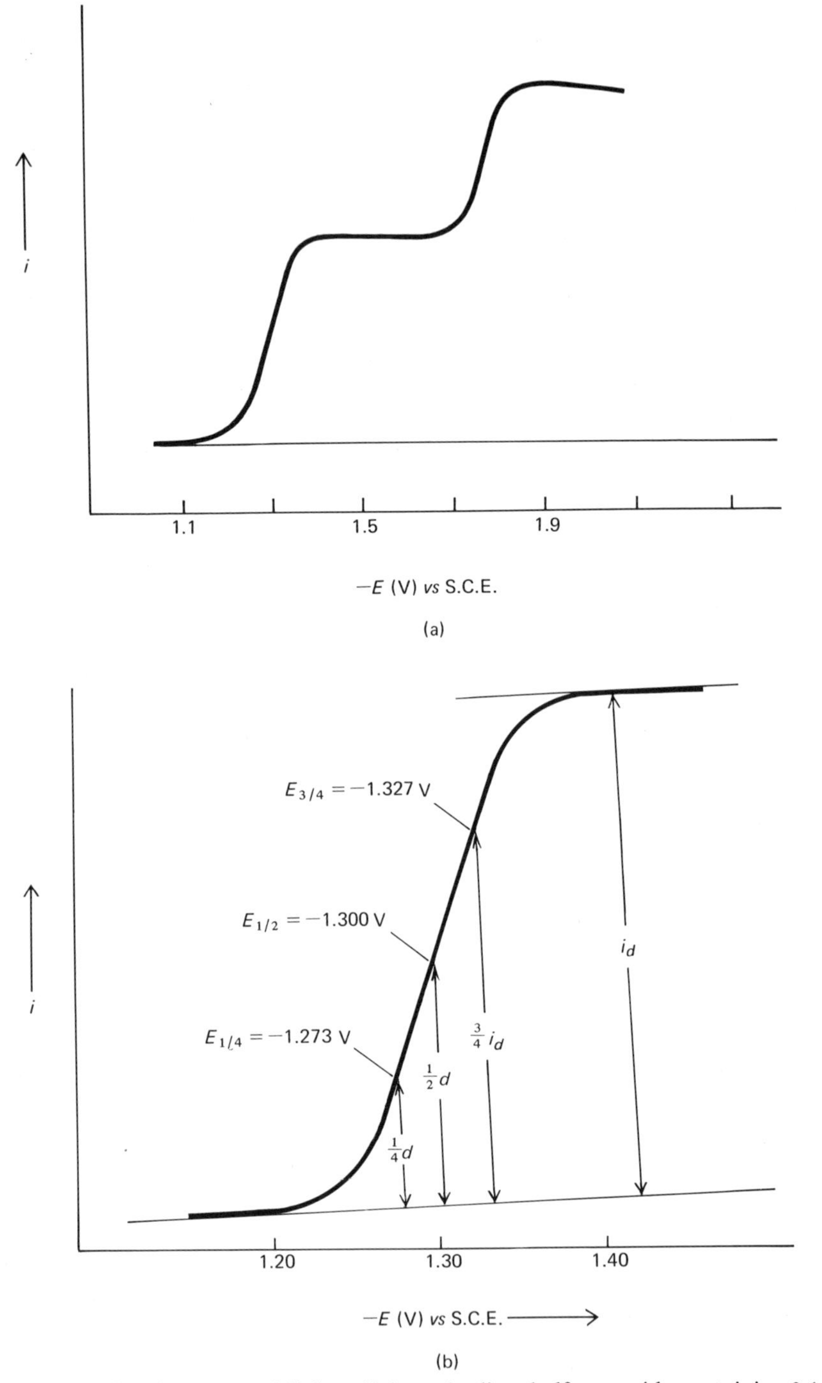

Figure 3.7 (a) Polarogram of diphenylfulvene in dimethylformamide containing 0.1-F tetraethylammonium bromide. (b) First wave of (a), expanded scale.

residual current, $E_{1/2}$ for the first wave was found graphically from Figure 3.7b to lie at -1.30 V *vs* S.C.E. A plot of i_{lim} *vs* $h_{\text{corr}}^{1/2}$ is a straight line; this can be established either graphically (Figure 3.2) or by noting that the quotient $i_{\text{lim}}/h_{\text{corr}}^{1/2}$ is a constant (Table 3.1). Measurement of m and t at -1.4 V (atop the first wave), where $i_d = 3.3\ \mu$A, afforded these values: m, 1.207 mg/sec; t, 3.53 sec. Using anthracene as a model compound, the diffusion coefficient D of diphenylfulvene in DMF is estimated as 1.19×10^{-5} cm^2/sec (Table 3.2). Computation of n from Eq. 3.7 gives a value of 0.99, that is, 1. From Figure 3.7b, $E_{1/4} - E_{3/4} = -1.273-(-1.327)$ V, or 54 mV. This datum and the n value suggest a reversible 1-electron reduction. A similar analysis applied to the second wave indicates that it represents an irreversible 1-electron reduction. (The second wave is smaller than the first because of the decreased polarographic drop times at potentials very negative of the pzc.) These conclusions will now be tested by other electrochemical techniques.

3.2 CYCLIC VOLTAMMETRY

Consider the implications of the polarographic results and interpretation just cited for diphenylfulvene. The compound is reduced in two discrete 1-electron steps; the first step is reversible on a polarographic time scale, and the second step is irreversible. What might one guess about the nature of the chemistry taking place at each polarographic wave? Initial 1-electron reduction of **1** should form the radical anion **8**, and transfer of a second electron to **8** should lead to dianion **9**. That the initial electron transfer should be rapid and reversible is in

$$\mathbf{1} \xrightarrow{e^-} (\ominus)\!-\!\dot{C}(C_6H_5)_2 \xrightarrow{-e^-} (\ominus)\!-\!\bar{C}(C_6H_5)_2$$

8 **9**

accord with both chemical intuition and electrochemical experience: many aromatic hydrocarbons are reversibly reduced to the corresponding radical anions via addition of an electron to the lowest unfilled molecular orbital of the hydrocarbon (*cf.* conversion of **4** to **6**, p. 68, and Chapter 7).[18] In the present case, reduction of **1** has the particular advantage that it leads to a species **8**, which is isoelectronic with, and indeed is a nonbenzenoid analog of, the highly stable triphenylmethyl radical. It would be unsurprising, furthermore, to find that reduction of **8** to **9** should be irreversible, for the dianion bears two negative charges in close proximity and should be highly reactive. Given these tentative proposals concerning the nature of the two 1-electron polarographic waves for **1**, what further evidence can be adduced to support them? In the remainder of this chapter there will be described a number of experimental techniques. Each will be shown to provide a different kind of information; specifically, each will be found to vindicate and strengthen the above deductions concerning formation of **8** and **9** in the polarographic reduction of diphenylfulvene.

Cyclic voltammetry, also called *cyclic triangular-wave voltammetry*, is, like polarography, a relatively simple technique, capable nevertheless of providing, with relatively little experimental effort, a great deal of useful information about electrochemical behavior. Cyclic voltammetry is itself a variant on a technique known as *stationary-electrode voltammetry*.[19] In the latter technique, the potential of a stationary electrode* is varied linearly, as in polarography, and the resulting current is measured on a recorder or oscilloscope. Typical stationary-electrode voltammograms are illustrated in Figure 3.8. Notice that the potential may be swept either anodically or cathodically, and that unlike polarographic waves, the curves are *peaked*. This is a consequence of the use of a stationary electrode: as the potential moves into the region where the substrate is reduced

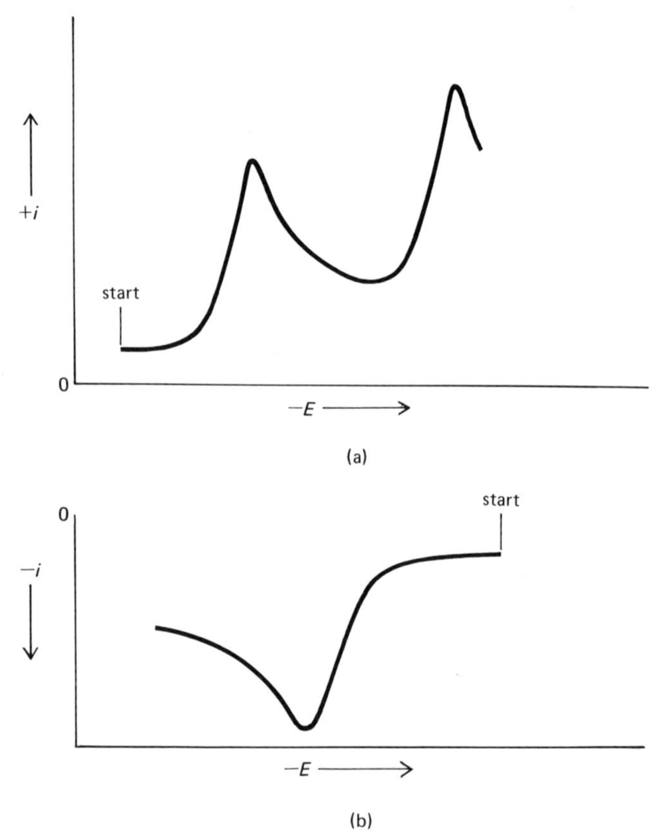

Figure 3.8 Stationary electrode voltammograms. (a) Cathodic sweep (two reduction steps). (b) Anodic sweep (one oxidation step).

* This is often a hanging mercury drop, but may also be a disk or wire composed of platinum, gold, carbon, tin oxide, or other substance, (or mercury coated on one of these). Mercury, of course, cannot be used in electrochemical oxidation studies.

(or oxidized), the region adjacent to the electrode becomes depleted of material, and the current decreases. The $t^{-1/2}$ dependence of the current predicted by the Cottrell equation (Eq. 2.22) is obeyed, in fact, at potentials past the peak (this may be shown by stopping the potential sweep after the peak and then measuring the current–time curve). This peaked behavior is masked at a D.M.E., because (1) the drop is expanding, (2) the drop continually falls off, making fresh solution available for the next drop, and (3) sweep rates are low in polarography. In 1948, Randles and Sevcik independently derived an equation relating the peak current, i_p, to the experimental parameters of the stationary-electrode polarographic experiment[20]:

$$i_p = 2.687 \times 10^5 n^{3/2} A D^{1/2} C v^{1/2} \tag{3.10}$$

where n, A, D, and C have their usual significance (Chapter 2). Observe, however, that as well as confirming the expected dependence of i_p on these parameters, Eq. 3.10, the *Randles–Sevcik equation*, also predicts that the peak current should be a function of the *potential sweep rate*, v (in volts per second). The Randles–Sevcik equation is based on the assumptions of a reversible process and linear diffusion (planar electrode), and when these conditions are fulfilled the ratio $i_p/v^{1/2}$ is indeed found to be a constant as predicted by Eq. 3.10.

The experimental feature of interest in cyclic voltammetry is this: after the stationary electrode polarogram of a substance has been measured by sweeping past the peak potential, E_p, *the direction of voltage scan is reversed.* This is done by causing the potential of the working electrode to vary as a cyclic triangular wave (Figure 3.9), hence the name of the technique. Either a single cycle (from

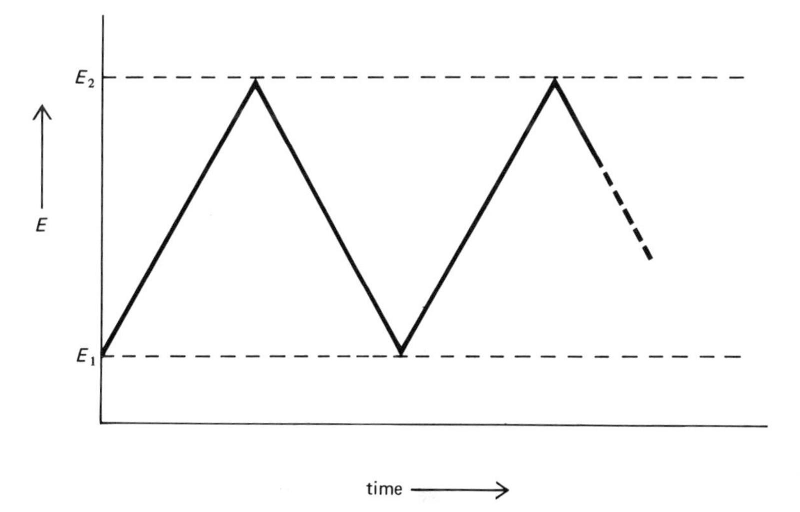

Figure 3.9 Triangular wave function.

E_1 to E_2 and back to E_1) or many cycles may be carried out. Typical *cyclic voltammograms* are illustrated in Figure 3.10. It can be seen from the figure that cyclic voltammetric behavior can exhibit a variety of forms. The shape of the cyclic voltammogram is highly dependent on the relative rates of electron transfer, mass transport,* and any chemical reactions occurring at the electrode surface; this enables one to deduce a great deal about the course of an electrode process from the cyclic voltammogram. Scan rates can be varied over a wide range (ca. 0.01–10,000 V/sec), providing, as will be seen, an extremely useful experimental parameter.**

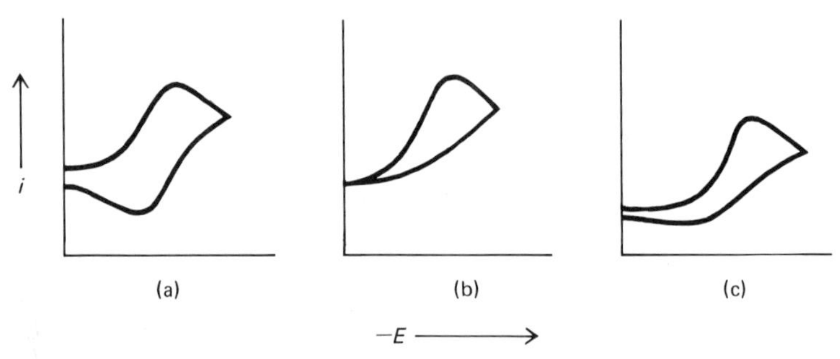

Figure 3.10 Representative cyclic voltammograms. (a) Reversible process. (b) Irreversible process. (c) EC process. All scans start in the cathodic direction.

Cyclic voltammetry serves nicely as an experimental test of reversibility. Consider, for example, the reversible reduction of a species O to R:

$$O + ne^- \rightleftharpoons R$$

(e.g., O and R might be **1** and **8**, respectively). The cyclic voltammogram for this process is illustrated in Figure 3.10a and expanded in Figure 3.11. Reduction of O to R takes place at ca. −1.3 V†; when the voltage scan direction is reversed at −1.6 V, an *anodic* peak due to oxidation of R back to O is observed, again at ca. −1.3 V. If E_{pa} and E_{pc} are the potentials corresponding to the anodic and cathodic peak currents, respectively, and $E_{pa/2}$ and $E_{pc/2}$ are the potentials corresponding to the half-peak currents, it may be shown that $E_{pa} - E_{pc}$ and $E_{pc/2} - E_{pa/2}$ will each be equal to $0.056/n$ V for a reversible process. Furthermore, within experimental error,

$$E_{pc/2} = E_{pa} = E^0 + 0.028/n \text{ V}$$

* Cyclic voltammetry is carried out in quiet solution, hence diffusion constitutes the sole means of mass transport.

** At very low scan rates, natural convection currents are set up in solution; the upper limit is determined by the fact that charging currents become excessive at high scan rates (the faradaic and charging currents increase with the one-half and first power of the scan rate, respectively).

† Observe that for a reversible process the current at E^0 is approximately $0.85 i_p$.

and

$$E_{pa/2} = E_{pc} = E^0 - 0.028/n \text{ V}$$

These relationships are all illustrated in Figure 3.11. (The rationale for using half-peak potentials is that $E_{p/2}$ is often easier to measure experimentally than E_p, because of the broadness of the peak.) Cyclic voltammetric criteria of

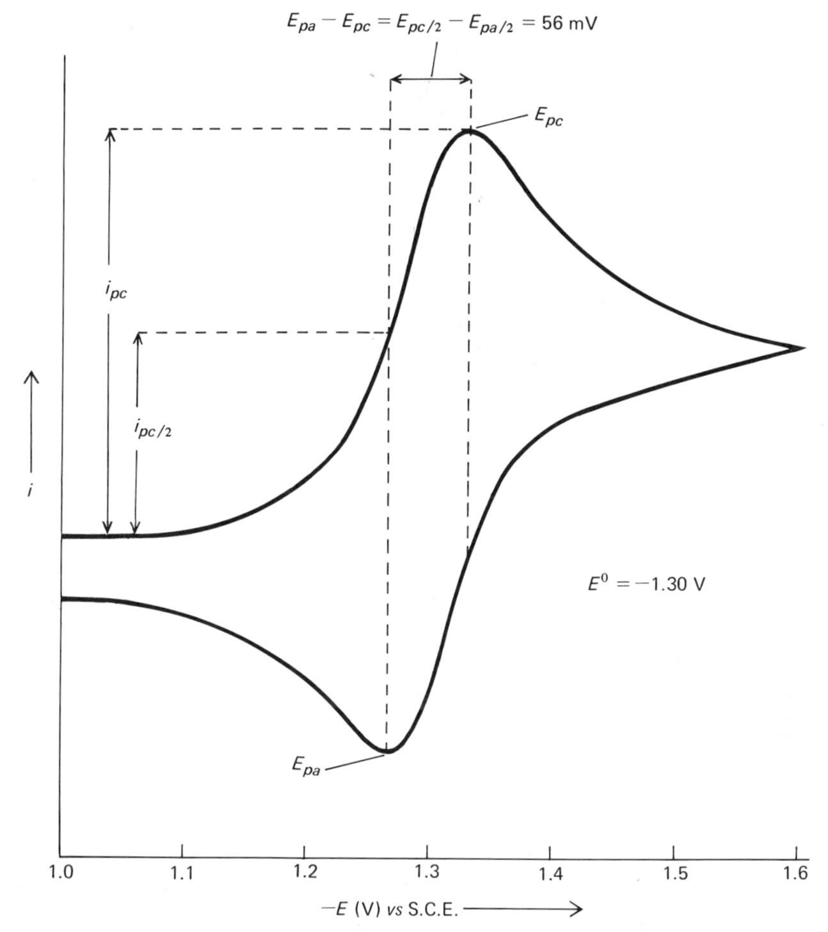

Figure 3.11 Cyclic voltammogram of a reversible 1-electron process, showing important features.

reversibility, then, include the correct spacing for peak and/or half-peak potentials, and, of course, the very fact that a reoxidation peak is actually observed for R. (Other criteria will be introduced shortly.) It will be apparent that a cyclic voltammogram resembling Figure 3.11 and satisfying these criteria is much better proof of electrochemical reversibility than either Tomes's criterion or a log plot.

Suppose that the electrode process of interest is totally irreversible:

$$\mathrm{O} + ne^- \longrightarrow \mathrm{R}$$

The cyclic voltammogram for this process is illustrated in Figure 3.10b. Note the complete absence of any measurable anodic current upon scan reversal past the peak potential. This is consistent with a very low rate of reoxidation of R and is conclusive experimental evidence of irreversibility.

A third frequently encountered type of behavior is shown in Figure 3.10c. An anodic peak is observed after scan reversal past the cathodic peak, but this peak is diminished considerably in height relative to the cathodic peak. In such cases it is often observed, furthermore, that at faster scan rates the relative height of the anodic peak increases until it is as large as the cathodic peak; at very slow scan rates the anodic peak may completely disappear. In other words, the cyclic voltammogram may be made to resemble either Figure 3.10a or 3.10b at very fast or very slow sweep rates, respectively. This behavior is usually associated with a chemical reaction subsequent to electron transfer*:

$$\mathrm{O} + ne^- \rightleftharpoons \mathrm{R}$$

$$\mathrm{R} \xrightarrow{k_1} \text{products}$$

such that a fraction of the R reacts chemically and is not available for reoxidation upon scan reversal. If the scan rate is very high relative to k_1, on the other hand, very little R will be lost to the succeeding chemical reaction, and the voltammograms will resemble the reversible case; conversely, if the scan rate is low relative to k_1, the chemical reaction will be essentially over before the voltage scan is reversed, and the electrode process will appear totally irreversible. Because k_1 might vary over a wide range, it is important to measure the voltammogram over as wide a range of scan rates as possible when a voltammogram is of the form shown in Figures 3.10a (or 3.11) and 3.10b. Otherwise, the existence of the coupled chemical reaction might not be discovered. Of course, if k_1 is very high, reversible oxidation of R may not be observed even at the highest scan rates used; if k_1 is very small, voltammograms will appear reversible at all scan rates. Examination of the cyclic voltammetric behavior of a substance

* The chemical reaction is said to be coupled to the electron transfer. According to a nomenclature system commonly used for multistep electrode reactions,[21] this is called an EC process, indicating that an initial electron transfer is followed by a chemical reaction. The reduction of the conjugate acid of a weak base (p. 66) would be a CE process in this system of nomenclature.

is of considerably greater value than simply for characterization of the reversibility of electron transfers, however. It is often possible to observe the formation and decay of reactive intermediates, and even to identify such intermediates and/or products. A few examples will serve to illustrate the point. A cyclic voltammogram for the electrochemical oxidation of aniline (**10**), reported by Bacon and Adams in 1968, is shown in Figure 3.12.[22] A well-developed oxidation

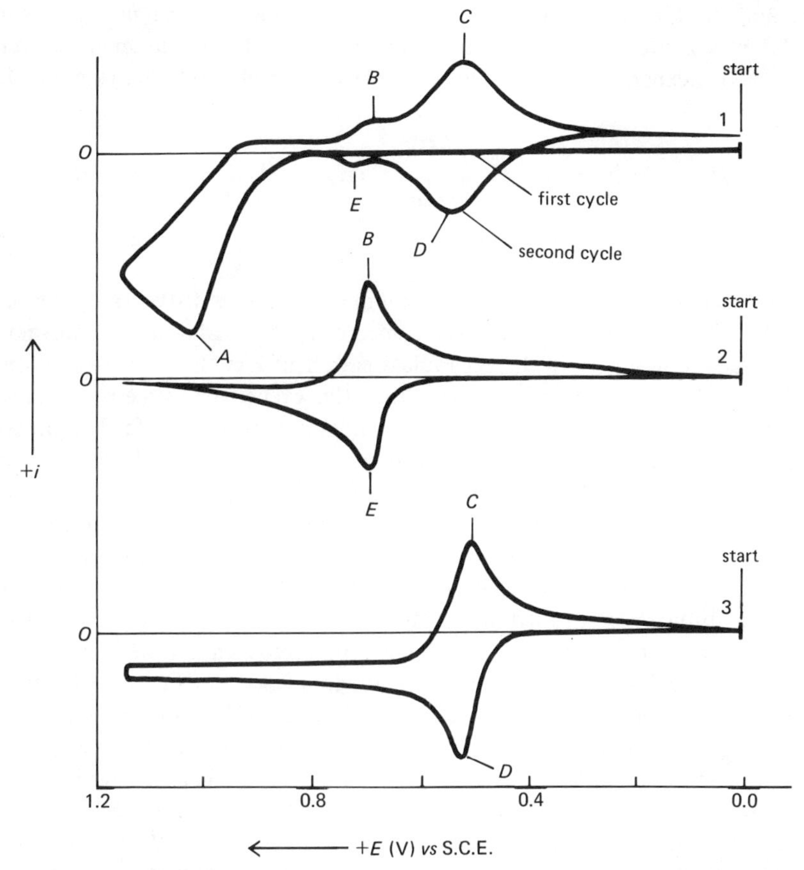

Figure 3.12 Cyclic voltammograms of aniline and its oxidation products (aqueous medium, pH 2.3). Curve 1, aniline (**10**); curve 2, benzidine (**11**), curve 3, 4-aminodiphenylamine (**12**). All scans start anodically from 0.0 V. (J. Bacon and R. N. Adams, *J. Amer Chem. Soc.*, **90**, 6598 [1968].)

peak (*A*) is observed at ca. +1.0 V (*vs* S.C.E.), but the absence of a corresponding cathodic peak at +1.0 V upon scan reversal indicates that the oxidation is irreversible. When the cathodic scan is continued past +1.0 V back toward the starting potential, however, a very interesting phenomenon is observed. Two new cathodic waves (*B* and *C*) are observed, and *on the second anodic cycle* two peaks (*D* and *E*) are observed. Peaks *B*, *C*, *D*, and *E* must be due to species that were formed *subsequent* to the oxidation of aniline, since these peaks do not appear until after the potential sweep has reached point *A* on the first cycle. It is also evident that two reversible couples are formed in the oxidation of aniline at *A*; these couples are represented by the pair of peaks at *B* and *E*, and the pair at *C* and *D*. Identification of the species responsible for the new reversible couples was made in the following manner. It was found in separate cyclic voltammetric experiments that benzidine (**11**) and *p*-aminodiphenylamine (**12**)

$C_6H_5NH_2$ $H_2N-C_6H_4-C_6H_4-NH_2$ $H_2N-C_6H_4-NHC_6H_5$

10   **11**   **12**

exhibit reversible redox couples (i.e., similar to Figure 3.10a), at potentials corresponding to peaks *B–E* and *C–D*, respectively. An oxidative mechanism for converting aniline to **11** or **12** via radical cation intermediates can be written (Chapter 8). The procedure of comparing the cyclic voltammetric behavior of known compounds suspected to be intermediates or products is fairly standard. Peak matching, of course, although of great value, provides only presumptive evidence of identity and requires further support where possible.

Michielli and Elving employed "wave clipping" and addition of a proton donor (phenol) to obtain information on intermediates in the electrochemical reduction of benzophenone (**13**) in pyridine.[23] A polarogram showed two 1-electron diffusion-controlled waves at -1.80 and -2.03 V (relative to the $Ag/AgNO_3$ reference electrode). The cyclic voltammogram shows in succession two cathodic peaks, Ic and IIc, and three anodic peaks, Ia, IIa, and IIIa, as the potential is scanned from 0 V to -2.3 V and back to 0 V (Figure 3.13a). On the second cycle, a new cathodic peak, IIIc, is observed. Wave clipping, that is, reversal of the scan direction at a potential (ca. -1.9 V) between peaks Ic and IIc, causes peaks IIa and IIIa to disappear; peak Ia remains (Figure 3.13b). This indicates very nicely that peaks IIa and IIIa are associated with oxidation of a substance or substances formed in the *second* reduction step (peak IIc), and also that peak Ia is due to reoxidation of the species formed in the first reduction step (Ic). Wave clipping between peaks IIa and IIIa does not cause peak IIIc to disappear, showing that peak IIIc is due to reduction of a species formed in peak IIa. Peak IIc, resulting from 2-electron reduction of benzophenone (*cf.* the polarographic results), must lead to a very reactive species, since it is totally irreversible.

Addition of increasing amounts of phenol to the solution enhanced peak Ic and diminished peaks IIc and Ia; at the same time the two 1-electron waves merged into one 2-electron wave at -1.76 V. These results are consistent with

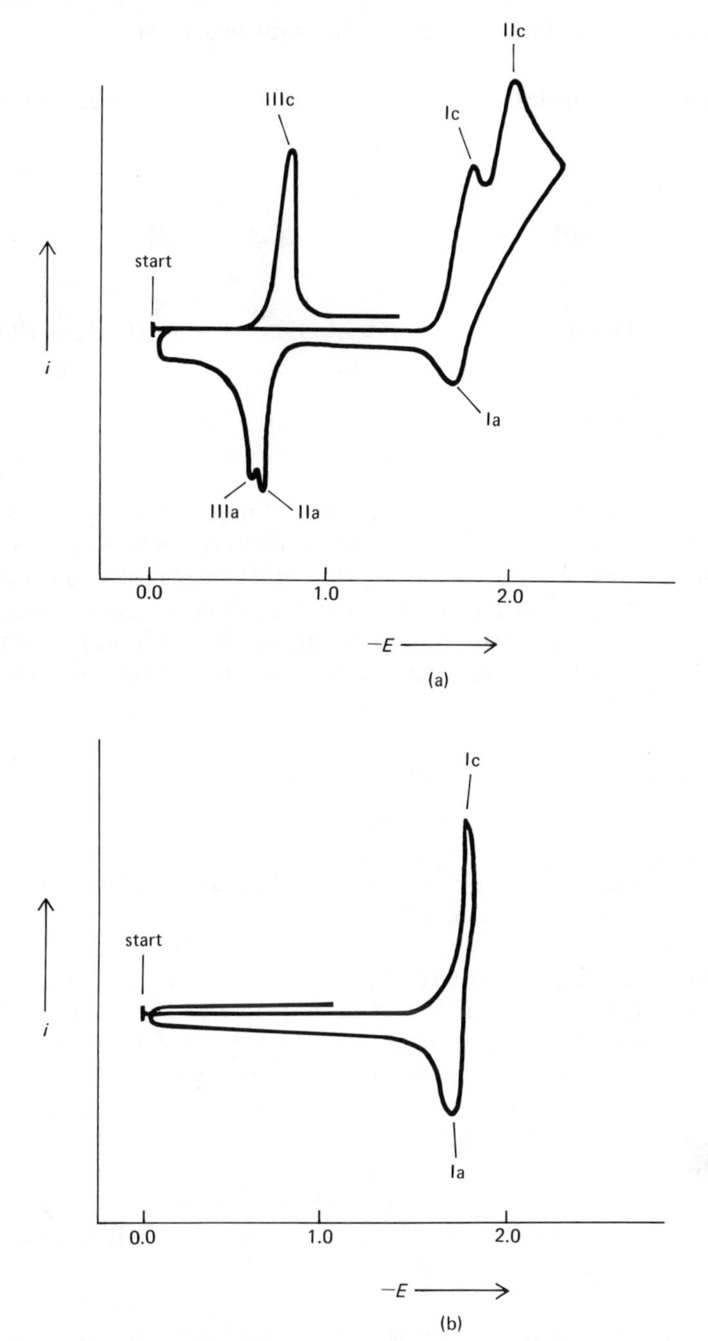

Figure 3.13 Cyclic voltammograms of benzophenone in pyridine. (a) Scan reversed after the second cathodic peak. (b) Scan reversed (clipped) after the first cathodic peak. Potential scans start cathodically from 0.0 V. (R. F. Michielli and P. J. Elving, *J. Amer. Chem. Soc.*, **90**, 1991 [1968].)

assignment of the couple Ic–Ia to reversible formation of benzophenone ketyl (**14**):

$$\underset{\mathbf{13}}{(C_6H_5)_2CO} + e^- \longrightarrow \underset{\mathbf{14}}{(C_6H_5)_2CO^{\dot{-}}}$$

$$\mathbf{14} + C_6H_5OH \longrightarrow C_6H_5O^- + \underset{\mathbf{15}}{(C_6H_5)_2\dot{C}OH}$$

$$\mathbf{14} + e^- \longrightarrow \underset{\mathbf{16}}{(C_6H_5)_2CO^{-2}} \xrightarrow{H^+} \underset{\mathbf{17}}{(C_6H_5)_2CHO^-}$$

$$\underset{\mathbf{17}}{(C_6H_5)_2CHO^-} \xrightarrow{H^+} \underset{\mathbf{18}}{(C_6H_5)_2CHOH}$$

In the presence of phenol, **14** is protonated, and the resulting radical **15** is then reduced further (Chapter 4).* Peak IIc must be associated with reduction of **14** to the highly reactive dianion **16**. Peaks IIa and IIIa may be due to oxidation of species such as **16**, **17**, or a pyridine-containing substance, any of which would be consistent with the fact that these waves are not formed if the potential is not scanned past peak IIc. It was suggested that peak IIIc is associated with formation of a pyridyl alcohol.

The Nicholson–Shain Criteria

In a now classic paper, Nicholson and Shain reported in 1964 the solution (by computer-assisted numerical methods) of the complex differential equations relating the current to rates of diffusion, coupled chemical reaction, and potential scan.[25] A number of common electrode processes, including coupled chemical reactions of several varieties, were investigated. A great deal of data was presented concerning the effects of coupled chemical reactions on the shapes of cyclic voltammograms, and methods of obtaining the rate constants for the chemical reactions from cyclic voltammetric data were suggested. For the present discussion, however, the special value of the Nicholson–Shain paper lies in the fact that it predicts qualitatively the way in which a number of characteristics of the cyclic voltammogram should depend on scan rate for a given electrode mechanism. One may, therefore, measure changes in these characteristics as a function of scan rate and hence deduce a great deal about the nature of the process occurring at the electrode. Using the notation of Nicholson and Shain, the various mechanisms considered by them include:

* A question is often raised concerning the nature of the proton source in aprotic solvents not containing additives such as phenol. Even after most drying treatments, water concentrations may be 10 m*M* or higher, as determined by Karl Fischer titration or gas chromatography.[24] Hoffmann eliminations on the tetraalkylammonium salts usually used as electrolyte are also generally fast in dipolar aprotic solvents.

Case I. Reversible Charge Transfer

$$\mathrm{O} + n\mathrm{e}^- \rightleftharpoons \mathrm{R}$$

Case II. Irreversible Charge Transfer

$$\mathrm{O} + n\mathrm{e}^- \xrightarrow{k} \mathrm{R}$$

Case III. Chemical Reaction Preceding a Reversible Charge Transfer

$$\mathrm{Z} \underset{k_b}{\overset{k_f}{\rightleftharpoons}} \mathrm{O}$$

$$\mathrm{O} + n\mathrm{e} \rightleftharpoons \mathrm{R}$$

Case IV. Chemical Reaction Preceding an Irreversible Charge Transfer

$$Z \underset{k_b}{\overset{k_f}{\rightleftharpoons}} \mathrm{O}$$

$$\mathrm{O} + n\mathrm{e}^- \xrightarrow{k} \mathrm{R}$$

Case V. Reversible Charge Transfer Followed by a Reversible Chemical Reaction

$$\mathrm{O} + n\mathrm{e}^- \rightleftharpoons \mathrm{R}$$

$$\mathrm{R} \underset{k_b}{\overset{k_f}{\rightleftharpoons}} \mathrm{Z}$$

Case VI. Reversible Charge Transfer Followed by an Irreversible Chemical Reaction

$$\mathrm{O} + n\mathrm{e}^- \rightleftharpoons \mathrm{R}$$

$$\mathrm{R} \xrightarrow{k_f} \mathrm{Z}$$

Case VII. Catalytic Reaction with Reversible Charge Transfer

$$\mathrm{O} + n\mathrm{e}^- \rightleftharpoons \mathrm{R}$$

$$\mathrm{R} + \mathrm{Z} \xrightarrow{k_f'} \mathrm{O}$$

Case VIII. Catalytic Reaction with Irreversible Charge Transfer

$$\mathrm{O} + n\mathrm{e}^- \xrightarrow{k} \mathrm{R}$$

$$\mathrm{R} + \mathrm{Z} \xrightarrow{k_f'} \mathrm{O}$$

Later treatments examined the effect of adsorption on cyclic voltammetric characteristics[26] and analyzed several variants* on the following scheme, known

* The cases examined included all four permutations of $\mathrm{A} \rightarrow \mathrm{B}$ or $\mathrm{C} \rightarrow \mathrm{D}$ as either reversible or irreversible; B was assumed to proceed irreversibly to C in all cases.

as the ECE process (see p. 84), which is very common in organic electrochemistry.[27]

$$A + n_1e^- \longrightarrow B$$
$$B \xrightarrow{k_f} C$$
$$C + n_2e^- \longrightarrow D$$

As indicated above, the response of several parameters of the cyclic voltammogram to changes in potential scan rate was shown by Nicholson and Shain to be useful in deciding to which of the above cases (if any) a given process belongs. These parameters, which are to be measured at each of a number of scan rates (over as wide a range of scan rates as possible), include (1) the ratio of anodic to cathodic peak* currents, i_{pa}/i_{pc}; (2) the cathodic *current function*, $i_{pc}/v^{1/2}$, where v is the scan rate in millivolts per second or volts per second; and (3) the change in cathodic half-peak potential per ten-fold change in scan rate, $\Delta E_{pc/2}/\Delta \log v$. The variation of each of these three parameters with scan rate for the eight cases treated by Nicholson and Shain is illustrated in Figures 3.14–3.16. The numbers on the abscissa of each figure have no absolute significance in the present context; they are used merely to serve as reminders that the

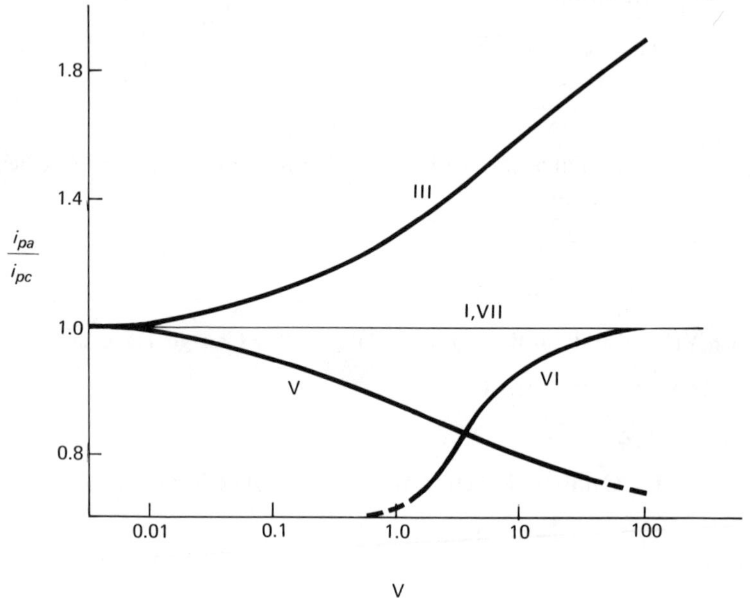

Figure 3.14 Plots of the ratio of anodic to cathodic peak currents, i_{pa}/i_{pc}, as a function of scan rate for various electrode processes. (R. S. Nicholson and I. Shain, *Anal. Chem.*, **36**, 722 [1964].)

* The peak current on the reverse sweep (i_{pa} here) cannot be measured directly from the cyclic voltammogram. Adams has described several procedures for evaluation of this quantity.[28]

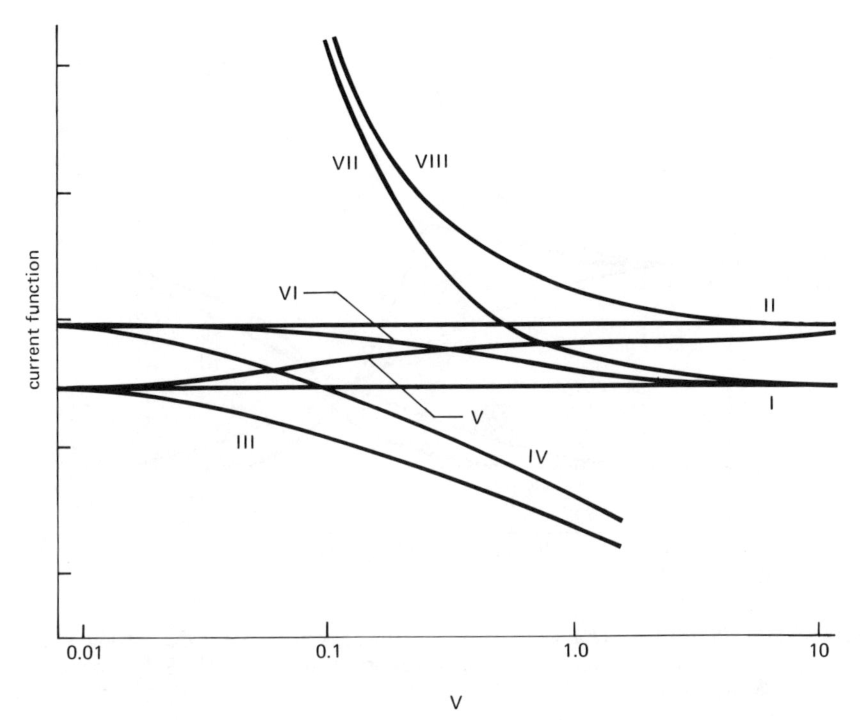

Figure 3.15 Plots of the cathodic current function, $i_{pc}/V^{1/2}$, as a function of scan rate for various electrode processes. (R. S. Nicholson and I. Shain, *Anal. Chem.*, **36**, 722 [1964].)

scan-rate scale is logarithmic. The behavior expected for a given case may be found by inspection of the figures. Since considerable detail concerning these criteria has been presented by Nicholson and Shain (and Adams),[28] discussion of the specific cases will be minimal here. It may, though, be worth pointing out a few salient features of the predictions implicit in Figures 3.14–3.16. These criteria predict, for example, that for a reversible process the current function and half-peak potential should be independent of scan rate.* The peak current ratio, i_{pa}/i_{pc}, should be equal to unity and also independent of scan rate. These predictions have been verified for reversible processes by a number of investigators.[23,29] Recalling the previous discussion of a chemical reaction following reversible charge transfer (Figure 3.10c), which may now be recognized as Case VI, it can be seen by inspection of Figures 3.14–3.16 that, as previously predicted, this mechanism resembles a reversible process (Case I) at high scan rates and an irreversible process (Case II) at low scan rates. Since no anodic

* Furthermore, the anodic and cathodic peak potentials and half-peak potentials should be separated by $56/n$ mV (p. 82).

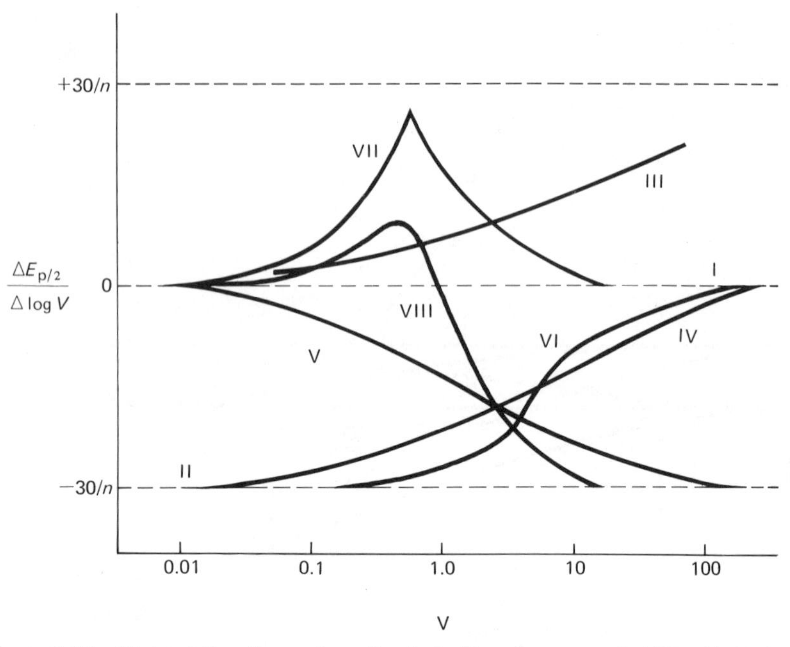

Figure 3.16 Plots of the change in cathodic half-peak potential, $\Delta E_{pc/2}/\Delta \log V$, as a function of scan rate for various electrode processes. (R. S. Nicholson and I. Shain, *Anal. Chem.*, **36**, 722 [1964].)

current is associated with irreversible reduction, Cases II, IV, and VIII are not pictured in Figure 3.14. For an irreversible process with no coupled chemical reactions (Case II), it may also be shown that the half-peak potential $E_{pc/2}$ should shift cathodically approximately $30/\alpha n$ mV for each ten-fold increase in the potential scan rate (Figure 3.16). This constitutes a method of determining αn; Nicholson and Shain, however, proposed a better method for obtaining αn as well as $k_{s,h}$ from the cyclic voltammetric data when E^0 is known or can be estimated.[24]

A few cautions must be stated with respect to the Nicholson–Shain criteria. The most obvious is, of course, the fact that there are quite a few electrode processes possible apart from those examined by Nicholson and Shain. However, the cases selected by them include some of the most common electrode processes, and because many of the more complex mechanisms amount to combinations of two or more of these simpler cases, the criteria will often be helpful in such situations. It should also be emphasized that the criteria were derived with the assumption of linear diffusion, that is, a planar electrode.* It is, on the other hand, often much more convenient to use a hanging mercury drop electrode

* Spherical diffusion correction factors were, however, supplied by Nicholson and Shain for Cases I and II.

(HMDE) (Chapter 9), because of the clean, reproducible nature of the mercury surface, the high overvoltage of hydrogen on mercury, and the fact that potentials measured at this electrode are easily related to polarographic data. Because of the sphericity of the HMDE, one might expect deviations from the predictions of Figures 3.14–3.16 at low scan rates. To estimate the magnitude of such deviations, Nicholson studied the effects of spherical diffusion on the cyclic voltammetric criteria for Case VI.[30] He found that only one of the three criteria, the behavior of the current function, $i_p/v^{1/2}$, with changes in scan rate (Figure 3.15), breaks down badly when a spherical electrode is used. Half-peak potentials (Figure 3.16) were found to be insensitive to sphericity, and the ratio of anodic to cathodic peak currents, i_{pa}/i_{pc}, behaves qualitatively similarly to the predictions of Figure 3.14. These findings are probably representative of the trends to be expected for coupled chemical reactions in general: this suggests that Figures 3.14 and 3.16 will be of most use in qualitative mechanistic analysis at the HMDE. Deviations from the Nicholson–Shain predictions are of more direct concern to those attempting to extract quantitative information, that is, rate constants of coupled chemical reactions, from cyclic voltammetric data. Nicholson found, for example, that for Case VI rate constants calculated from theory for a planar electrode, using measurements taken at the HMDE, will be smaller than the true rate constants; however, he also showed how corrections for sphericity could be introduced into the rate-constant calculations.[30] The spherical diffusion error is not encountered, of course, in studies of electrochemical oxidations of organic compounds, which may be carried out at shielded planar solid electrodes. Incidentally, notice that although the Nicholson–Shain criteria were computed for cathodic processes, application to oxidations is straightforward; for example, the ordinate of Figure 3.14 would become i_{pc}/i_{pa}, and the current function (Figure 3.15) would be $i_{pa}/v^{1/2}$.

Example: Diphenylfulvene

On pp. 78 and 79 there were presented polarographic data and conclusions, respectively, concerning the electrochemical reduction of diphenylfulvene (**1**) in DMF. The cyclic voltammetric behavior of **1** is illustrated in Figures 3.17 and 3.18. In the first voltammogram, the scan was reversed (clipped) at -1.4 V, and an anodic wave for oxidation of radical anion **8** back to the fulvene is clearly visible. The presumption of reversibility previously made from polarographic data is vindicated completely, furthermore, by application of the Nicholson–Shain criteria to Figure 3.17 (Table 3.4).[17] Notice, however, Figure 3.18, in which the voltage scan is not reversed until after the second reduction process, corresponding to the second polarographic wave of the fulvene. The anodic waves corresponding to the first and second 1-electron reduction steps are considerably diminished relative to the cathodic peak currents. This is in accord with the irreversibility previously noted for the

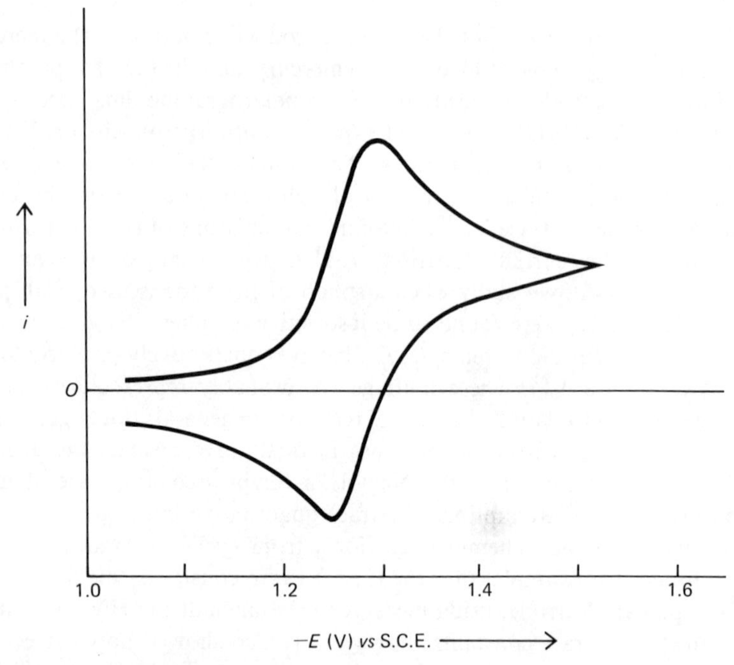

Figure 3.17 Cyclic voltammogram of 6,6-diphenylfulvene (**1**) in dimethylformamide, with scan reversal following the first peak.

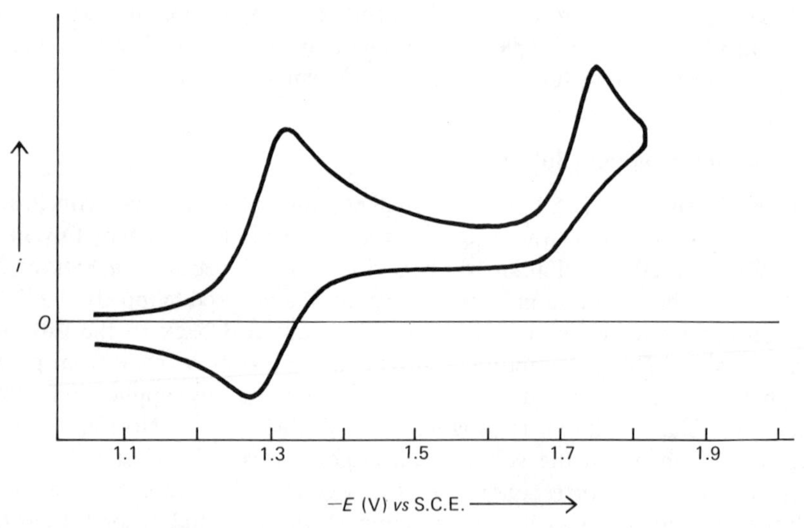

Figure 3.18 Cyclic voltammogram of 6,6-diphenylfulvene (**1**) in dimethylformamide, with scan reversal following the second peak.

Table 3.4 Cyclic voltammetric behavior of diphenylfulvene[a]

ν (mV/sec)	E_{pc} (V)[b]	E^{0} (V)[b, c]	$E_{pc} - E_{pa}$ (mV)	i_{pc}/i_{pa}	$i_{pc}/\nu^{1/2}$
20	−1.33	−1.30	56	1.04	0.59
50	−1.33	−1.30	59	1.03	0.60
100	−1.32	−1.30	58	0.99	0.58
200	−1.32	−1.30	56	1.00	0.57

Source: A. J. Fry and W. E. Britton, unpublished results.

[a] First voltammetric wave in dimethylformamide containing 0.1-F tetraethylammonium bromide.

[b] *vs* S.C.E.

[c] Potential at which $i = 0.85\ i_{pc}$.

second polarographic wave and with chemical intuition, which predicts that dianion **9** should be highly reactive. [This behavior is very similar to that already noted for benzophenone in pyridine (p. 86)[23] and is, in fact, representative of the behavior of many aromatic hydrocarbons and ketones in aprotic media.] One might suppose that a reasonable pathway for disappearance of **9** is reaction with the supporting electrolyte, tetraethylammonium bromide. Further evidence supporting this conclusion will be adduced in the next section. At this point, it is enough to point out that the cyclic voltammetric data fully support and extend the mechanistic postulates originally made from polarographic data.

3.3 CONTROLLED-POTENTIAL ELECTROLYSIS

With information about the behavior of a new system at hand as a result of polarographic and cyclic voltammetric experiments, it is possible to proceed to what is for the synthetic chemist the most interesting aspect of electrode behavior. This consists of electrochemical synthesis of a sufficiently large quantity of the product of the electrode reaction to allow its isolation and identification by the usual chemical and spectroscopic methods of organic chemistry, as a prelude to later large-scale synthesis. The methods used for preparative-scale electrolysis fall into two categories: the *controlled-current* and *controlled-potential* methods. The former was used almost exclusively until relatively recently because of its experimental simplicity.[31] However, the advantages of controlled-potential electrolysis over controlled-current methods have been so well recognized in the last two decades, and the apparatus for the former has been so simplified, that controlled-potential electrolysis is now almost universally employed. This is particularly true when one is examining the mechanism of a multistep electrode process. As will be seen in the next chapter, controlled-current electrolysis has occasional applications for simple electrochemical preparations, but the disadvantages of this technique will be apparent when it is discussed there.

Controlled-potential electrolysis, as its name implies, consists simply of an experiment in which an electrochemical reaction is allowed to proceed to completion at an electrode whose potential is held constant. Consider again the polarogram of diphenylfulvene (Figure 3.19). If the potential of a large mercury cathode in contact with a solution of diphenylfulvene is maintained at potential *A* and the system is allowed to come to electrochemical equilibrium, the diphenylfulvene will be converted cleanly to radical anion **8**, since only the first electron transfer can take place at this potential (hence the term *controlled-potential electrolysis*). If, on the other hand, the cathode is maintained at potential *B*,

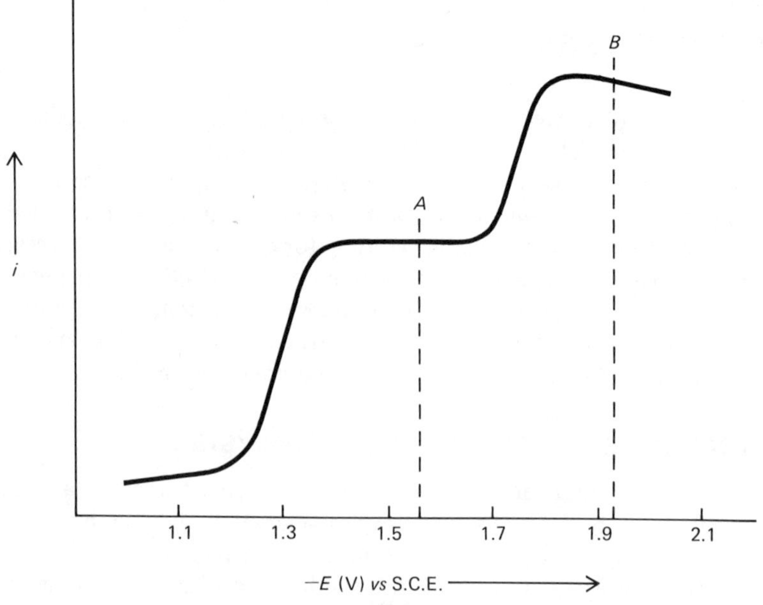

Figure 3.19 Polarogram of 6,6-diphenylfulvene (**1**) in dimethylformamide.

complete 2-electron reduction to **9** will occur, followed of course by any subsequent reactions of **9**. It remains only to examine the solution after electrolyses at potentials *A* and *B* for evidence that the products of the individual electron-transfer steps indeed have the structures **8** and **9** (or a product thereof), respectively, already postulated for them from other evidence. The experimental results are as follows.[17] Electrolysis of **1** at potential *A* affords a stable, deep green solution. Electrolysis at potential *B*, either of this green solution or of **1**, affords a yellow solution and is accompanied by gas evolution (ethylene). An attempt to measure the nmr spectrum of the green solution is quite instructive: one observes only very broad unstructured absorption in the aromatic region, as well as considerably broadened solvent absorption (*cf.* Figure 3.27). This phenomenon is characteristic of paramagnetic species (pp. 113–114). The yellow color of the solution formed through complete reduction of **1** is quenched by addition of water, indicating the presence of a strongly basic species. Finally, measurement of the total amount of electricity consumed in electrolysis (Section 3.4) indicates that, in agreement with the polarographic *n* values, 1 electron/molecule of **1** is required to generate the green paramagnetic species, while 2 electrons/molecule of **1** are required to produce the yellow solution. All of these results are in complete accord with the deductions made from polarographic and cyclic voltammetric experiments on diphenylfulvene (pp. 79 and 93), respectively: the green substance is, reasonably enough, the stable radical anion **8**, and reduction of **8** to **9** must be followed rapidly by Hoffmann elimination on the supporting electrolyte to produce the yellow basic cyclopentadienide **19** and ethylene, previously inferred (p. 95).

$$\mathbf{9} + Et_4N^+ \longrightarrow [C_5H_4^{\ominus}]\text{—}CH(C_6H_5)_2 + CH_2{=}CH_2 + Et_3N$$

19

Notice the outstanding feature of controlled-potential electrolysis, the fact that the potential of the working electrode may be adjusted to any desired value. It is common organic chemical practice to arrange a set of agents, such as metal hydrides, according to their relative reducing power toward various functional groups, in order to decide which to use in a specific instance. Controlled-potential electrolysis allows one to use a single agent (the electrode) whose potency is continuously variable over a very wide range. One may, therefore, carry out oxidations or reductions with a fineness of selectivity unmatched by any set of chemical reagents. Furthermore, because relative rates of electrochemical reduction or oxidation of different functional groups do not necessarily parallel the corresponding relative reactivity toward chemical agents, it is possible to effect electrochemically many conversions not accessible by any other means. A few examples were given in Chapter 1, and this statement will be amplified in Chapter 4.

What are the practical requirements for the experiments just described? First, it is clear that electrolysis currents must be substantially higher than in polarography or cyclic voltammetry (in the range of 1–1000 mA, rather than microamperes) if one wishes sufficient quantities of electrolysis products for structural analysis.* A simple calculation from Faraday's law will indicate, for example, that electrolysis of 1 g of a substance of molecular weight 100, undergoing 1-electron reduction, would require 965 C, or, for example, 300 mA for 54 min. Since larger currents must be passed than in polarography or cyclic

* All of the usual methods of organic structure determination, such as gas chromatography, nmr, ir and uv spectroscopy, may, of course, be used in analysis. Discussion of this subject is unnecessary here.

voltammetry, the microelectrodes used in the latter techniques—the D.M.E. or HMDE—must be replaced by an electrode of considerably greater area. [This is usually (in reductions) a mercury pool, but may also be a large gauze or sheet of another metal.] For the same reason, solutions are stirred, and often higher concentrations of electroactive substance are used in electrolysis than in voltammetry, although this may sometimes be unwise (see below).

The electrical circuitry for controlled-potential electrolysis is basically the same as that used in polarography or cyclic voltammetry: a 3-electrode configuration (anode, cathode, and reference electrode) is used, and the potential of the working electrode* relative to the reference is adjusted to the desired value by a simple potentiometric circuit, with an external reference voltage source[32] (Figure 3.20).** The instrument used to effect controlled-potential electrolysis is called a *potentiostat*. The function of the potentiostat requires that it continually adjust the total voltage difference between the anode and cathode, E_{tot}, in order to keep the working-electrode potential constant throughout electrolysis. This will be seen from the following considerations.[32] The total

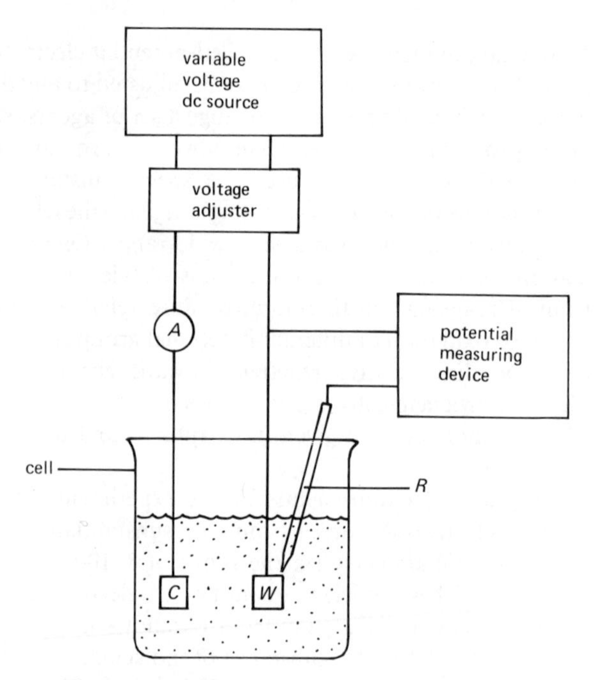

Figure 3.20 Potentiostat (schematic diagram).

* For the definition of *working* and *counter* electrodes, see p. 5.

** In polarography and cyclic voltammetry, an electronic or mechanical voltage scan replaces this manual adjustment of the working electrode potential, and, of course, faster circuit response times are required at high scan rates.

voltage output of the potentiostat at any time during electrolysis, that is, the applied voltage E_{tot} between the anode and cathode, is partitioned in the following manner,

$$E_{tot} = E_{anode} - E_{cathode} + iR$$

where E_{anode} and $E_{cathode}$ are the potentials of the anode and cathode, respectively, relative to the reference electrode, i is the electrolysis current, and R is the resistance of the solution between the anode and cathode (Figure 3.20). All of the terms on the right side of the equation are subject to change as electrolysis proceeds: $E_{cathode}$ and E_{anode} because the Nernst equation requires that electrode potentials change as the concentrations of electroactive species change, R because ions will be either formed or consumed in an electrolysis, and i because it is dependent on the instantaneous bulk concentration of electroactive material (Eq. 2.32). The potentiostat keeps the potential of the working electrode ($E_{cathode}$ in a reduction, E_{anode} in an oxidation) constant by continually adjusting E_{tot} to compensate for changes in electrode potentials and/or the iR term during electrolysis.* Potentials E_{anode} and $E_{cathode}$ are on the order of at most a few volts, but since the resistance of organic solvents is usually rather high, the iR term can be 100 V or more.

To appreciate certain features of experimental design in a controlled-potential electrolysis, it will be instructive at this point to examine quantitatively the way in which the current varies with time in a large-scale electrolysis in stirred solution. The steady-state current was shown previously to be given by

$$i_{lim} = \frac{n\mathcal{F}DAC^*}{\delta} \tag{2.32}$$

The limiting current is, of course, directly proportional to the bulk concentration of electroactive species, and hence voltammograms in stirred solution (Figure 1.6) must be measured sufficiently rapidly that the bulk concentration of electroactive substance is not appreciably diminished during the voltammogram. Of course, an electrolysis to completion does not conform to this stricture—it is not a steady-state experiment. Equation 2.32 may be rewritten to describe the instantaneous current at time t:

$$i_{lim,t} = \frac{n\mathcal{F}DAC^{*,t}}{\delta} \tag{3.11}$$

where $C^{*,t}$ is the bulk concentration at time t. Equation 2.32 is a special case of Eq. 3.11 for the condition $t = 0$, and hence i_{lim} and C^* in Eq. 2.32 are the *initial* current and concentration of electroactive substance, respectively, in the

* Schwarz and Shain have described the operation of a potentiostat in terms of modern *operational amplifier* methodology.[33]

preparative electrolysis. From Faraday's law considerations, the instantaneous current at time t is also given by

$$i_{\lim, t} = n\mathscr{F} \frac{dN}{dt} \tag{3.12}$$

where dN/dt represents the number of moles dN of electroactive substance electrolyzed in the infinitesimal interval dt. Since $N = VC^{*,t}$, where V is the volume of the electrolysis solution in liters and $C^{*,t}$ is the bulk concentration of electroactive substance in moles per liter,

$$i_{\lim, t} = n\mathscr{F} V \frac{dC^{*,t}}{dt} \tag{3.13}$$

Combining Eqs. 3.11 and 3.13,

$$-n\mathscr{F} V \frac{dC^{*,t}}{dt} = \frac{n\mathscr{F} DAC^{*,t}}{\delta} \tag{3.14}$$

where the negative sign is introduced to account for the fact that $C^{*,t}$ diminishes with time. Equation 3.14 can be rearranged to

$$\frac{dC^{*,t}}{C^{*,t}} = -\frac{DA}{V\delta} dt \tag{3.15}$$

Integrating both sides,

$$\ln C^{*,t} = -\frac{DA}{V\delta} t + \text{constant} \tag{3.16}$$

The constant must be $\ln C^*$, since $C^{*,t} = C^*$ when $t = 0$. Then

$$C^{*,t} = C^* \exp\left(-\frac{DAt}{V\delta}\right) \tag{3.17}$$

where C^* is the initial concentration of electroactive material. Also, since the current at any point during electrolysis, i_t, is proportional to the instantaneous bulk concentration, $C^{*,t}$,

$$\log i_t = -\frac{0.43 DA}{V\delta} t + \log i_{\lim} \tag{3.18}$$

where $i_{\lim}$ is the initial current (again given by Eq. 2.32). Equations 3.16 and 3.18 are completely equivalent. Taking Eq. 3.16 as an example, notice that it may be rewritten as

$$\log C^{*,t} = -\beta t + \log C^*$$

where

$$\beta = \frac{0.43DA}{V\delta} \tag{3.19}$$

That is, the decay of current and concentration should be kinetically a simple first-order process. This has been verified experimentally by many investigators.[34] One corollary of this feature of controlled-potential electrolysis is that the time necessary to effect electrolysis to a given point, such as 99% completion, is independent of the initial concentration of electroactive species, within certain inherent limitations of the particular potentiostat being used. The nature of these limitations will be clear if we first examine the significance of β, the *electrochemical rate constant*. (The latter term is used because the rate of electrolysis is a function only of D, A, V, and δ.) The electrochemical rate-constant concept, although based on the admittedly crude Nernst diffusion-layer theory, is very useful in the practical design of electrochemical cells. For example, Eq. 3.19 suggests that efficient convection will decrease the time necessary to effect electrolysis by decreasing δ and hence increasing β. Rates of convection adequate for synthetic applications are easily attainable by stirring, but for analytical applications where very short times are important, other expedients, such as pumping of the solution at or past the electrode surface have been employed. In a particularly elegant application, Bard reported extremely short electrolysis times by using ultrasonic energy to stir the electrode–solution interface.[35]

Inspection of Eq. 3.19 will reveal that shorter electrolysis times are also favored by a high ratio of electrode surface area A to solution volume V. Again, in most synthetic applications this ratio need not be extremely high—a typical (cylindrical) cell might have an $A : V$ ratio in the range 0.5–1.5 cm^{-1} (Chapter 9). For special applications, however, electrolysis can be considerably speeded by using a very high ratio of electrode area to solution volume. This principle reaches its extremum in the so-called thin-layer electrode (TLE), which amounts to electrolysis of a film of solution of 50–100 μ in thickness.[36] Electrolysis times are very short in a TLE, even though diffusion is usually the only means of mass transport, because the cell thickness is actually comparable with typical diffusion-layer thicknesses, but the rather minute quantities of solution used limit the technique to mechanistic, rather than synthetic, applications.* The other term involved in β is the diffusion coefficient, which is not particularly amenable to

* Considerable ingenuity has gone into the design of thin-layer electrodes, in view of the rapidity with which experiments in them can be carried out. A particularly attractive area recently has been development of the so-called *optically transparent thin-layer electrode* (OTTLE), in which the walls of the cell are transparent, and often of a conducting material, in which case they also serve as the electrodes.[37] An OTTLE may be placed directly in the sample beam of a spectrophotometer, so the buildup and decay of electrochemically generated intermediates may easily be monitored. The OTTLE promises to be increasingly valuable in the analysis of electrode mechanisms.

manipulation, since it is characteristic of the electroactive substance. The rather large temperature coefficient of most diffusion coefficients (2% per degree) does mean, though, that higher electrolysis rates can be attained if necessary by raising the solution temperature. In view of the fact that β can be adjusted readily by changes in A, V, or δ, this procedure is rarely necessary. In a complex electrode process involving coupled chemical reactions, one might even inadvertently change the course of the reaction through temperature effects on the rate of the chemical reactions.

It was mentioned on p. 101 that, within the limitations of a given potentiostat, the length of time required to electrolyze a given substance is independent of its initial concentration. It should likewise be pointed out here that the limitations of the potentiostat also militate against overly slavish attention to Eq. 3.19 in the design of cells for preparative electrolysis. One does not necessarily strive for the largest possible β, that is, the shortest electrolysis times. The output voltage (E_{tot}) and current capabilities of the specific potentiostat being used limit both the amounts that can be electrolyzed and the rate at which electrolysis can be allowed to proceed. In fact, to a certain degree these two parameters have a reciprocal relationship, as can be seen in the following way. Examine Figure 3.21,

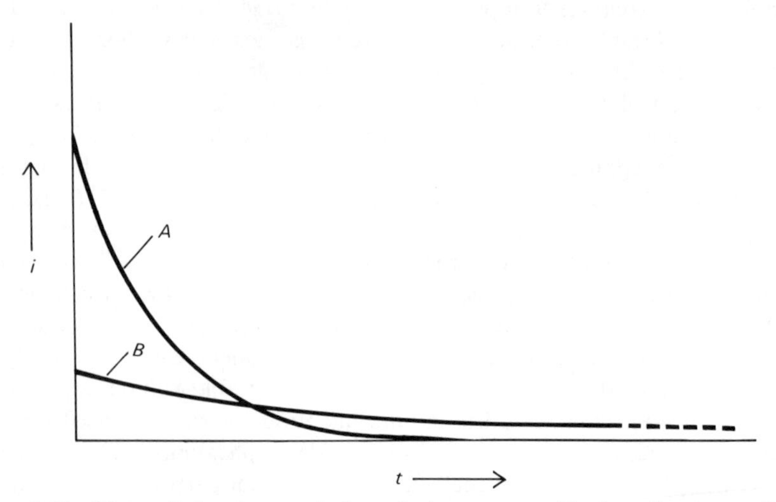

Figure 3.21 Plots of the decay of electrolysis current with time for two cells of differing efficiency.

which represents the decay of electrolysis current with time (Eq. 3.18) for electrolysis of equal amounts of the same substance in two cells whose electrolytic rate constant β is considerably different. Curve A represents a cell of high β and consequently short electrolysis time; curve B represents the slower electrolysis associated with a smaller value of β. The nature of the instrumental problem will be clear from an examination of the initial current, i_{lim}, for the two cases. Curve *A*, naturally, has a high initial current associated with its rapid electrolysis time. Suppose that i_{lim} happens to be 1 A and 0.1 A for curves *A* and *B*, respectively. If the maximum current capability of the potentiostat is, say, only 0.5 A, it cannot supply the high currents demanded by cell *A* early in the electrolysis. Another limitation on the magnitude of currents that can be passed in an electrolysis is associated with solutions of high resistance. Recall (p. 99) that essentially all of the voltage output of the potentiostat across the anode and cathode, E_{tot}, is associated with the *iR* drop through the solution. Suppose that one has a potentiostat with a current capability of 1 A and a maximum voltage capability, E_{tot}, of 100 V. If the cell resistance *R* is 1000 Ω, Ohm's law states that the maximum current *available for electrolysis* is 0.1 A (100 V/1000 Ω). Thus, the output *voltage* capability of the potentiostat can prevent it from utilizing its maximum *current* capability. Only if the cell resistance were lowered to 100 Ω could this potentiostat supply the 1 A demanded by cell *A*. Because of the rather high resistances associated with organic solvents, it is the voltage output capability of the potentiostat, rather than its current capability, that is usually the limiting factor in electrolysis. A potentiostat of at least 100 V and 1 A capabilities is normally required for electrolysis of organic compounds in useful amounts. (Other features to be desired in a potentiostat are discussed in Chapter 9.) Notice that the electrolysis represented by curve *B* in Figure 3.21 is within both the voltage and current capabilities of such a potentiostat, as long as $R \leq 1000\ \Omega$. In this case one could maintain satisfactory potential control only by using the less efficient cell, at the expense of longer time necessary to effect complete electrolysis.

A number of alternatives may be employed when one encounters an experimental situation where either the voltage or the current demands of an electrolysis cannot be supplied by the potentiostat. The trivial solution is simply to increase the current and/or voltage capacity of the potentiostat. With certain exceptions, to be encountered in the section on instrumentation (Section 9.1.2), this is normally not possible. A more realistic response to the problem may be deduced from the hypothetical electrolyses of Figure 3.21. One may move from an untenable situation such as in cell *A* to one whose demands can be met by the potentiostat merely by adjusting any of the factors that determine the electrolytic rate constant β. Prominent among these is δ, and the cell efficiency can usually be adjusted to a satisfactory level simply by slowing the stirring rate. In addition one might, if necessary, decrease the electrode area or increase the solution volume by dilution. Given the rather high resistances characteristic of organic solvents, the initial *iR* drop through a cell of extremely high β (electrolysis times < 5 min or so) cannot be handled by most potentiostats, but it is usually not difficult to find cell parameters such that electrolyses are complete in 10–60 min. It is desirable to work near the limits of the potentiostat to achieve maximum electrolysis efficiency, and some adjustment of cell parameters is usually necessary in a new system until an approximate β compatible with both short electrolysis times and instrumental limitations is found.

3.4 COULOMETRY

It should be clear by now that polarography, cyclic voltammetry, and controlled-potential electrolysis complement each other very nicely in an electrochemical investigation. The first two techniques provide a substantial amount of mechanistic information, and the last can vindicate mechanisms proposed from voltammetric evidence by providing ample quantities of electrode products for characterization by the usual methods. Adequate information for synthetic design is usually available by combined use of these three techniques. Some caution must be exercised, however, when comparing data taken in polarographic and cyclic voltammetric experiments with products obtained in a preparative-scale electrolysis. Although the same electrode process *usually* is observed in both kinds of experiment, divergent behavior is occasionally observed. For example, the initial electron-transfer step might lead to a product that could react by either bimolecular or unimolecular paths. At the low concentrations normally used in polarography, the bimolecular route might well be unimportant, but if a higher concentration of substrate is used in the preparative-scale electrolysis, as is often the case, the bimolecular path could compete more favorably. This sort of behavior could be diagnosed, naturally, through a product dependence on the initial concentration of electroactive material. A more insidious situation can arise, however, when the product of initial electrochemical reduction undergoes a slow chemical reaction to form a new species that is itself electroactive:

$$\mathrm{O} + \mathrm{e}^- \rightleftharpoons \mathrm{R}$$

$$\mathrm{R} \xrightarrow{k_d} \mathrm{X}$$

$$\mathrm{X} + \mathrm{e}^- \xrightarrow[\text{fast}]{} \mathrm{Z}$$

Suppose that k_d is of a magnitude such that R has a half-life of several minutes. On the relatively short time scale (seconds) of polarography and cyclic voltammetry, conversion of R to X will not occur to an appreciable degree, and in these experiments the reduction will appear a simple reversible process (Nicholson–Shain Case I, p. 89). On the other hand, since controlled-potential electrolysis may take from 10 min to several hours, ample time will be available for the conversion of R to X and thence to Z. This process would, then, appear to consume 1 electron/molecule of O by polarography and cyclic voltammetry, and 2 electrons/molecule of O when carried on a preparative scale. An extreme case of such behavior has been reported in the electrochemical reduction of fluoranthene, which exhibits a 1-electron polarographic wave in Cellosolve but affords tetrahydrofluoranthene, a 4-electron reduction product, upon preparative-scale electrolysis.[38]

The general behavior described here, that is, slow secondary electrochemical processes not revealed by polarography or cyclic voltammetry but observed on the time scale of preparative electrolysis, may be diagnosed by *coulometry*.[39] This technique, as its name implies, involves measurement of the *amount* of electricity (in coulombs) consumed in the electrolysis. This may then be related via Faraday's constant to the number of electrons consumed per mole of electroactive substance. The latter may, in turn, be compared to the n value determined polarographically from Ilkovic's equation. Secondary kinetic complications, competing pathways, and so on, can give rise to values of n that are nonintegral or otherwise differ from values determined polarographically. Although it is actually relatively uncommon to encounter an electrode process that takes a different course on a preparative scale than at a microelectrode, it is nevertheless always wise to supplement the polarographic n value with a measurement of the corresponding n value in the large-scale electrolysis. This requires, in fact, very little extra experimental effort, since the measurement is actually made *during* the preparative-scale electrolysis. Hence, from one experiment one learns both the amount of electricity consumed per mole of reactant and the product of electrolysis.

3.4.1 Principles of Coulometry

Faraday's law states that 9.65×10^4 C are consumed per equivalent of substance electrolyzed. One coulomb (an ampere-second) is actually a rather large amount of electricity, and it is possible to measure Q, the number of coulombs consumed in an electrolysis, with considerable accuracy and precision. Q is obtained through integration of the current–time curve (e.g., Figure 3.21):

$$Q = \int_0^t i \, dt \tag{3.20}$$

where t is the time at which the electrolysis current has decayed to a level indistinguishable from the background current.* The number of electrons n consumed per mole of electroactive substance is, of course, equal to $Q/x \cdot 96{,}500$, where x is the number of moles of material electrolyzed. A number of methods have been developed for integration of the electrolysis current–time curve. All have in common the feature that a secondary electrical element** is placed *in series* with the electrolysis cell, so that the same quantity of electricity must pass through both elements. The secondary element might be a second electrolysis

* A measurable background current (50–100 μA) is usually observed in most systems, generally due to slow faradaic processes, e.g., electrolysis of impurities in the solvent or electrolyte. It is often convenient to preelectrolyze the solvent for 10–30 min to reduce the background to a minimum before adding the electroactive substance. The quantity Q should be corrected for background; for this purpose the latter may be assumed to be constant during the electrolysis.

** This element must have a relatively low resistance, since the iR drop across it is part of E_{tot}.

cell in which a known reaction is allowed to proceed. In this case the amount of electrolysis product in the second cell could be related to Q. For example, one might carry out the electrolysis of water in such a cell, collect the electrochemically generated hydrogen and oxygen, and compute from Faraday's law the quantity of electricity consumed. This technique is capable of high accuracy but is cumbersome where large amounts of gas would be generated. Much more frequently, the secondary electrical element is simply a standard resistor in series with the cell (Figure 3.22). The voltage drop V across this known resistance,

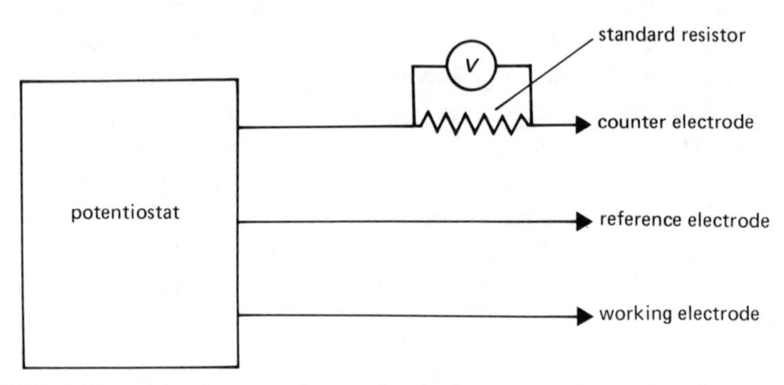

Figure 3.22 Schematic diagram for coulometric measurements involving measurement of current passed through a standard resistor in series with cell.

which will be proportional to the current i passing through it ($V = iR$), may be integrated by any of a number of methods. Probably the most convenient method for the purposes discussed here is recording of the voltage–time curve on an ordinary strip-chart recorder. The area under the resulting curve may be integrated by any of a number of methods (disk integrator, planimetry, etc.). This method gives entirely satisfactory ($\pm$1–3%) results when electrolysis is carried out on a preparative scale (>10 mg) and when the electrolysis times are not too short for the response time of the recorder. Precise and accurate coulometric data can be obtained by electronic integration. This usually involves the charging of a capacitor (the method usually used for integration of nmr spectra). Bard and Solon, however, converted the voltage drop across a standard resistor into digital form by use of a voltage-to-frequency converter and fed the output of the latter directly to an electronic counter.[40] This method is precise, uses commercially available components, and is fast enough for high-speed coulometry using ultrasonically stirred cells (p. 101).[35]

Not only is current integration using a strip-chart recorder and standard resistor the most convenient procedure for coulometry on a preparative scale, but it also has the added advantage over electronic or chemical integration methods that a permanent graphic record of the electrolysis current–time curve is produced, while the other methods only provide the total quantity of electricity consumed. This graph may itself, on occasion, provide valuable information about the mechanism of the electrode process. The straightforward exponential decay of current with time (Eq. 3.18 and Figure 3.21) is only true for the extreme situations where (1) electron transfer occurs without coupled chemical reactions (Cases I and II in the Nicholson–Shain nomenclature), or (2) the rates of any preceding coupled chemical reactions are much higher than the electrochemical rate constant β, or the rates of any catalytic reactions are much lower than β. In either of these situations, mass transport by convection is rate determining, and Eq. 3.18 is valid. When slow secondary chemical reactions are involved, the current–time plot may differ markedly from exponential behavior; this information is lost in those integration procedures that provide only the final value of Q. A number of examples will serve to illustrate the value of recording the current–time relationship. Buchta and Evans investigated the electrochemical reduction of dibenzoylmethane (**20**) in dimethylsulfoxide.[41] The polarographic behavior of this substance is shown in Figure 3.23. The limiting current for the first wave at -1.38 V *vs* S.C.E. is proportional to neither $h_{corr}^{1/2}$ nor the concentration of **20**,

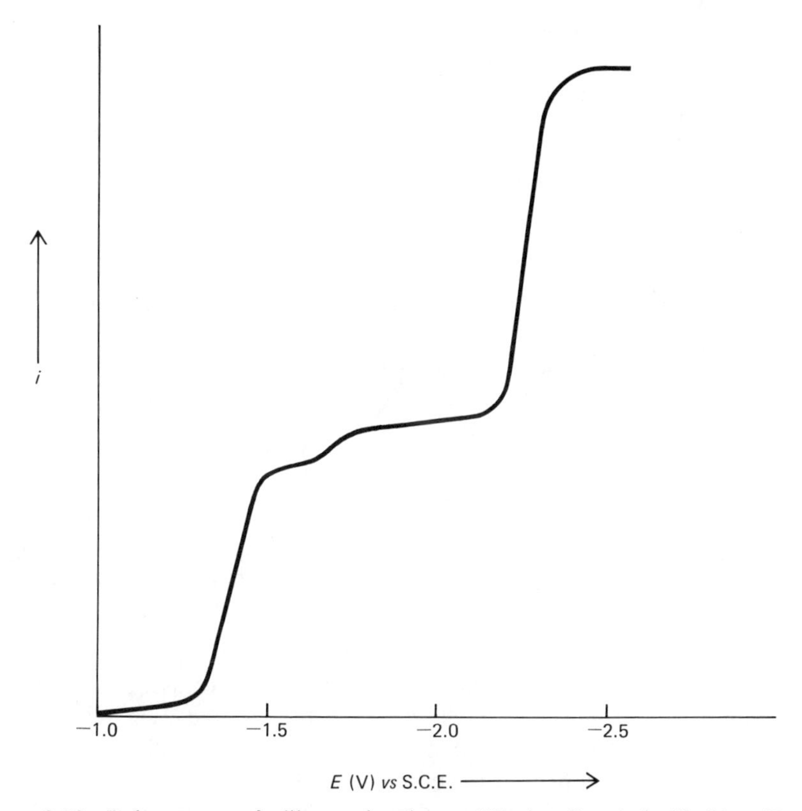

Figure 3.23 Polarogram of dibenzoylmethane (**20**) in dimethylsulfoxide. (R. C. Buchta and D. H. Evans, *Anal. Chem.*, **40**, 2182 [1968].)

showing that reduction is not diffusion controlled. The log plot for this wave has a slope of 70 mV, indicating irreversibility. The small wave at −1.69 V and the larger one at −2.25 V appear at the same potentials for dibenzoylmethane pinacol (**21**) and the enolate of dibenzoylmethane (**22**), respectively. Because of this coincidence and the fact that these two waves are disproportionately smaller when the concentration of **20** is decreased, it was suggested by Buchta and Evans that they must be due not to direct reduction of **20**, but instead to reduction of **21** and **22** formed in chemical reactions subsequent to initial reduction of **20**. Coupled chemical reactions could also account for the fact that the first wave is not diffusion controlled. When a controlled-potential electrolysis was carried out at a potential corresponding to the first polarographic wave, the current–time curve took the unusual form shown in Figure 3.24. After the

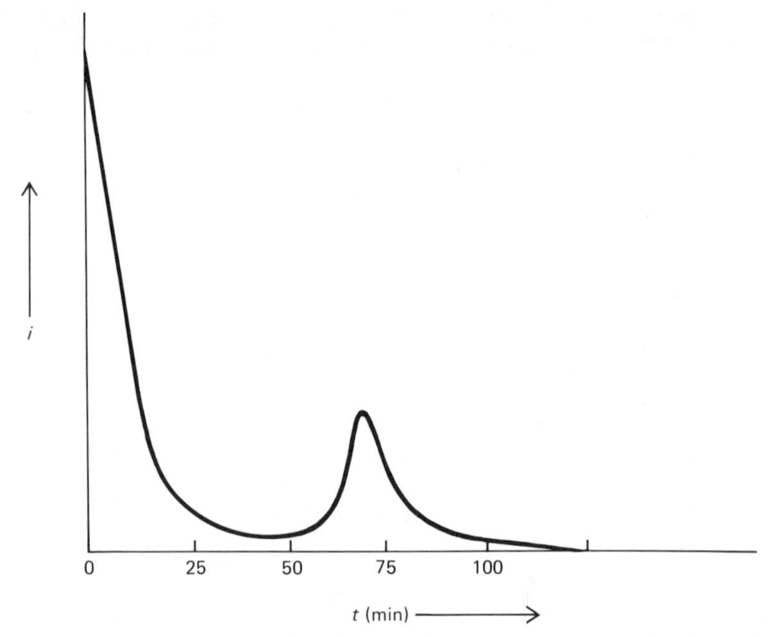

Figure 3.24 Plot of current *vs* time for the electrolysis of dibenzoylmethane (**20**) in dimethylsulfoxide at −1.50 V (*vs* S.C.E.). (R. C. Buchta and D. H. Evans, *Anal. Chem.*, **40**, 2183 [1968].)

expected initial decrease, the current was observed to increase, pass through a maximum at ca. 70 min, and finally decay to background. Integration of the current–time curve showed that 0.55 electrons/molecule of dibenzoylmethane had been consumed by the time the minimum at 40 min had been reached; from this point until the end of the electrolysis, another 0.24 electrons/molecule was consumed. This striking current–time behavior was accompanied by equally striking color changes: the solution was yellow in the early stages of electrolysis (<40 min) but became deep blue in the region of the current maximum at ca. 70 min. Finally, the color faded through green to yellow as the current decayed to background after 100 min. The solution was shown by chemical and polarographic experiments to contain only **21** and the yellow enolate **22** (independently synthesized) after 40 min. The later transient blue color, interestingly, was shown through electron spin resonance spectroscopy (esr) to be due to the *benzil radical anion* **24**. No other paramagnetic species could be detected during electrolysis. Much supporting evidence, which need not be detailed here, was adduced to support the following mechanism:

$$\underset{\mathbf{20}}{4(C_6H_5CO)_2CH_2 + 2e^-} \longrightarrow \underset{\mathbf{21}}{\begin{array}{c} OH \\ | \\ C_6H_5CCH_2COC_6H_5 \\ | \\ C_6H_5CCH_2COC_6H_5 \\ | \\ OH \end{array}} + \underset{\mathbf{22}}{2(C_6H_5CO)_2\bar{C}H}$$

$$\mathbf{21} \xrightarrow[\substack{retro\text{-aldol} \\ k_d}]{\text{base}} \underset{\mathbf{23}}{C_6H_5COCOC_6H_5} + 2CH_3COC_6H_5$$

$$\mathbf{23} + e^- \longrightarrow \underset{\mathbf{24}}{[C_6H_5COCOC_6H_5]^{\dot{-}}}$$

The unusual form of the current–time curve was shown to arise from the fact that the "base" catalyst for the *retro*-aldol reaction is actually the benzil radical anion; the reaction is thus catalyzed by a reaction product,* and this behavior was termed "electrolytic autocatalysis." By appropriate analysis of the current–time curve it was possible to estimate that k_d lies in the range 0.4–0.6 min^{-1}.

The preceding example, it will be noted, involved correlation of the unusual current–time curve with simultaneous measurements of other properties (in this case, color and esr spectra) of the solution during the course of electrolysis. Correlations of this sort are generally useful in situations where kinetic complications result in nonexponential decay. A study by Webb, Mann, and Walborsky illustrates this point very nicely.[42] The reaction under investigation was the electrochemical reduction of 1-bromo-1-methyl-2,2-diphenylcyclopropane (**25**) at a mercury cathode in DMF containing TEAB.

* **22** is probably the catalyst for the *retro*-aldol reaction early in the electrolysis, until appreciable quantities of the much more strongly basic anion **24** build up in solution.

C_6H_5 C_6H_5 Br CH_3 → C_6H_5 C_6H_5 H CH_3

25 **26**

The reaction was shown by coulometry to consume 2 electrons/molecule of **25**, and **26** was the only product isolated at the conclusion of the electrolysis. Plots were made of the decrease in both electrolysis current and starting material, and of the increase in **26**, with time (Figure 3.25). It was observed that **25** disappears faster than **26** appears. This indicates the existence of a relatively longlived intermediate in the reduction. Inspection of the shape of the current–time curve in Figure 3.25 suggests further that this intermediate is formed in a 1-electron reduction of **25**, since starting material has mostly disappeared when relatively less current has passed. The intermediate must then be reduced in a second 1-electron step to fulfill the overall coulometric data. Webb, Mann, and Walborsky suggested the following ECE mechanism for the reduction:

$$\mathrm{RBr} + e^- \longrightarrow [\mathrm{RBr}]^{\dot{-}} \longrightarrow \mathrm{R}\cdot + \mathrm{Br}^-$$
$$\mathrm{R}\cdot + e^- \longrightarrow \mathrm{R}^-$$
$$\mathrm{R}\cdot + n\mathrm{Hg}^0 \longrightarrow \mathrm{RHg}_n\cdot$$
$$\mathrm{RHg}_n\cdot + e^- \longrightarrow n\mathrm{Hg}^0 + \mathrm{R}^-$$
$$2\mathrm{RHg}_n\cdot \longrightarrow \mathrm{R}_2\mathrm{Hg} + (n-1)\mathrm{Hg}^0$$
$$\mathrm{R}_2\mathrm{Hg} + e^- \longrightarrow \mathrm{RHg}\cdot + \mathrm{R}^-$$
$$\mathrm{RHg}\cdot + n\mathrm{Hg}^0 \longrightarrow \mathrm{RHg}_{n+1}\cdot$$
$$\mathrm{R}^- + \mathrm{Et}_4\mathrm{N}^+ \longrightarrow \mathrm{RH} + \mathrm{Et}_3\mathrm{N} + \mathrm{CH}_2{=}\mathrm{CH}_2$$

The dialkylmercury intermediate (**27**) was preferred as the transient intermediate over the cyclopropylmercuric bromide **28**, since cyclic voltammetry showed that

$[C_6H_5$ C_6H_5 $CH_3]_2Hg$ C_6H_5 C_6H_5 CH_3 HgBr

27 **28**

28 is reduced considerably more easily than is **25**, the starting material. This being the case, significant concentrations of **28** could not accumulate in solution during electrolysis. The dicyclopropylmercury compound **27**, on the other hand, is somewhat harder to reduce than is the starting material; furthermore, peaks observed during cyclic voltammetry on **25** match those observed for authentic **27**. Both of the latter facts clearly support the assignment of **27** as the intermediate whose existence was inferred from Figure 3.25.

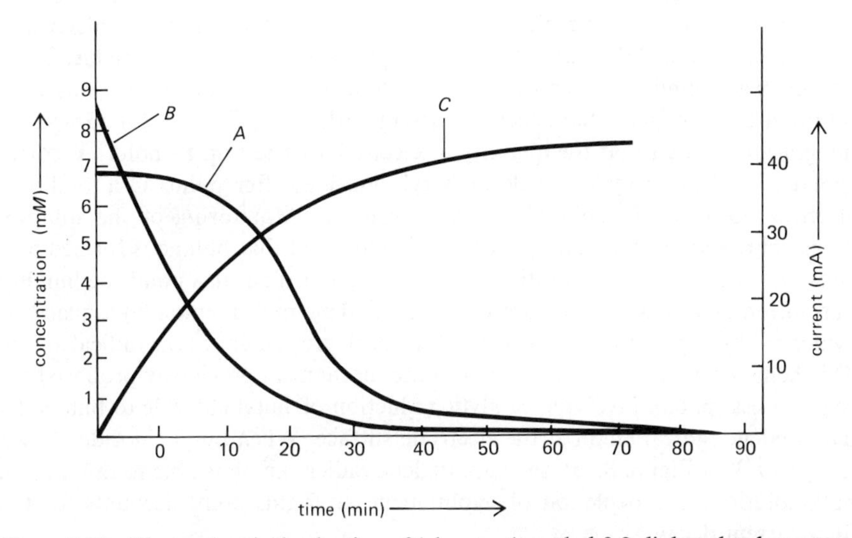

Figure 3.25 Electrochemical reduction of 1-bromo-1-methyl-2,2-diphenylcyclopropane (**25**) in acetonitrile. Curve *A*, electrolysis current *vs* time; curve *B*, concentration of **25** *vs* time; curve *C*, concentration of 1-methyl-2,2-diphenylcyclopropane (**26**) *vs* time. (J. L. Webb, C. K. Mann, and H. M. Walborsky, *J. Amer. Chem. Soc.*, **92**, 2044 [1970].)

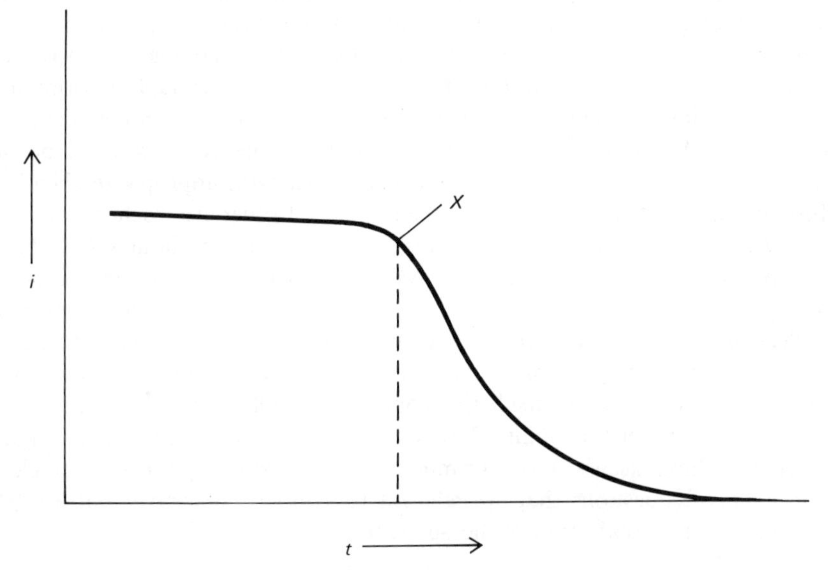

Figure 3.26 Plot of electrolysis current *vs* time for the controlled-potential electrolysis of a mixture of *n*-butyl chloride and naphthalene in dimethylformamide. (J. W. Sease and R. C. Reed, private communication.)

Another informative current–time curve was observed in a reaction involving a catalytic step.[9a] During electrolysis of a 1:4 mixture of naphthalene and *n*-butyl chloride in DMF at a potential where only naphthalene is reduced, the current was found to be constant over a long initial period, which was then followed by an exponential decay of current with time (Figure 3.26). Graphical integration showed that the quantity of electricity passed up to point *X* corresponded to 2 electrons/molecule of butyl chloride; after point *X*, a total of 1 electron/molecule of naphthalene was consumed. Monitoring of the solution by gas chromatography during electrolysis indicated that butane is formed at a constant rate until it ceases rather abruptly at point *X*; at this point the buildup in solution of appreciable quantities of naphthalene radical anion takes place, as evidenced by appearance and growth of the dark green color of the radical anion. This behavior is wholly consistent with the mechanism previously proposed for this process (p. 68), involving catalytic reduction of butyl chloride to butane by naphthalene radical anion at the electrode surface. When butyl chloride is used up (point *X* in Figure 3.26), the naphthalene radical anion is able to build up in bulk solution, and depletion of naphthalene after this point accounts for the final current decay.*

3.5 OTHER INSTRUMENTAL TECHNIQUES

It should now be clear that each electrochemical technique provides a different sort of information about the mechanism of an electrode process. Taken together, the four discussed so far (polarography, cyclic voltammetry, controlled-potential electrolysis, and coulometry) usually provide enough information about a reaction for intelligent experimental design, and it is not unusual to encounter in the literature rather detailed electrode mechanisms based on no more than these four types of experiment together with appropriate chemical information. There are, however, a large number of other electrochemical techniques that have not been discussed thus far. Most of these, though, have been developed either for quantitative analysis, for measurement of the rates of heterogeneous electron transfer (e.g., $k_{s,h}$) or of chemical reactions coupled to the electron-transfer step, or for detailed examination of double-layer structure. Although such techniques constitute a valuable and fascinating area of electrochemical research, they lie outside the scope of this book. It may be well, though, to mention briefly at this point a few techniques that are of less general significance than those discussed. One or more of the following might be used to clear up mechanistic questions that are relevant to synthetic objectives but are left unanswered after application of the above four.

* The current–time curve in Figure 3.26 differs from that calculated for a catalytic process by Geske and Bard, who predicted an initial current decay before the plateau.[43] These authors, however, assumed both slow to moderate regeneration of O ($k_f = 1\ M^{-1}\ \text{sec}^{-1}$ or so) and a large excess of species Z [for the definition of k_f and Z, see Cases VII and VIII, p. 89]. Neither of these conditions is fulfilled for the experiment illustrated in Figure 3.26.

3.5.1 Electron Spin Resonance

As the reader is no doubt already aware, electron spin resonance (esr) is useful not only for detection of species containing unpaired electrons (*cf.* the work of Buchta and Evans, discussed on p. 109), but also for measurement of the distri-

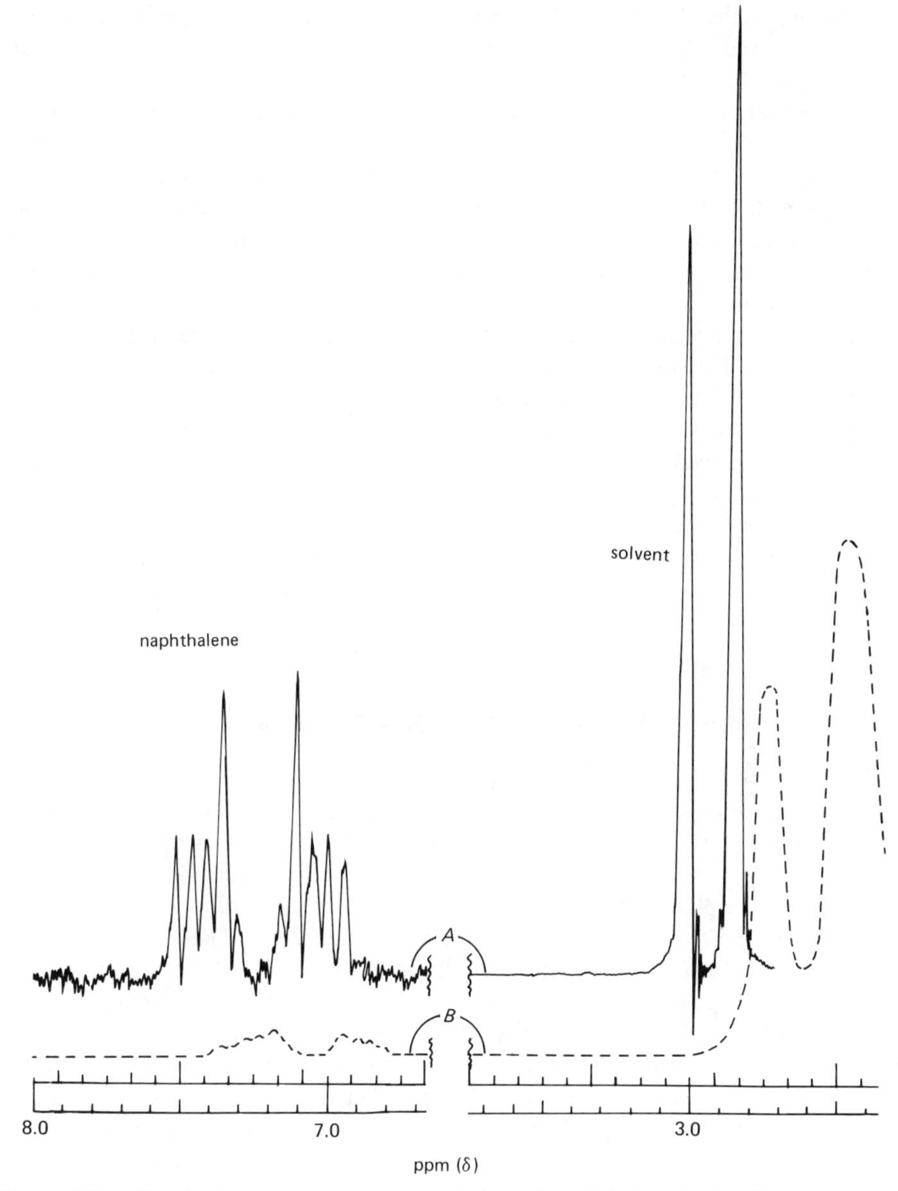

Figure 3.27 Curve *A*: nmr spectrum of a solution of naphthalene in 1,2-dimethoxyethane. Curve *B*: nmr spectrum of a solution of naphthalene radical anion in 1,2-dimethoxyethane.

bution of spin density in free radicals, through analysis of esr hyperfine coupling.[44] One-electron reduction of many aromatic hydrocarbons, ketones, and so on in aprotic solvents affords moderately stable radical anions; this, coupled with the fact that an electrochemical cell may be located directly in the cavity of the esr spectrometer, has made *in situ* electrochemical generation of such species routine practice. Because of the sensitivity of esr spectroscopy, electrochemically generated radicals may be observed and identified even when highly reactive or in low steady-state concentration.

If qualitative evidence for the presence of a free radical is all that is required and esr facilities are not available, one may examine the nmr spectrum of the electrolysis solution. If a paramagnetic species is present, its spectrum will be broadened, often to such an extent that it is not observable at all, due to rapid spin-lattice relaxation effects induced by the electrical field associated with the unpaired electron.[45a] Furthermore, when a proton-containing solvent is used, this same field will broaden and shift the *solvent* resonances through *inter*-molecular relaxation effects.[45b] Both of these effects are illustrated in Figure 3.27. The solid line represents a solution of naphthalene in 1,2-dimethoxy-ethane; the dashed line, broadened and shifted relative to the neutral species, represents the same solution after conversion of naphthalene to its radical anion **6**. The great disadvantage of this nmr method is, of course, the fact that, since electrolyses presently cannot be carried out in an nmr spectrometer, the radical must be sufficiently stable to survive preparation and transfer to the nmr cavity. This limits the technique to the detection of fairly longlived paramagnetic species. Likewise, quantitative or even qualitative structural information is not provided.

3.5.2 The Rotated-Disk and Ring–Disk Electrodes

It was stated in Chapter 2 that it is generally impossible to obtain an exact analytical solution for the complex hydrodynamic relationships involved in stirred solution. There is one exception to this rule, however. A rigorous solution was obtained by Levich for the so-called rotated-disk electrode (RDE), which consists of a planar electrode (generally a circular disk) rotated about an axis perpendicular to the plane of the electrode (Figure 3.28).[46] The direction of mass transport, which is shown schematically in Figure 3.28, has been discussed by Adams,[46b] who also provides a good introduction to the construction and use of the RDE.[47] Observe that rotation of the electrode imparts both tangential and radial motion to the solution in contact with the disk (containing products formed at the electrode surface). To maintain mass balance as this material leaves the electrode, the solution containing the electroactive substance is continually transported to the electrode surface in a direction perpendicular to the plane of the electrode. (Adams has provided a dramatic photographic illustration of this mass-transport phenomenon.[48])

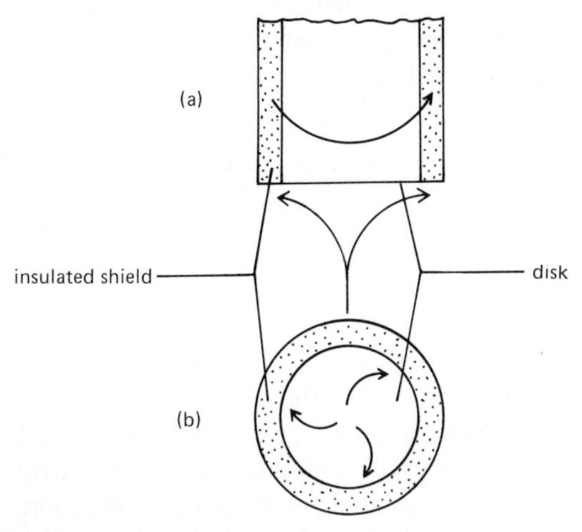

Figure 3.28 Motion of solution with respect to the surface of a rotated-disk electrode. (a) Side view. (b) Bottom view. (L. Meites, *Polarographic Techniques*, 2nd ed., New York, Wiley, 1965, p. 422.)

The limiting current at the RDE has been shown to be given by[49]

$$i_{\text{lim}} = 0.620 n\mathscr{F} A D^{2/3} C^* V^{-1/6} \omega^{1/2} \tag{3.21}$$

where n, $\mathscr{F}$, A, and D have their usual significance, V is the kinematic viscosity in square centimeters per second, ω is the angular velocity of the disk in radians per second ($\omega = 2\pi N$, where N is the rotation rate in revolutions per second), i_{lim} is in microamperes, and C^* is in millimoles per liter. Equation 3.21 can be experimentally verified fairly readily for N in the range 1–40 rps, and is obeyed at higher rotation rates if the electrode is properly designed. The rotated-disk electrode is capable of responding to considerably lower concentrations of electroactive material than are stationary electrodes such as the D.M.E. or the HMDE, but for qualitative purposes has no particular advantages over the latter. The *ring–disk electrode*,[50] however, does represent a useful variant of the RDE. A bottom view of this electrode is shown in Figure 3.29. The inner

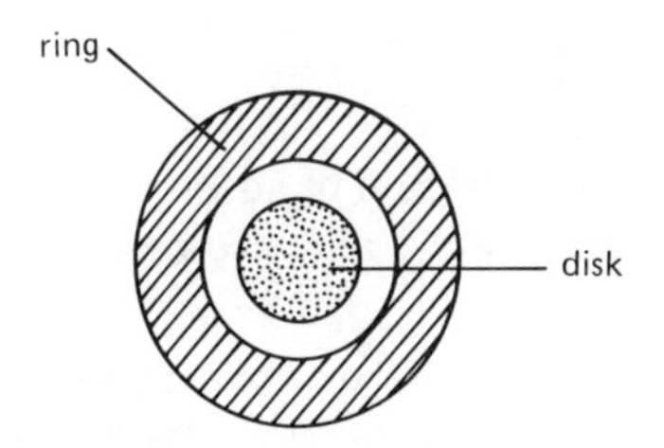

Figure 3.29 Ring–disk electrode, bottom view. (L. Meites, *Polarographic Techniques*, 2nd ed., New York, Wiley, 1965, p. 427.)

darkened area consists of an ordinary RDE. The outer darkened area is an annular ring electrode, whose potential is adjustable independently from that of the disk. The light-colored area merely serves to insulate the ring from the disk.

Suppose that one has a reaction of the ECE type (p. 84):

$$O + e^- \rightleftharpoons R$$
$$R \xrightarrow{k} X$$
$$X + e^- \longrightarrow Y$$

If the disk is maintained at a potential sufficient to reduce O, and the ring–disk ensemble is rotated, the R formed at the disk will have partly decomposed to X by the time the electrolyzed solution reaches the ring electrode. One may then carry out at the ring either cyclic voltammetry for qualitative identification of X or controlled-potential electrolysis of X to Y. Furthermore, one may determine k by monitoring the relative concentrations of R and X as a function of ω. The ring–disk electrode promises to be of increasing value, particularly when practical problems involved in design for high rotation rates are overcome.

3.5.3 Adsorption

A number of methods have been developed to study the role of adsorption in electrode processes. This complex subject cannot be treated adequately here, but it may be helpful to describe a few experiments that may be used to study adsorption.

Surface-Tension Measurements

In the late nineteenth century, Lippmann developed the *capillary electrometer*,* an instrument for measuring mercury–solution interfacial tension as a function of applied potential.[51] Soon thereafter he discovered the existence of electrocapillary maxima, such as shown in Figure 2.17. Adsorbed materials lower the mercury–solution interfacial tension (p. 53), and hence the capillary electrometer is used extensively in adsorption studies. One convenient experimental approach is suggested by the *Gibbs equation*[52]:

$$\Gamma_{\text{ads}} = -\frac{1}{RT}\left(\frac{\partial \gamma}{\partial \ln C_{\text{ads}}}\right)_{E,T} \tag{3.22}$$

where Γ_{ads} is the surface concentration of adsorbed material in moles per square centimeter at a specified temperature and electrode potential, γ is the interfacial

* Heyrovsky has provided a description of both the operation of this device and how he discovered polarography, for which he received the Nobel Prize in chemistry in 1959, during experiments on electrocapillary phenomena using the capillary electrometer.[53]

tension in dynes per centimeter, and C_{ads} is the bulk concentration of adsorbed substance in moles per liter. A plot of the change in interfacial tension with adsorbent against $RT \ln C_{ads}$ will therefore be a straight line of slope Γ_{ads}. From Avogadro's number, one may compute from Γ_{ads} the area taken up by a single molecule at the electrode surface (assuming a unimolecular film of adsorbed substance). This information may then be of use in determining the *orientation* of the adsorbed material at the electrode surface and can thus explain phenomena resulting from a special adsorption geometry (Chapter 4).

A variant on direct measurement of γ involves measurement of the weight of a drop emerging from the D.M.E., since the drop weight is directly proportional to γ.[54] Furthermore, if the height of the D.M.E. is greater than a meter or so, the mercury flow rate from the capillary is constant, and one may measure instead the drop time t, which is more convenient than weighing drops. Corbusier and Gierst,[54] and later, Bard and Herman,[55] described automatic drop-timing devices in which falling drops trigger a photocell to start and stop an electronic counter. Such a device lends itself to reasonably rapid measurement of electrocapillary data. Corbusier and Gierst showed how to convert drop-time measurements into Γ_{ads} by calibrating the apparatus with a solution of known surface tension.[54]

Capacitance

It has already been pointed out that the double layer has a capacitance associated with it (Section 2.4). This capacitance, like the surface tension, is dependent on factors, including adsorption, that affect double-layer structure. There are a number of methods, both static (ac impedance bridge)[56] and dynamic (ac polarography),[57] for measuring double-layer capacitance. Changes in this property with concentration of adsorbent also allow one to obtain Γ_{ads},[58] along with a great deal of other detailed information on double-layer structure.

References

[1] L. Meites, *Polarographic Techniques*, 2nd ed., New York, Wiley, 1965.
[2] Meites, *op. cit.*, pp. 125–140.
[3a] K. Wiesner, *Z. Elektrochem.*, **49**, 164 (1943); [b] K. Wiesner, *Coll. Czech. Chem. Commun.*, **12**, 64 (1947).
[4] S. G. Mairanovskii, *Catalytic and Kinetic Waves in Polarography*, New York, Plenum, 1968, pp. 7–55.
[5a] C. L. Perrin, *Progr. Phys. Org. Chem.*, **3**, 165 (1965); [b] P. Zuman, *Progr. Phys. Org. Chem.*, **5**, 81 (1967).
[6] J. Heyrovsky and J. Kuta, *Principles of Polarography*, New York, Academic, 1966, p. 368.
[7] J. E. Crooks and R. S. Bulmer, *J. Chem. Educ.*, **45**, 725 (1968).
[8] G. A. Rechnitz and H. A. Laitinen, *Anal. Chem.*, **33**, 1473 (1961).
[9a] J. W. Sease and R. C. Reed, private communication; [b] A. J. Fry and W. E. Britton, unpublished experiments.

[10a]J. F. Garst, P. W. Ayers, and R. C. Lamb, *J. Amer. Chem. Soc.*, **88**, 4260 (1966); [b]G. D. Sargent, J. N. Cron, and S. Bank, *J. Amer. Chem. Soc.*, **88**, 5363 (1966).
[11a]I. G. Markova and L. G. Feoktistov, *J. Gen. Chem. U.S.S.R.*, **38**, 933 (1968); [b]L. G. Feoktistov, M. M. Goldin, and I. G. Markova, *Extended Abstracts, 21st Meeting*, Comite International de Thermodynamique et de Cinetique Electrochimiques, Prague, Czechoslovakia, September 1970, p. 13.
[12a]A. Walkley, *J. Amer. Chem. Soc.*, **63**, 2278 (1941); [b]J. J. Lingane, *Ind. Eng. Chem. Anal. Ed.*, **15**, 583 (1943).
[13]R. G. Reed, M.A. thesis, Wesleyan University, 1969.
[14]Meites, *op. cit.*, p. 145.
[15]K. S. V. Santhanam, L. O. Wheeler, and A. J. Bard, *J. Amer. Chem. Soc.*, **89**, 3386 (1967).
[16]Meites, *op. cit.*, pp. 138–140.
[17]A. J. Fry and W. E. Britton, unpublished experiments.
[18]M. J. Peover, in A. J. Bard, ed., *Electroanalytical Chemistry*, New York, Marcel Dekker, 1967, vol. 2, pp. 1–51.
[19]Meites, *op. cit.*, pp. 413–420.
[20a]J. E. B. Randles, *Trans. Faraday Soc.*, **44**, 327 (1948); [b]A. Sevcik, *Coll. Czech. Chem. Commun.*, **13**, 349 (1948).
[21]A. C. Testa and W. H. Reinmuth, *Anal. Chem.*, **33**, 1320 (1961).
[22]J. Bacon and R. N. Adams, *J. Amer. Chem. Soc.*, **90**, 6596 (1968).
[23]R. F. Michielli and P. J. Elving, *J. Amer. Chem. Soc.*, **90**, 1989 (1968).
[24]J. W. Sease, R. C. Reed, and W. L. Morgan, unpublished research.
[25]R. S. Nicholson and I. Shain, *Anal. Chem.*, **36**, 706 (1964).
[26a]R. H. Wopschall and I. Shain, *Anal. Chem.*, **39**, 1514, 1527 (1967); [b]M. H. Hulbert and I. Shain, *Anal. Chem.*, **42**, 162 (1970).
[27a]R. S. Nicholson and I. Shain, *Anal. Chem.*, **37**, 178, 190 (1965); [b]M. Mastragostino, L. Nadjo, and J.-M. Saveant, *Electrochim. Acta*, **13**, 721 (1968).
[28]R. N. Adams, *Electrochemistry at Solid Electrodes*, New York, Marcel Dekker, 1969, chap. 5.
[29a]J. L. Sadler and A. J. Bard, *J. Amer. Chem. Soc.*, **90**, 1979 (1968); [b]J. Phelps, K. S. V. Santhanam, and A. J. Bard, *J. Amer. Chem. Soc.*, **89**, 1752 (1967); [c]M. E. Peover and B. S. White, *J. Electroanal. Chem.*, **13**, 93 (1967).
[30]M. L. Olmstead and R. S. Nicholson, *J. Electroanal. Chem.*, **14**, 133 (1967).
[31]S. Swann, Jr., in A. Weissberger, ed., *Technique of Organic Chemistry*, 2nd ed., New York, Wiley, 1956, vol. 2, pp. 385–523.
[32]J. J. Lingane, *Electroanalytical Chemistry*, 2nd ed., New York, Wiley, 1958, chap. 13.
[33]W. M. Schwarz and I. Shain, *Anal. Chem.*, **35**, 1770 (1963).
[34]Meites, *op. cit.*, p. 516.
[35]A. J. Bard, *Anal. Chem.*, **35**, 1125 (1963).
[36a]R. S. Nicholson, *Anal. Chem.*, **42**, 131R (1970); [b]C. N. Reilley, *Rev. Pure Appl. Chem.*, **18**, 137 (1968); [c]J. C. Sheaffer and D. G. Peters, *Anal. Chem.*, **42**, 430 (1970); [d]A. T. Hubbard and F. C. Anson, in A. J. Bard, ed., *Electroanalytical Chemistry*, New York, Marcel Dekker, 1970, vol. 4, p. 129ff.
[37a]W. R. Heineman, J. N. Burnett, and R. W. Murray, *Anal. Chem.*, **40**, 1974 (1968); [b]P. T. Kissinger and C. N. Reilley, *Anal. Chem.*, **42**, 12 (1970).
[38]I. Bergman, in G. J. Hills, ed., *Polarography—1964*, New York, Wiley, 1966, vol. 2, p. 925.
[39]G. W. C. Milner and G. Phillips, *Coulometry in Analytical Chemistry*, London, Pergamon, 1967.
[40]A. J. Bard and E. Solon, *Anal. Chem.*, **34**, 1181 (1962).

[41]R. C. Buchta and D. H. Evans, *Anal. Chem.*, **40**, 2181 (1968).
[42]J. L. Webb, C. K. Mann, and H. M. Walborsky, *J. Amer. Chem. Soc.*, **92**, 2042 (1970).
[43]D. H. Geske and A. J. Bard, *J. Phys. Chem.*, **63**, 1057 (1959).
[44a]P. B. Ayscough, *Electron Spin Resonance in Chemistry*, London, Methuen, 1967; [b]R. N. Adams, *J. Electroanal. Chem.*, **8**, 151 (1964).
[45a]J. W. Emsley, J. Feeney, and L. H. Sutcliffe, *High Resolution Nuclear Magnetic Resonance Spectroscopy*, New York, Pergamon, 1965, vol. 2, p. 25; [b]J. A. Pople, W. G. Schneider, and H. J. Bernstein, *High-Resolution Nuclear Magnetic Resonance*, New York, McGraw-Hill, 1959, pp. 207–212.
[46a]Meites, *op. cit.*, pp. 421–428; [b]Adams, *op. cit.*, pp. 80–102.
[47]Adams, *op. cit.*, pp. 283–286.
[48]Adams, *op. cit.*, p. 82.
[49]Meites, *op. cit.*, p. 424.
[50a]Adams, *op. cit.*, pp. 96–102; [b]Meites, *op. cit.*, pp. 427–428; [c]R. H. Sonner, B. Miller, and R. E. Visco, *Anal. Chem.*, **41**, 1498 (1969); [d]J. D. E. McIntyre and W. F. Peck, *Anal. Chem.*, **41**, 1713 (1969).
[51a]G. Lippmann, *Compt. Rend.*, **76**, 1407 (1873); [b]G. Lippmann, *Pogg. Ann.*, **149**, 561 (1873).
[52]D. M. Mohilner, in A. J. Bard, ed., *Electroanalytical Chemistry*, New York, Marcel Dekker, vol. 1, pp. 241–409.
[53]J. Heyrovsky and J. Kuta, *Principles of Polarography*, New York, Academic, 1966, chap. 1.
[54]P. Corbusier and L. Gierst, *Anal. Chim. Acta*, **15**, 254 (1956).
[55]A. J. Bard and H. B. Herman, *Anal. Chem.*, **37**, 317 (1965).
[56a]Mohilner, *op. cit.*, p. 285ff.; [b]R. Parsons and P. C. Symons, *Trans. Faraday Soc.*, **64**, 1077 (1968); [c]D. C. Grahame, *J. Phys. Chem.*, **61**, 701 (1957), and references therein.
[57a]B. Breyer and H. H. Bauer, in P. J. Elving and I. M. Kolthoff, eds., *Chemical Analysis*, New York, Wiley, 1961, vol. 13; [b]D. E. Smith, in A. J. Bard, ed., *Electroanalytical Chemistry*, New York, Marcel Dekker, 1966, vol. 1, pp. 1–155.
[58]P. W. Board, D. Britz, and R. V. Holland, *Electrochim. Acta*, **13**, 1633 (1968).

4 MANIPULATION OF EXPERIMENTAL PARAMETERS

Even though the essential mechanistic features of a given process may have been elucidated with the aid of the experimental techniques outlined in the preceding chapter, it is still normally of great interest to learn the response of that system to changes in experimental parameters. With some knowledge of the latter the chemist might, for example, learn how to divert the reaction into a new channel, maximize the yield of one product from an electrolysis that produces several, or otherwise exercise that element of creativity characteristic of the best synthetic experiments. There are a number of parameters—potential, solvent, pH, supporting electrolyte, added trapping or surface-active agents—that might be manipulated in an exhaustive study of a given process, although it is unlikely that all of these would be examined in a single study. The object of the following discussion is to suggest the kinds of experimental response that may be encountered, both as a guide to possible manipulation of electrode reactions and as an aid to further understanding of phenomena that may be met during electrochemical studies.

The effects of changes in certain electrochemical parameters, such as electrode potential or added proton donors, are now quite well understood. Others, such as the effects of double-layer structure, added surface-active agents, or nature of the supporting electrolyte on the course of electrode reactions, at present tend to be largely empirical. The latter is true mainly because such effects have never been studied specifically with synthetic applications in mind, rather than because they are intrinsically unimportant or uninteresting. A number of examples of such "surface" phenomena have been selected to point out the use to which these may be put on occasion, or the situations in which they may arise.

4.1 POTENTIAL

As was indicated in the previous chapter, the principal advantage in the use of controlled-potential electrolysis for oxidation or reduction of organic compounds as opposed to chemical oxidants or reductants arises from the fact that one may vary the electrode potential continuously over a very wide range. This allows one to effect an extremely wide range of conversions using a single apparatus (the potentiostat plus electrolysis cell) and to carry out these reactions with a selectivity unmatched by any set of chemical reagents of graded reducing or oxidizing power. One may, for example, reduce a molecule to any of several possible products simply by selection of the appropriate electrode potential (*cf.* diphenylfulvene, p. 96) or electrolyze one component of a mixture while leaving another untouched. This last feature is convenient when one wishes to carry out a reaction between the inert component and an electrochemically generated intermediate.

4.1.1 Relative Ease of Reduction of Functional Groups

To best utilize the synthetic power of controlled-potential electrolysis for selective modification of a complex polyfunctional molecule, one should know approximately the electrochemical behavior of each individual functional group in the molecule. Knowing, for example, that aromatic imines (Schiff bases) may be reduced to the corresponding amines under conditions [dimethylformamide (DMF) containing tetraethylammonium bromide (TEAB)] where aliphatic imines are electrochemically inert[1] allows one to predict that the following conversions would proceed in straightforward fashion:

C_6H_5N NCH_3 ⟶ C_6H_5NH NCH_3

CH_3N NCH_3 (C_6H_5) ⟶ CH_3NH NCH_3 (C_6H_5)

These selective reductions would, on the other hand, be rather difficult to do using conventional chemical reducing agents. From this simple example, one might extrapolate that it would be quite desirable from the synthetic point of view to have available a list of functional groups in order of relative ease of electrochemical reduction or oxidation. A compilation of useful data, from the extensive organic electrochemical literature, is presented in Tables 4.1 and 4.2.

Table 4.1 Reduction potentials of representative organic compounds

Substrate	$E_{1/2}$ (V)[a]	Solvent[b]	References[c]
	Organic Halogen Compounds		
Phenacyl bromide	−0.16	10% EtOH, pH 7.1	1
Bromoacetone	−0.34	H_2O, pH 4.6	1
Carbon tetrachloride	−0.78	75% D–W	1
Chloroacetone	−1.15	H_2O, pH 4.6	1
Benzyl bromide	−1.27	DMF	2
Allyl bromide	−1.29	75% D–W	1
Iodobenzene	−1.62	75% D–W	1
Chloroform	−1.67	75% D–W	1
Ethyl iodide	−1.67	75% D–W	1
3-Iodo-3-hexene	−1.80	DMF	3
Benzyl chloride	−1.94	75% D–W	1
Benzyl chloride	−2.1	DMF	2

Table 4.1 – *Continued*

Ethyl bromide	−2.08	75% D−W	1
n-Butyl bromide	−2.23	75% D−W	1
Bromobenzene	−2.32	75% D−W	1
Methylene chloride	−2.33	75% D−W	1
Benzyl fluoride	−2.45	DMF	2
Vinyl bromide	−2.47	75% D−W	1
Chlorobenzene	−2.54	DMF	1
Vinyl chloride	N.R.	75% D−W	1
Ethyl chloride	N.R.	75% D−W	1
	Carbonyl and Related Compounds		
Benzoquinone	+0.04	H_2O, pH 7.0	1
Benzoquinone	−0.40	DMSO	1
Anthraquinone	−0.83	DMF	1
Benzalacetophenone	−0.89	50% EtOH, pH 7.2	1
Benzophenone anil	−0.96	40% EtOH, pH 7.4	4
Benzalacetone	−1.11	50% EtOH, pH 7.2	1
Methyl vinyl ketone	−1.20	10% EtOH, pH 7	1
Methyl vinyl ketone	−1.42	H_2O, pH 7.0	5
Mesityl oxide	−1.6	50% EtOH, pH 7.2	5
Benzophenone	−1.31	50% EtOH, pH 8.5	1
Benzophenone	−1.72	DMF	1
Benzalaniline	−1.83	DMF	6
Acetophenone	−1.49	10% EtOH, pH 7.1	1
Acetophenone	−1.99	DMF	1
Acetone	−2.46	75% D−W	1
Camphor anil	−2.6	DMF	7
Camphor alkylimines	N.R.	DMF	7
	Alkenes and Related Compounds		
Azobenzene	−1.36	DMF	8
Benzalaniline	−1.83	DMF	6
trans-Stilbene	−2.14	DMF	1
trans-Stilbene	−2.20	75% D−W	1
	Acetylenes		
Diphenylacetylene	−2.19	75% D−W	1
Phenylacetylene	−2.37	75% D−W	1
Acetylene	N.R.	75% D−W	1
	Miscellaneous		
Anthracene	−1.94	75% D−W	1
Phenanthrene	−2.46	75% D−W	1
Naphthalene	−2.47	75% D−W	1
Diethyl sulfate	−2.6	AN	9
Acetic anhydride	−2.8	AN	10

[a] Potentials relative to S.C.E.

[b] D−W = dioxane−water; DMF = dimethylformamide; DMSO = dimethylsulfoxide; AN = acetonitrile. Electrolytes are various tetraalkylammonium salts.

Table 4.1 – *Continued*

[c] References:

1. L. Meites, *Polarographic Techniques*, 2nd ed., New York, Wiley, 1965, pp. 671–711.
2. R. C. Reed, Ph.D. thesis, Wesleyan University, 1971.
3. A. J. Fry and M. A. Mitnick, *J. Amer. Chem. Soc.*, 91, 6207 (1969).
4. H. Lund, *Acta Chem. Scand.*, 13, 249 (1959).
5. I. M. Kolthoff and J. J. Lingane, *Polarography*, 2nd ed., New York, Wiley, 1952, vol. 2, p. 665. (Reprinted by permission of John Wiley and Sons, Inc.)
6. J. M. W. Scott and W. H. Jura, *Can J. Chem.*, 45, 2375 (1967).
7. A. J. Fry and R. G. Reed, *J. Amer. Chem. Soc.*, 91, 6448 (1969).
8. J. L. Sadler and A. J. Bard, *J. Amer. Chem. Soc.*, 90, 1979 (1968).
9. D. Lipkin, *Symposium on Radical Anions*, 155th American Chemical Society National Meeting, San Francisco, April 1968.
10. T. J. Curphey, C. W. Amelotti, T. P. Layloff, R. L. McCartney, and J. H. Williams, *J. Amer. Chem. Soc.*, 91, 2817 (1969).

Table 4.2 Oxidation potentials of representative organic compounds

Substrate	$E_{1/2}$ (V)[a]	Solvent[b]	References[c]
	Aromatic Hydrocarbons		
Azulene	0.91	HOAc	1
Pyrene	1.20	HOAc	1
Anthracene	1.20	HOAc	1
Biphenylene	1.30	HOAc	1
Acenaphthene	1.36	HOAc	1
Hexamethylbenzene	1.52	HOAc	1
1-Methylnaphthalene	1.53	HOAc	1
Acenaphthylene	1.53	HOAc	1
2-Methylnaphthalene	1.55	HOAc	1
Durene	1.62	HOAc	1
Fluorene	1.65	HOAc	1
Phenanthrene	1.68	HOAc	1
Naphthalene	1.72	HOAc	1
Mesitylene	1.90	HOAc	1
Biphenyl	1.91	HOAc	1
Toluene	2.29	AN	2
Benzene	2.39	AN	2
	Alkenes		
Tetrakis(dimethylamino)-ethylene	−0.75	AN	2
1,1-*Bis*(dimethylamino)-2-methylpropene	0.05	AN	3
1,1-*Bis*(dimethylamino)-ethylene	0.48	AN	3
α-Dimethylaminostyrene	0.70	AN	3
Bitropyl	1.29	AN	2
Tropilidene	1.39	AN	2
Cyclooctatetraene	1.42	HOAc	1
trans-Stilbene	1.51	HOAc	1
1,1-Diphenylethylene	1.52	HOAc	1
Acenaphthylene	1.53	HOAc	1
Cyclohexene	2.31	AN	4

Table 4.2 – *Continued*

2-Butene	2.52	AN	4
Ethylene	3.16	AN	4
	Hetero-Substituted Aromatics		
N,N-Dimethylaniline	0.71	AN	3
Diphenylamine	0.83	AN	2
Triphenylamine	0.92	AN	2
Aniline	0.85, 1.01	AN	2
Phenol	1.35	AN	2
Anisole	1.67	AN	2
Thioanisole	1.82	AN	2
	Miscellaneous		
2,5-Dimethylfuran	1.20	HOAc	1
Furan	1.70	HOAc	1
Thiophene	1.91	HOAc	1
Biphenyl-2-carboxylic acid	1.71	HOAc	1
N,N-Dimethylformamide	1.90	HOAc	1

[a] Potential relative to S.C.E.: literature values measured against other reference electrodes have been corrected as follows: Ag/0.1-*M* $AgNO_3$ = +0.31 V *vs* S.C.E.; Ag/0.01-*M* $AgNO_3$ = +0.26 V *vs* S.C.E. (see reference 5).

[b] HOAc = acetic acid; AN = acetronitrile; electrolytes were generally tetraalkylammonium salts.

[c] References:
1. L. Eberson and K. Nyberg, *J. Amer. Chem. Soc.*, 88, 1686 (1966).
2. N. L. Weinberg and H. R. Weinberg, *Chem. Rev.*, 68, 449 (1968).
3. J. M. Fritsch, H. Weingarten, and J. D. Wilson, *J. Amer. Chem. Soc.*, 92, 4038 (1970).
4. M. Fleischmann and D. Pletcher, *Tetrahedron Lett.*, 6255 (1968).
5. C. K. Mann and K. K. Barnes, *Electrochemical Reactions in Nonaqueous Systems,* New York, Marcel Dekker, 1970, p. 26. (Reprinted by permission of Marcel Dekker, Inc.)

(Data for any given functional group must be considered to be only approximate, however, due to possible shifts in half-wave potential as a result of inductive or resonance interaction of that group with others in the molecule.) Compilations such as those in Tables 4.1 and 4.2 can lead to surprising and intriguing conclusions. For example, notice that a number of functional groups often thought of as more or less chemically reactive, that is, acid anhydrides, esters, alkenes, and alkyl chlorides, are difficult or impossible to reduce electrochemically, while chemically less reactive groups such as aromatic bromine or iodine and geminal polyhalides are often easily reduced electrochemically.[2] The fact that the relative ease of electrochemical reduction of a series of functional groups does not necessarily parallel the relative reducibility of those groups by chemical agents means that one may not only reduce a complex molecule selectively, but may readily carry out transformations that would be impossible by another method. Examples that may be cited to support this statement include those in Chapter 1 and the following:

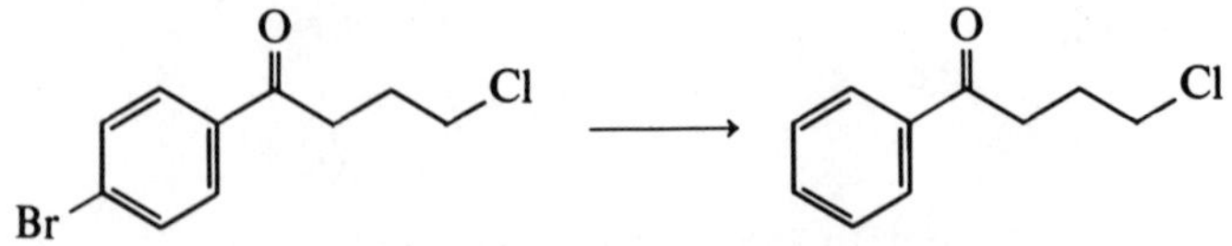

O, Cl, Br — **1** → O, Cl — **2** *(Ref. 2)*

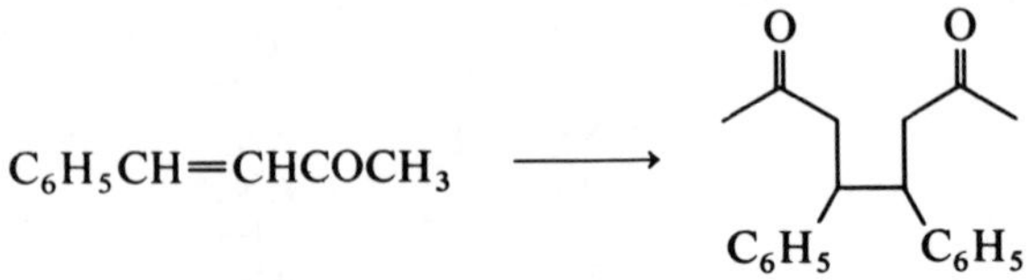

$C_6H_5CH{=}CHCOCH_3$ → O, O, C_6H_5, C_6H_5 *(Ref. 3)*

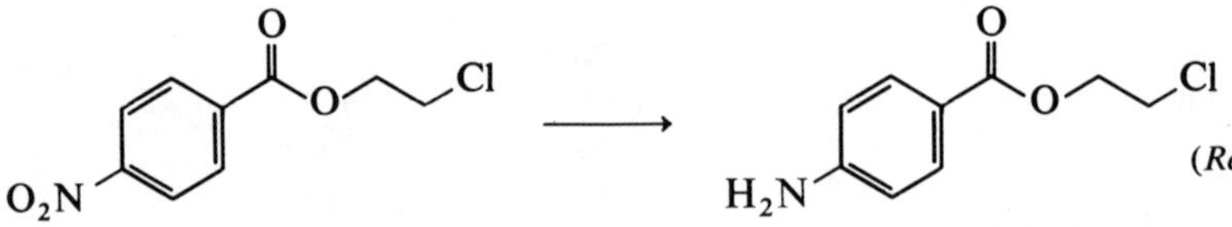

O, O, Cl, O_2N → O, O, Cl, H_2N *(Ref. 4)*

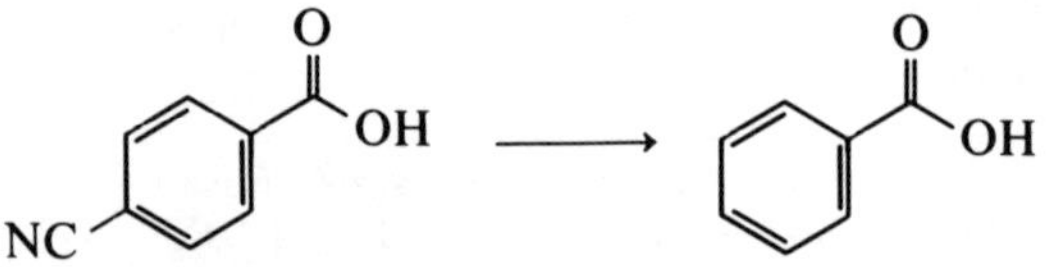

O, OH, NC → O, OH *(Ref. 5)*

The conversion of **1** to **2** is particularly interesting in view of the other potentially reducible groups, that is, carbonyl, chlorine atom, and benzene ring, that survive electrolysis intact. There is no need to block the carbonyl group by ketalization before electrolysis. One might imagine any number of attractive hypothetical conversions from inspection of Tables 4.1 and 4.2. There always exists, naturally, the possibility of intramolecular reaction of one moiety of a molecule with an intermediate generated electrochemically elsewhere in the molecule; for example,

Radical

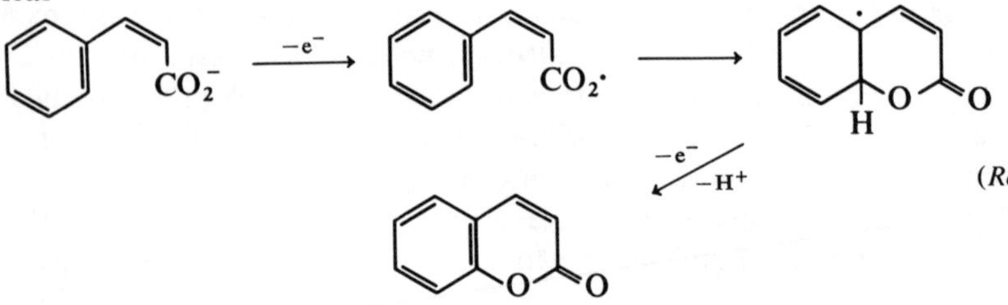

CO_2^- $\xrightarrow{-e^-}$ $CO_2\cdot$ → H, O, O; $\xrightarrow{-e^-,\ -H^+}$ O, O *(Ref. 6)*

Carbanion

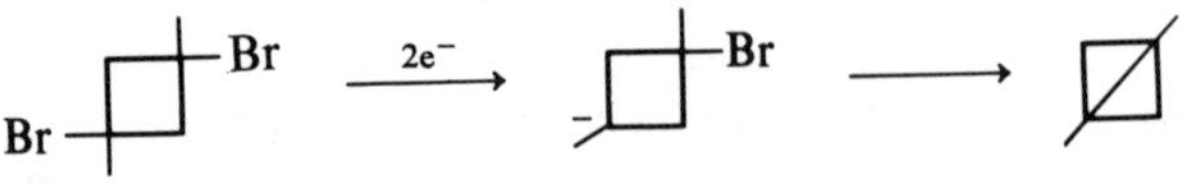

Br, Br $\xrightarrow{2e^-}$ Br → *(Ref. 7)*

Carbonium Ion

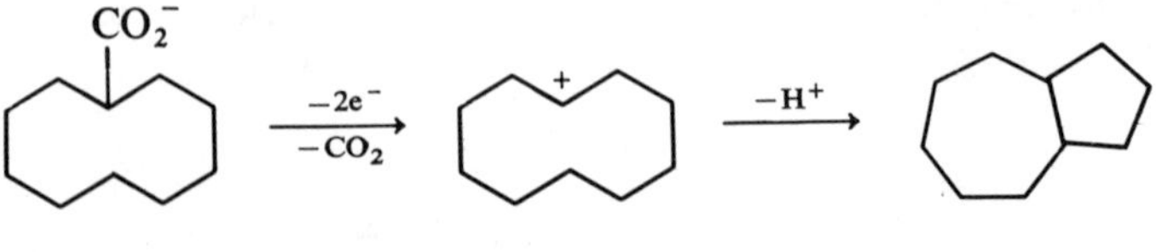

CO_2^- $\xrightarrow[-CO_2]{-2e^-}$ + $\xrightarrow{-H^+}$ *(Ref. 8)*

An application of controlled-potential electrolysis of obvious promise lies in selective oxidation or reduction of the complex intermediates encountered during the synthesis of natural products. One is often faced in such work with the problem of operating on one functional group of the molecule to the exclusion of several others. This is, of course, precisely the forte of the controlled-potential technique. Also, because relative chemical and electrochemical reactivities of functional groups do not always coincide, the two methods often complement one another nicely:

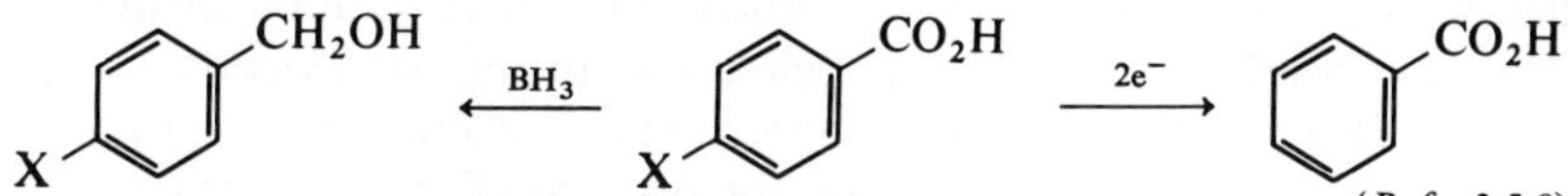

CH_2OH $\xleftarrow{BH_3}$ CO_2H $\xrightarrow{2e^-}$ CO_2H

X X

(Refs. 2,5,9)

(X = Br or CN)

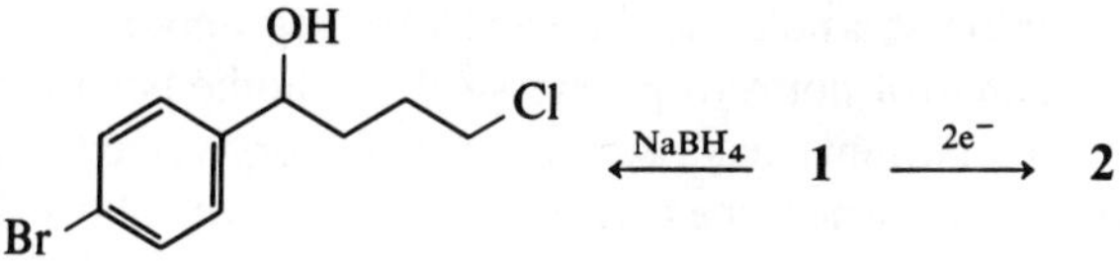

OH Cl $\xleftarrow{NaBH_4}$ **1** $\xrightarrow{2e^-}$ **2** *(Ref. 2)*

Br

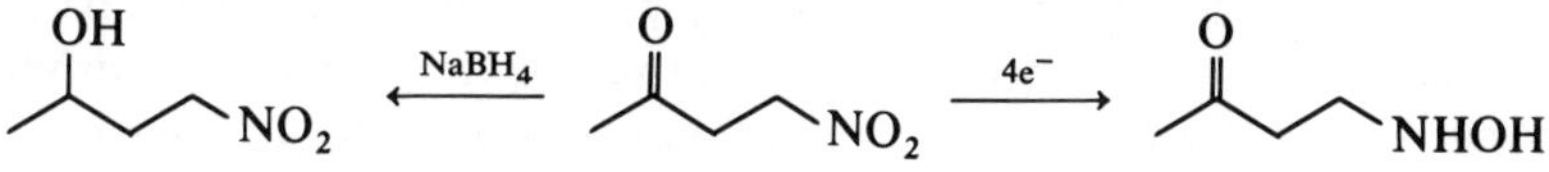

OH NO_2 $\xleftarrow{NaBH_4}$ O NO_2 $\xrightarrow{4e^-}$ O NHOH

(Refs. 10,11)

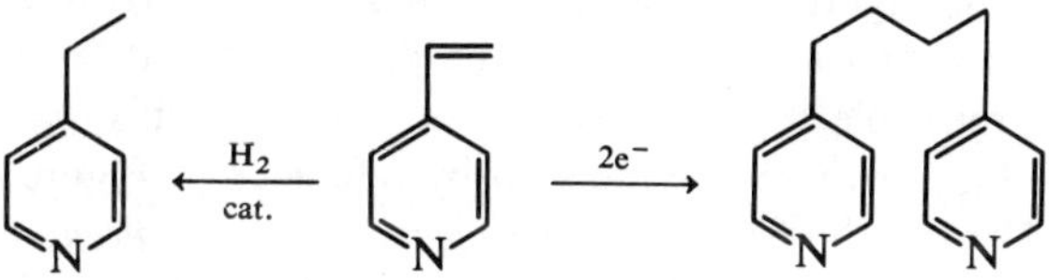

N $\xleftarrow[cat.]{H_2}$ N $\xrightarrow{2e^-}$ N N *(Ref. 3)*

4.1.2 Methodology

In many, probably most, electrochemical investigations it suffices to set the electrode potential for large-scale electrolysis at a point corresponding to a potential anywhere on the plateau of the polarographic wave of interest. This procedure may lead one astray if followed slavishly, however. The nature of the problem of selecting the control potential from polarographic data is conveniently divided into consideration of the respective behavior of reversible and irreversible processes.

Reversible Processes

Since half-wave potentials for a reversible process are identical (or nearly so) both in polarography and stirred solution (Chapter 2), one will indeed experience no difficulty if a potential on the plateau but just past the top of the desired polarographic wave (i.e., potential A in Figure 3.19) is selected as the large-scale control potential,* as long as the resistance of the solution has been properly compensated for (Chapter 9).

Irreversible Processes

Most organic electrode processes are irreversible for one reason or another, and hence it is of particular interest to consider the behavior of such processes. Since half-wave potentials for irreversible processes in stirred solution are quite dependent on experimental conditions, such as the stirring rate (Section 2.3), polarographic data can provide only an approximate value for the control potential. Fortunately, for an irreversible reduction one may effect complete conversion to the reduced form by electrolysis *at any point on the rising portion* of the voltammetric wave, as well as on the plateau. It is therefore often possible to select the control potential simply by beginning at a potential known to be well anodic of the desired wave, and setting the control point to progressively cathodic potentials until the electrolysis rate (i.e., current) has risen to a satisfactory value.** Electrolysis times are, of course, longer when one is carrying out an electrolysis at a point part way up the wave. There is no reason in principle, other than possibly because the output capability of the potentiostat may be insufficient to meet the current demand on the plateau, why one would not wish to electrolyze on the plateau, *for a compound containing only one voltammetric wave*, or when one wishes to electrolyze at the most cathodic of a series of waves, such as potential B in Figure 3.19. When one wishes to electrolyze at the first of two closely spaced waves there are, on the other hand, at least two good reasons why it is preferable to electrolyze at a potential only part way up the first wave. The first of these may be seen by reference to Figure 4.1. Notice that at potential E_1 atop the first wave, *the second wave has already begun*. Electrolysis at E_1 will result, therefore, in overreduction. It would be wiser to select a more anodic potential, such as E_2, where the second wave can be considered negligible, even though one must accept a longer electrolysis time at E_2. The other reason for picking a point only part way up the wave arises from the fact that half-wave potentials in stirred solution may sometimes be several tenths of a volt anodic of the polarographic value. One possible consequence of this fact is illustrated in

* It should be obvious at this point that one would never electrolyze on the rising portion of the wave for a truly reversible process. If one were to electrolyze at, e.g., $E_{1/2}$, an equimolar mixture of starting material and product would result.

** One may, in fact, using this method construct a complete voltammogram on the stirred solution, either point by point or with the aid of mechanical or electronic voltage scan. This is the method of choice in complex situations, such as when products of electrolysis at a single wave are potential dependent.[12]

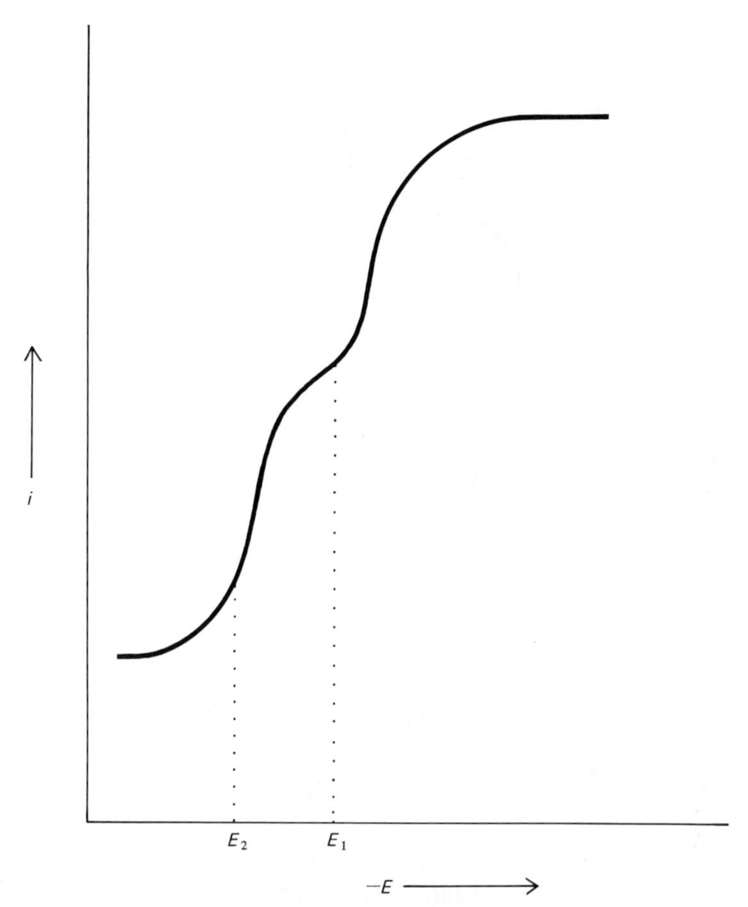

Figure 4.1 Polarogram of a substance exhibiting two closely spaced reduction steps.

Figure 4.2, which represents polarographic (dashed lines) and stirred (solid line) voltammograms for a substance exhibiting two closely spaced irreversible waves. Observe that in this (admittedly extreme) example, E_1, a potential atop the first polarographic wave, corresponds to a potential atop the *second* wave in stirred solution. Here it is clear that only by actually measuring the stirred solution voltammogram or, what is almost the same, by increasing the potential cathodically only until a satisfactory current flows will one be able to carry out the electrolysis with the desired selectivity.

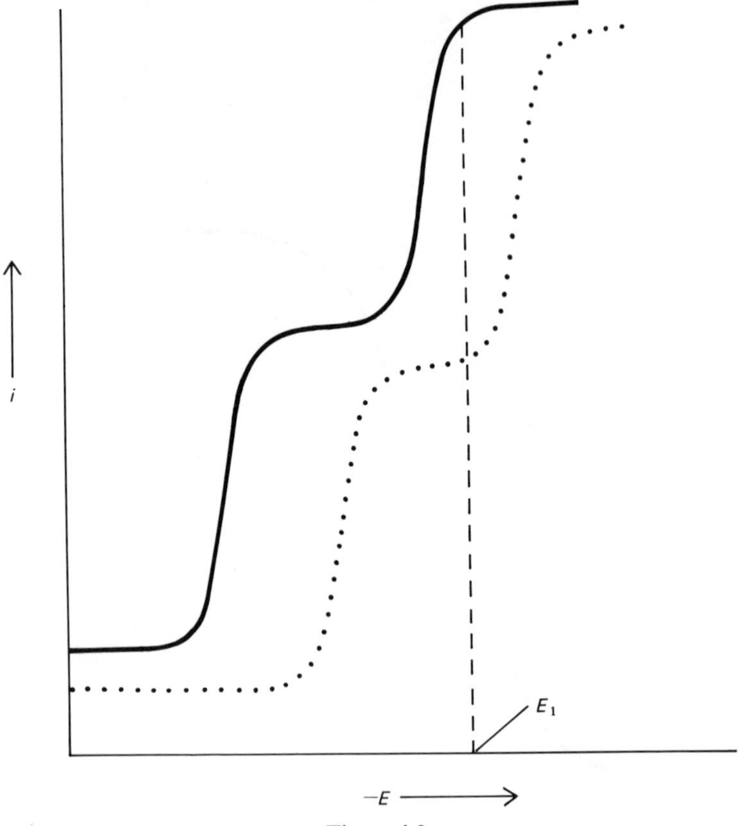

Figure 4.2

4.1.3 Nonpotential-Dependent Electrolysis

Up to this point heavy emphasis has been placed on control of the electrode potential during electrolysis, because of the close relationship between electron-transfer rates and electrode potential, which constitutes the fundamental basis of voltammetry and electrolysis. It is necessary, however, to describe here a few electrolysis techniques that are not tied as closely to electrode potential as is controlled-potential electrolysis itself (and that, though useful, are consequently of less general interest).

Constant-Current Electrolysis

As was mentioned in Chapter 3, controlled-potential electrolysis requires adjustment of the output voltage of the electrolysis source to correct changes in the potential of the working electrode relative to the reference electrode. All modern potentiostats incorporate feedback circuitry to perform this operation automatically, but until the development of the first potentiostat incorporating this feature by Hickling in 1942,[13] it was necessary to carry out the continual adjustment of output voltage manually during the course of electrolysis.* The tedium associated with this manual procedure meant, in fact, that until the relatively recent development of the potentiostat, controlled-potential electrolysis was rarely employed.** The procedure generally used for electrolysis before the availability of the potentiostat (and still used to some extent despite its disadvantages) consists of *constant-current electrolysis*. Constant-current electrolysis is extremely simple experimentally. It is merely necessary to apply a direct current across the anode and cathode. It is desirable to be able to adjust the current to a desired value; from Ohm's law, this may be done by using as current source either a variable voltage in series with the cell (as resistor) (Figure 4.3a) or a fixed voltage source and variable resistor (also in series with the cell) (Figure 4.3b).

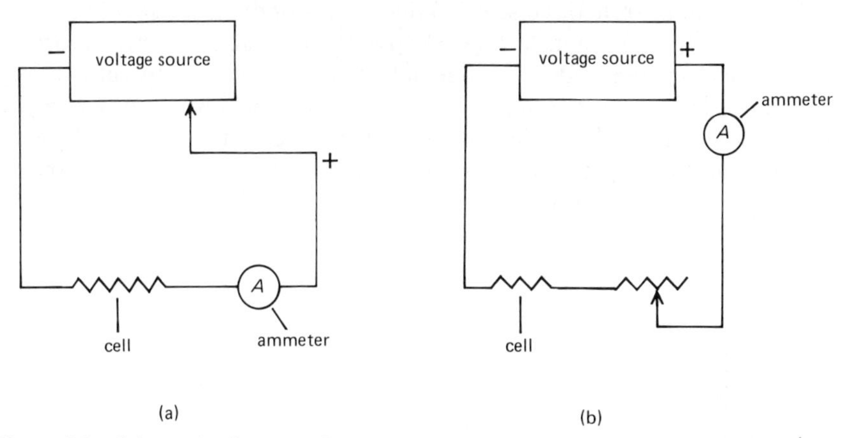

Figure 4.3 Schematic diagrams for constant-current electrolysis. (a) Variable voltage power supply. (b) Fixed voltage source in series with a variable resistor.

What happens when a constant current is applied across the electrolysis cell? Let us illustrate with respect to diphenylfulvene, whose behavior under controlled-potential electrolysis has already been discussed (Section 3.3). At the outset of electrolysis, the cathode will assume a potential E_1 given by the Nernst equation

$$E_1 = E^0 - \frac{RT}{n\mathscr{F}} \ln \frac{[\text{Red}]}{[\text{Ox}]}$$

* By contrast, a controlled-potential electrolysis employing a potentiostat proceeds to completion unattended after one has adjusted the electrode potential at the outset.

** As early as 1897, Haber had recognized the potential of the working electrode as the electrochemical parameter most critical in determining the course of electrolysis,[14] and had carried out manually a number of controlled-potential electrolyses.

where [Red] and [Ox] correspond to the concentrations of the diphenylfulvene radical anion and diphenylfulvene itself, respectively. As electrolysis proceeds, the ratio [Red]/[Ox] will increase, but because this ratio appears as its logarithm in the Nernst equation, the electrode potential will drift cathodically relatively slowly until reduction of the fulvene to its radical anion is almost complete. Near the end of this electrolysis, however, the ratio of [Red] to [Ox] will change rapidly, and hence the electrode potential will undergo a rather rapid cathodic excursion. The electrode potential will then quickly reach a potential E_2 where *reduction of the fulvene radical anion to the dianion is possible*, and once again the electrode potential will remain near E_2 until this second reduction is nearly complete. At the latter point, the electrode potential will again shift rapidly cathodic until a potential corresponding to the next most easily reduced species, in this case the solvent itself, is reached. This behavior is illustrated in Figure 4.4. It follows naturally from this discussion that the major disadvantage of constant-current electrolysis lies in its lack of selectivity: all possible reduction processes will take place in turn as the electrode potential continually drifts cathodically. (In the case just cited, for example, the radical anion could be produced only by constant monitoring of the electrolysis mixture to detect the point where the potential shift from E_1 to E_2 occurs.) The synthetic chemist desiring to carry out selective reduction of polyfunctional molecules ought, therefore, to employ

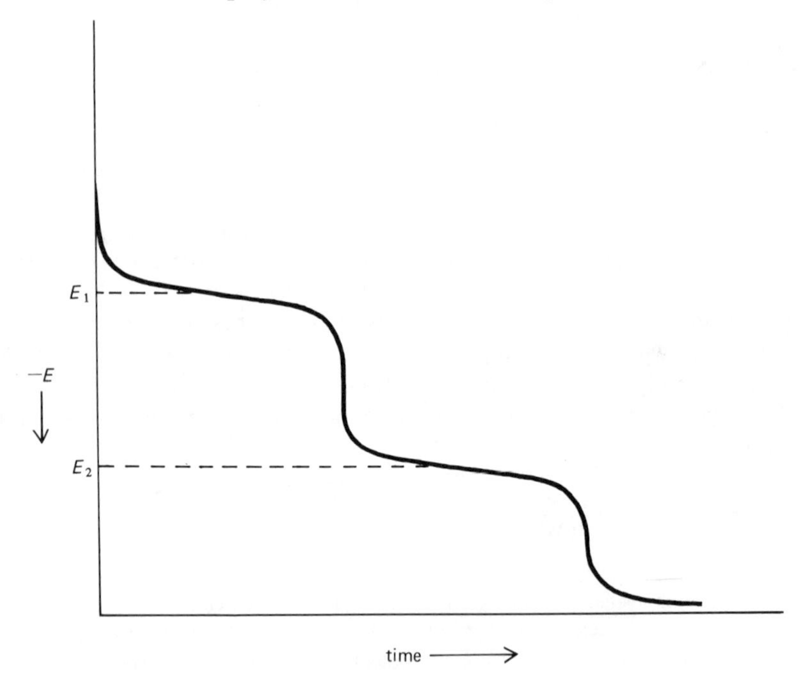

Figure 4.4 Plot of cathode potential *vs* time for the constant-current electrolysis of diphenylfulvene in an aprotic solvent.

controlled-potential electrolysis. There are, however, some situations in which constant-current electrolysis can be used. For example, a species of interest may have only one electrochemically active group, or one may wish to electrolyze at its most cathodic (or anodic, in oxidations) wave. In these situations, one cannot overreduce (or oxidize) the molecule, so control of the electrode potential is unnecessary. Another area where constant-current electrolysis is often used is for the electrochemical *in situ* generation of reagents. For example, suppose one oxidizes a solution containing an olefin and bromide ions. The latter are easily oxidized to molecular bromine, which will then quickly react with the olefin. In this case, one could use a large excess of bromide (as supporting electrolyte), so that the ratio $[Br_2]/[Br^-]$ would remain essentially constant during the electrolysis (appearance of the color of bromine would, incidentally, signal the endpoint of the bromination). In this kind of experiment, one carries out a complete conversion of the organic material even though the electroactive substance (bromide) is hardly consumed.

An area that received much attention prior to invention of the potentiostat involves adjustment of the solvent decomposition potential through variations in solvent, electrolyte, or electrode material.[15] This may be demonstrated in the following manner. Figure 4.5 represents the voltammogram in ethanol of a hypothetical species containing two reducible functional groups. Potential A represents the decomposition potential (at which hydrogen is evolved) of ethanol at a

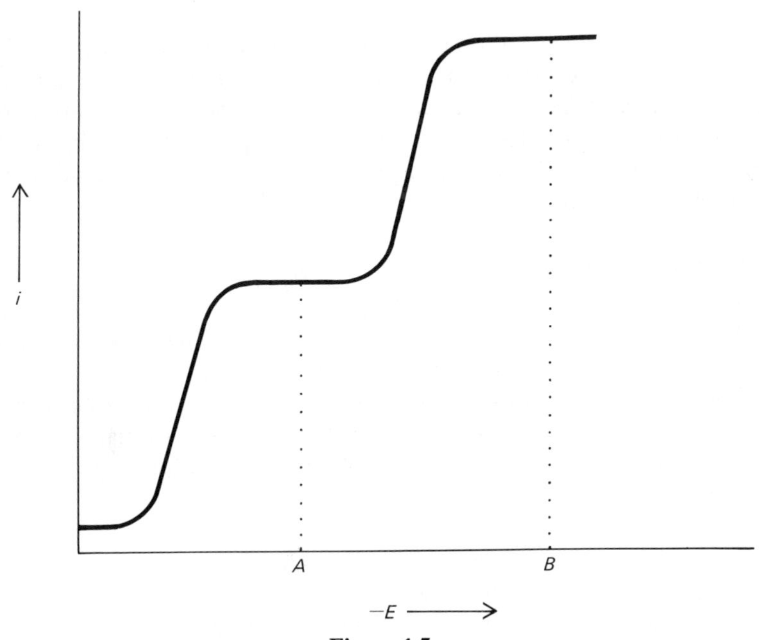

Figure 4.5

copper cathode; *B* represents the decomposition potential of ethanol at mercury. If one carried out constant-current electrolysis at a mercury cathode, complete reduction of the hypothetical substance would take place. If, however, one used a copper cathode, the potential could never become cathodic enough for the second process to take place, and one would thus have carried out a selective reduction at the first voltammetric wave. This procedure can be of only extremely limited use to anyone wishing to electrolyze a wide variety of molecules, and cannot be generally recommended. Despite the existence of a rather broad range of applicability of constant-current electrolysis, one may carry out, using the controlled-potential method, any conversion possible with constant current, as well as many others impossible or quite inconvenient with the latter.

Catalytic Processes

It will be recalled (Section 3.1) that overall indirect reduction of a species Z is sometimes possible, via a catalytic pathway, at potentials where Z is electroinactive:

$$\mathrm{O} + e^- \longrightarrow \mathrm{R}$$

$$\mathrm{R} + \mathrm{Z} \longrightarrow \mathrm{O} + \mathrm{Y}$$

The overall process amounts to

$$\mathrm{Z} + e^- \longrightarrow \mathrm{Y}$$

Looking at the example cited on pp. 68 and 112 from a synthetic point of view, it will be noted that addition of a catalytic quantity of naphthalene enables one to effect the efficient indirect reduction of alkyl chlorides, which are normally inert toward electrochemical reduction but are rapidly reduced by the naphthalene radical anion. This catalytic process could likewise be used to reduce other functional groups satisfying the latter criteria.

Intramolecular Electron Transfer

A process similar in some respects to the preceding catalytic type has been recognized by a number of investigators in recent years. This process consists of addition of an electron to one electrophore* of a complex molecule, followed by *intramolecular* electron transfer and subsequent reaction of another part of the molecule. This behavior has been recognized for a number of years as characteristic of the halonitrobenzenes, for example, for the *p*-nitro compounds (**4**) in acetonitrile[16]:

* Miller has introduced the term *electrophore* to describe any functional group or molecular subunit that is independently electroactive[17]; the word was coined by analogy to the term *chromophore*, commonly used to designate any such subunit capable of undergoing spectroscopic excitation.

$$\text{NO}_2\text{-C}_6\text{H}_4\text{-X}\ (\mathbf{4}) \xrightarrow{e^-} [\text{NO}_2\text{-C}_6\text{H}_4\text{-X}]^{\dot{-}}\ (\mathbf{3}) \xrightarrow[-X^-]{k_1} \text{NO}_2\text{-C}_6\text{H}_4^{\bullet} \longrightarrow \text{products}$$

The nitroaromatic electrophore is undoubtedly the initial electron acceptor in these reductions, even where X is iodine, as may be seen from polarographic data (Table 4.3) and from cyclic voltammetric experiments, which demonstrate the intermediacy of radical anions **3**, whose decomposition rate k_1 depends on X. The carbon–halogen bond cleavage represented by k_1 formally represents leakage of the added electron from the pi system to the orthogonal sigma system. It is

Table 4.3 Reduction potentials of halonitrobenzenes and related compounds[a]

Substrate	$E_{1/2}$ (V), *vs* S.C.E.	References[c]
Nitrobenzene	−1.18	1
p-Iodonitrobenzene	−1.09	1
p-Bromonitrobenzene	−1.15	1
p-Chloronitrobenzene	−1.08	1
Iodobenzene	−1.62[b]	2
Bromobenzene	−2.38	2
Chlorobenzene	−2.58	2

[a] In DMF containing 0.1-*F* tetraethylammonium salts, 25°C.

[b] 75% dioxane–water.

[c] References:

1. T. Kitagawa, T. P. Layloff, and R. N. Adams, *Anal. Chem.*, **35**, 1086 (1963).
2. L. Meites, *Polarographic Techniques*, 2nd ed., New York, Wiley, 1965, pp. 681–682. (Reprinted by permission of John Wiley and Sons, Inc.)

known from esr experiments, however, that there exists some coupling between the two systems (i.e., they are not fully orthogonal),[18] and this coupling presumably provides the mechanism by which bond cleavage takes place. In this experiment one has, in effect, used the pi electrophore (the nitro group) as an intramolecular catalyst for removal of the halogen atom. Other examples of this phenomenon have been recognized during electrochemical reduction of bromotriphenylethylene,[17] halonaphthalenes,[19] nitrobenzyl halides,[20] and halobenzophenones.[21]

4.2 INTERCEPTION OF INTERMEDIATES

4.2.1 Proton Availability

It has been known for some time that the electrochemical behavior of many organic compounds is profoundly affected by proton availability in the medium in which investigations were carried out. Most early polarographic investigations of organic substances were carried out in proton-rich solvents, such as water, ethanol, or aqueous dioxane, since the conductivity of these solvents is relatively high, but even in these solvents the course of an electrode reaction was often observed to differ depending on whether the solution was acidic, neutral, or alkaline. Substantial progress in understanding the role of the proton donor in organic electrode reactions has been achieved through mechanistic investigations in aprotic solvents, such as DMF, acetonitrile (AN), and dimethyl sulfoxide (DMSO).[22,23] Such solvents readily lend themselves to systematic investigation of the effect of the proton on the course of an electrochemical reaction, since one may add varying known amounts of a proton donor, such as phenol* or benzoic acid, to the solvent while monitoring the resulting effects on electrochemical behavior.[25,26] It is often possible to relate the electrochemical behavior of a substance in aprotic solvents containing varying amounts of an added proton donor to the pH dependence of its reactivity in protic solvents. For example, the behavior of a substance in alkaline aqueous or ethanolic solution often parallels its behavior in aprotic media. It would be a mistake to attempt to draw such analogies too closely, however, in view of the many physical and chemical differences between, for example, DMF and alkaline ethanol.

The electrochemical behavior of anthracene (**5**) in DMF containing increasing amounts of phenol is illustrated in Figure 4.6.[27] Observe that in the aprotic solvent two waves are observed, corresponding to stepwise reduction first to the radical anion **6** and then to the dianion **7**. As the phenol concentration is increased, the first wave grows at the expense of the second, until the first wave

* Phenol is considerably superior to either benzoic acid or water as an added proton donor in DMF.[24] The polarogram of a solution of benzoic acid in DMF shows a wave due to reduction of protons to hydrogen (p. 65).[24] No such wave is observed for phenol, even at 1-*M* concentration; phenol is therefore electroinactive, and hence apparently not appreciably dissociated in DMF. Molecular phenol must be the proton donor in this solvent. Water is, on the other hand, a poor proton donor in DMF and DMSO, because the high degree of structure in water–DMF and water–DMSO mixtures drastically decreases the proton-donating ability of water.[25] Water–acetonitrile mixtures are not highly structured, and water is a good proton donor in this solvent.[25]

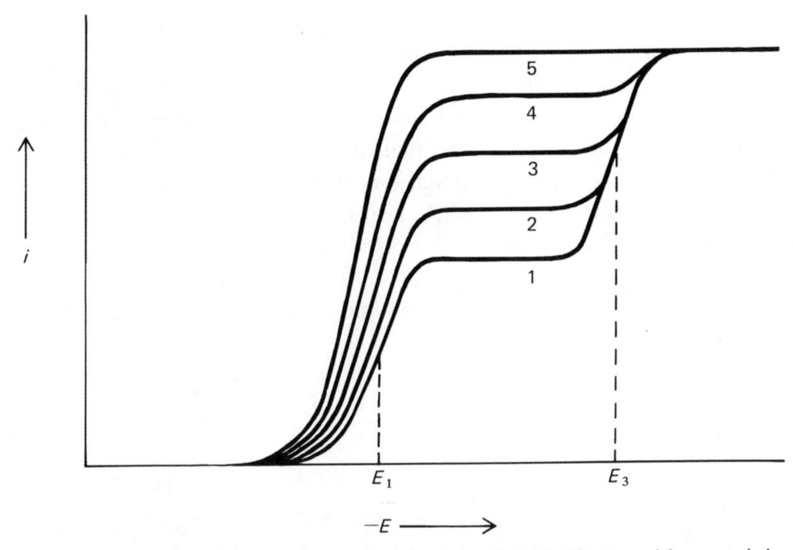

Figure 4.6 Polarogram of anthracene (**5**) (mM) in dimethylformamide containing increasing amounts of phenol. Concentration of phenol in curve 1, 0; curve 2, 10^{-3} M; curve 3, 3×10^{-3} M; curve 4, 8×10^{-3} M; curve 5, 12×10^{-3} M.

has doubled in height and the second has disappeared. The total wave height remains *constant* during these changes, indicating that the single wave at high phenol concentration corresponds to a 2-electron reduction. Hoijtink first observed such behavior during the polarographic reduction of aromatic hydrocarbons in dioxane whose water content ranged from 4 to 25%.[28,65] He attributed the phenomenon to reduction according to the following scheme:

$$R + e^- \xrightleftharpoons{E_1} R^{\dot{-}}$$

I and/or II

I:

$$R^{\dot{-}} + HX \xrightarrow{k_1} RH\cdot$$

$$RH\cdot + e^- \xrightleftharpoons{E_2} RH^-$$

II:

$$R^{\dot{-}} + e^- \xrightleftharpoons{E_3} R^{-2}$$

$$R^{-2} + HX \xrightarrow{k_2} RH^-$$

$$HX + RH^- \longrightarrow RH_2 + X^-$$

In solvents of low proton availability, such as aprotic solvents, 96% dioxane, or alkaline ethanol, the radical anion $R^{\dot{-}}$ formed at potential E_1 will not normally be protonated on the time scale of the experiment; it will instead diffuse away from the electrode into bulk solution. A second 1-electron wave for reduction to the dianion R^{-2} will be observed at potential E_3. (For anthracene in DMF, $R^{\dot{-}}$ and R^{-2} correspond, of course, to **6** and **7**, respectively.) However, in solvents of greater proton donating ability, such as 75% dioxane, neutral ethanol, or DMF containing added phenol, the radical anion $R^{\dot{-}}$ will be protonated rapidly on the polarographic time scale to produce a neutral free radical RH·. Hoijtink showed from Hückel molecular orbital calculations that RH· will, in general, be *easier* to reduce than the starting material R; that is, the potential E_2 at which RH· is reduced will generally be *anodic* of E_1.[29] Reduction of RH· to the anion RH^- is illustrated by a dashed line in Figure 4.7. (In the example cited above, RH·, RH^-, and the final product RH_2 correspond to structures **8**, **9**, and **10**.)

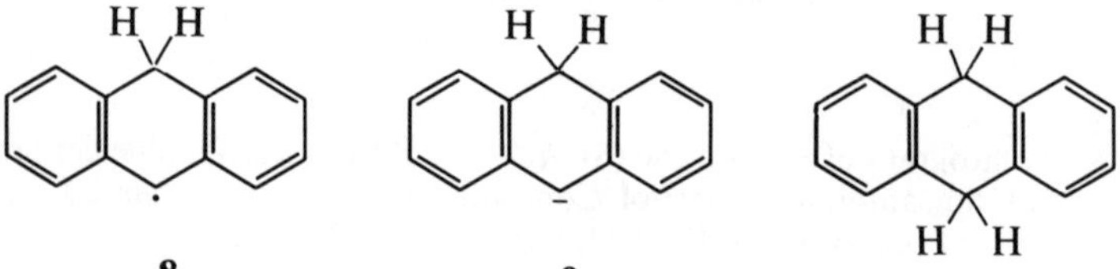

H H H H H H H H

8 **9** **10**

Since RH· is therefore formed by protonation at a potential cathodic of E_2, it will be reduced as rapidly as it is formed, and an overall 2-electron reduction will take place in proton-donating solvents. The change in the polarogram of anthracene with increasing proportions of phenol corresponds to the increasing importance of pathway I, an ECE process (p. 116), relative to pathway II. It is not strictly true that the proton-donor concentration alone determines which reduction path is followed. The latter actually depends on the magnitude of the pseudo-first-order protonation rate constant $\{k_1[HX]\}$, and hence pathway I might even be followed at low proton-donor concentrations whenever $R^{\dot{-}}$ is abnormally basic, that is, when k_1 is very high. This prediction has been confirmed during a study of the electrochemical behavior of several Schiff bases (imines), such as **11**, which follow the ECE pathway even in DMF containing no added proton donor

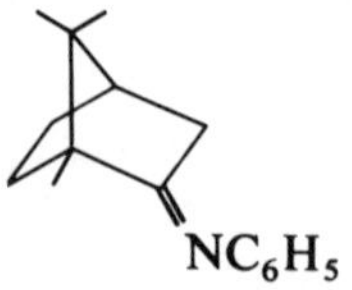

NC_6H_5

11

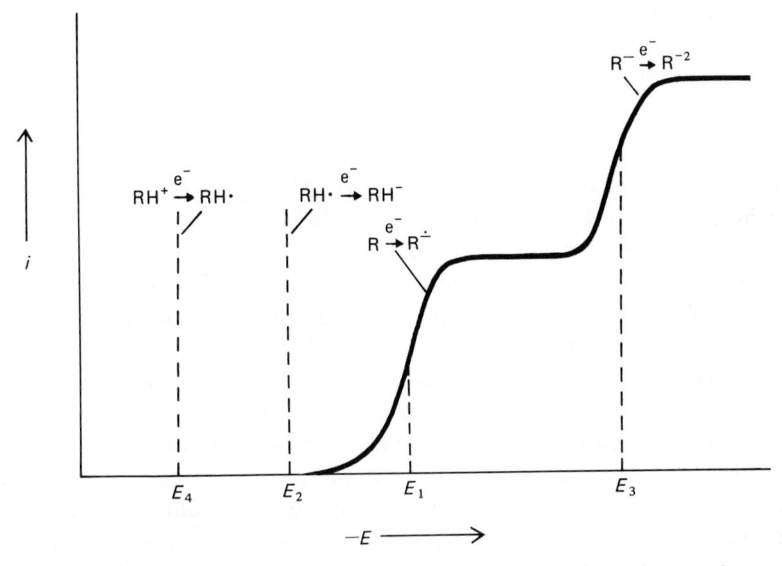

Figure 4.7 Solid line: typical polarogram of an aromatic hydrocarbon under aprotic conditions.

other than TEAB (see footnotes, pp. 88 and 147).[1] It should also be noted that even in the presence of proton donors a highly reactive radical anion $R^{\dot{-}}$ may undergo other reactions, such as disproportionation or dimerization, *faster* than protonation. When this occurs, it can usually be identified polarographically.[22,24] Occasionally, the dihydro derivative RH_2 is itself electroactive, and waves for its reduction are observed.[30]

A further component of this mechanistic scheme is observed in media sufficiently acidic that the starting material is protonated, so that the electroactive species is actually the conjugate acid RH^+ (or equivalently, a complex between R and HX).[22] Because of the positive charge on the latter, its reduction potential E_4 will normally be anodic of E_1 and even E_2, and a typical polarogram (Figure 4.7) will consist of two 1-electron waves, just as in aprotic media, but corresponding instead to the two steps

$$RH^+ + e^- \xrightarrow{E_4} RH\cdot$$

$$RH\cdot + e^- \xrightarrow{E_2} RH^-$$

Aromatic hydrocarbons are not sufficiently basic to exhibit this behavior in the solvents normally used in organic electrochemistry, but it has been detected for a number of substrates containing more or less basic heteroatoms, such as aromatic ketones and Schiff bases. Lund,[31] for example, reported that benzophenone anil (**12**)

$$(C_6H_5)_2C{=}NC_6H_5$$

12

exhibits two 1-electron waves in ethanol in the pH range 5–7.4, a single 2-electron wave from pH 7.4 to 10.75, and two 1-electron waves again at pH 13.* These three modes of behavior are then, in order: (a) stepwise reduction of the conjugate acid of **12** in acidic media, (b) 2-electron reduction by the ECE pathway in neutral media (pH 7.4–10.75), and (c) stepwise reduction to the radical anion and dianion of **12** in alkaline media.

The use of polarography in selecting conditions for controlled-potential preparative electrolysis is clearly illustrated in the preceding example. If one wished to reduce **12** to the corresponding amine **13**, one would select any set of conditions (pH 7.4–10.75) where a single 2-electron wave is observed, while electrolysis under either very acidic or very alkaline conditions would be selected if one desired dimeric product **14**, formed via 1-electron reduction. Lund did, in fact,

$$(C_6H_5)_2CHNHC_6H_5$$

13

$$\begin{array}{c}(C_6H_5)_2CNHC_6H_5\\ |\\ (C_6H_5)_2CNHC_6H_5\end{array}$$

14

isolate **13** in 91% yield by electrolysis at pH 10.[31] Electrolysis at the first wave at pH 6 was impossible, however, because of difficulty in buffering the solution, and electrolysis was not attempted at pH 13. Electrolysis of aryl ketones may likewise be guided by polarographic behavior: electrolysis at the first wave under acidic conditions, where two 1-electron waves are observed, affords pinacols in high yields, and electrolysis at either the second wave under such conditions or at the single 2-electron wave observed in neutral or weakly alkaline media affords the corresponding carbinols.[32]

The same generalizations concerning the choice of electrolysis conditions may be made with respect to electrolysis in aprotic solvents, with the added possibility, not available in protic media, of preparing more or less stable solutions of $R^{\bar{\cdot}}$ in the absence of proton donors. Thus, given the preceding discussion of the effects of proton availability, it may be seen that electrolysis of typical substrates may be diverted in a desired direction by the proper choice of conditions as revealed by polarographic experiments (Table 4.4). The reduction of

* Since the solvent used by Lund was not water, the term "pH" is not correct. This, together with the fact that the basicity of **12** is undoubtedly enhanced in the field of the double layer (pp. 156 and 209), requires that only qualitative significance be given to these pH values.

Table 4.4 Product dependence on experimental conditions in the electrochemical reduction of representative aromatic compounds

Substrate	Desired Product	Experimental Conditions
Anthracene (**5**)	Anthracene radical anion (**6**)	First wave aprotic medium[a]
	9,10-Dihydroanthracene (**10**)	Second wave, aprotic medium[b] or Neutral protic medium[a] or Aprotic medium containing lithium ion[b] or Aprotic medium containing a weak acid[a]
	9,9-Dihydro-10,10-dianthracene (**15**)	First wave in "super acid"[b]
Benzophenone	Benzophenone radical anion	First wave, aprotic medium [a]
	Benzhydrol	Second wave, aprotic medium[a] or Neutral protic medium (pH 5–11)[a] or Second wave in acid (pH < 5)[a] or Aprotic medium containing lithium ion[b] or Second wave in alkali (pH > 11)[b] or Aprotic solvent containing a weak acid[a]
	Benzpinacol (**16**)	First wave, acidic medium (pH < 5)[a] or First wave, aprotic solvent containing a strong acid[a]

[a] Observed experimentally; see Chapters 4, 6, and 7.

[b] Predicted.

anthracene to **15** in highly acid media has not been reported but is clearly implied

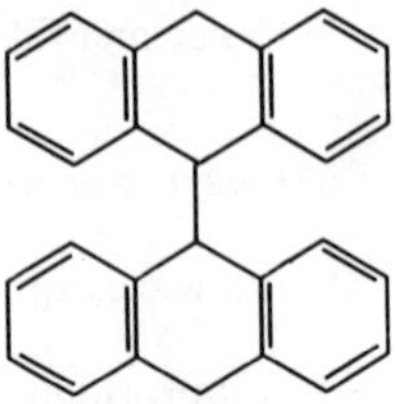

15

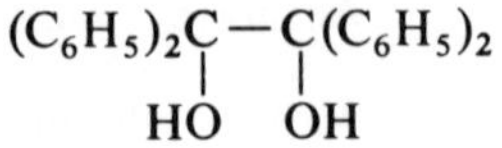

$(C_6H_5)_2C(OH)-C(OH)(C_6H_5)_2$

16

by the previous discussions. One would require an acid strong enough to form at least small quantities of the conjugate acid of the hydrocarbon, while not effecting electrophilic attack upon it. Several such acids have been reported in recent years, including fluorosulfonic acid and other "super acids."[33] Conceivably, reduction of the acid itself (to hydrogen) could take place before reduction of anthracene; in that case, an alternative would be electrolysis in the presence of a strong aprotic Lewis acid, such as boron trifluoride.

4.2.2 Trapping Agents

It is generally possible to intercept electrochemically generated intermediates whenever the trapping agent used (1) does not react with the electroactive substance prior to electrolysis, and (2) is itself electrochemically inert at the potentials used. A prominent example of the use of trapping agents is, of course, the addition of proton donors to aprotic solvents to achieve the effects just discussed. One might wish to add other agents, however, either to stabilize highly reactive electrochemically generated species, to divert a reaction down a new pathway, or perhaps to demonstrate the intermediacy of a putative species by actually intercepting it. Wawzonek and coworkers, for example, demonstrated the intermediacy of carbanionic intermediates during the electrochemical reduction of diphenylacetylene in aprotic solvents by isolation of diphenylmaleic anhydride (**17**), diphenylfumaric acid (**18**), and *meso*-diphenylsuccinic acid (**19**) when carbon

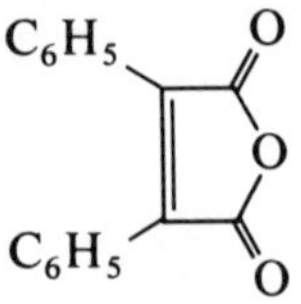

C_6H_5, C_6H_5, O, O, O

17

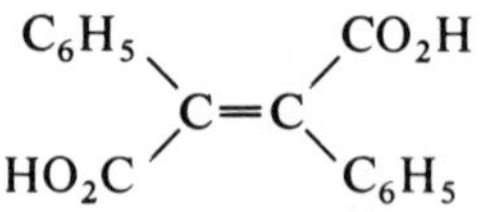

$(C_6H_5)(HO_2C)C{=}C(CO_2H)(C_6H_5)$

18

$C_6H_5CH(CO_2H)CH(CO_2H)C_6H_5$

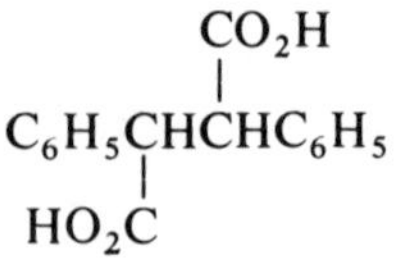

19

dioxide was bubbled through the solution during electrolysis.[34] The precise structure of such intermediates is, however, unclear. Wawzonek and coworkers suggested the following mechanism for the reaction:

$$C_6H_5C{\equiv}CC_6H_5 + e^- \longrightarrow \underset{\mathbf{20}}{C_6H_5\dot{C}{=}\bar{C}C_6H_5} \xrightarrow{e^-} \underset{\mathbf{21}}{(C_6H_5)\bar{C}{=}\bar{C}(C_6H_5)\ (cis)} + \underset{\mathbf{22}}{(C_6H_5)\bar{C}{=}\bar{C}(C_6H_5)\ (trans)}$$

$$\mathbf{21} \xrightarrow{2CO_2} (C_6H_5)(^-O_2C)C{=}C(C_6H_5)(CO_2^-) \xrightarrow[-H_2O]{2H^+} \mathbf{17}$$

$$\mathbf{22} \xrightarrow{2CO_2} (C_6H_5)(^-O_2C)C{=}C(CO_2^-)(C_6H_5) \xrightarrow{2H^+} \mathbf{18}$$

It was suggested that dianions **21** and **22** were the precursors of **17** and **18**, respectively, and that **19** arose through reduction of **17** or **18**. The acetylene exhibits two 1-electron polarographic waves in DMF, however, and formation of either **21** or **22** would require reduction at the potential of the *second* wave. Unfortunately, electrolyses were carried out under constant-current conditions, so formation of dianions cannot rigorously be excluded. Nevertheless, the following ECE mechanism is much more likely, by analogy to the role of proton donors in intercepting radical anions (pathway I, p. 137):

$$\mathbf{20} + CO_2 \longrightarrow \underset{\mathbf{23a}}{(C_6H_5)\dot{C}{=}C(C_6H_5)(CO_2^-)} \rightleftharpoons \underset{\mathbf{23b}}{(C_6H_5)\dot{C}{=}C(C_6H_5)(CO_2^-)}$$

$$\mathbf{23a} \xrightarrow{e^-} (C_6H_5)\bar{C}{=}C(C_6H_5)(CO_2^-) \xrightarrow[2H^+,\ -H_2O]{CO_2} \mathbf{17}$$

$$\mathbf{23b} \xrightarrow{e^-} (C_6H_5)\bar{C}{=}C(C_6H_5)(CO_2^-) \xrightarrow[2H^+]{CO_2} \mathbf{18}$$

Isomeric free radicals, whether generated chemically or electrochemically, are known to interconvert rapidly, so loss of stereochemistry in radicals **23a** and **23b** would appear to be reasonable.* This mechanism could be distinguished from that proposed by Wawzonek and coworkers by carrying out electrolysis at the potential of the first reduction wave of diphenylacetylene in the presence of carbon dioxide. Isolation of **17** and **18** in the same ratio (1 : 2.3) as in the constant-current electrolysis would consitute very substantial evidence in favor of the ECE mechanism, although not rigorously disproving dianion intermediates under constant-current conditions. Likewise, formation of 1,4-dicarboxy-1,4-dihydronaphthalene (**26**) of unspecified stereochemistry in 43% yield by electrolysis of naphthalene in the presence of carbon dioxide quite likely also involves an ECE mechanism, as suggested by Wawzonek and Wearring[34]:

e⁻ → **24** —CO_2→ H CO_2^- **25**

25 —e⁻→ H CO_2^- —CO_2, $2H^+$→ H CO_2H, H CO_2H **26**

rather than the dianion **27**.

27

These examples have been introduced not only to illustrate the use to which carbon dioxide may be put as a trap for carbanions (and therefore in syntheses of carboxylic acids and diacids), but also to indicate the difficulty that typically arises when one attempts to assign a mechanism to an electrode reaction carried out under constant-current conditions. Since all possible reduction paths may

* In fact it is now believed that phenyl-substituted vinyl radicals are linear.[35]

take place, it will not always be clear on which step in the reduction process the trapping agent is operating. The use of controlled-potential electrolysis, in conjunction with voltammetric data, is preferable here as elsewhere.

Acetic anhydride is, like carbon dioxide and proton donors such as phenol, a useful anion trap in aprotic solvents.[36, 37] This substance, although difficult or impossible to reduce electrochemically, is quite reactive toward nucleophiles. Curphey et al. took advantage of these properties to intercept labile electrochemically generated cyclopropane diols as their acetates[38]:

$$\xrightarrow[2Ac_2O]{2e^-} \quad + \; 2OAc^-$$

$$\xrightarrow[2Ac_2O]{2e^-} \quad + \; 2OAc^-$$

Acetic anhydride (and dialkyl sulfates, which are also very hard to reduce) have not been used as frequently as their synthetic potentialities as carbanion traps would appear to warrant.

Neutral radical intermediates in electrochemical reactions are not as easily intercepted as are anions, since the radicals themselves, as well as many molecules often used as radical traps, are often easily reduced. There are a few reports of radical trapping, however. Sugino and Nonaka, and others, employed activated olefins to intercept ketyl radicals generated during reduction of acetone under acidic conditions.[39,40] In these reactions a carbon–carbon bond is formed in synthetically useful yields:

$$(CH_3)_2C{=}O \xrightarrow[e^-]{H^+} (CH_3)_2\dot{C}OH \xrightarrow{CH_2{=}CHCN}$$

$$(CH_3)_2\underset{\displaystyle OH}{\underset{|}{C}}CH_2\dot{C}HCN \xrightarrow[H^+]{e^-} (CH_3)_2\underset{\displaystyle OH}{\underset{|}{C}}CH_2CH_2CN$$

A number of examples of interception of anodically generated cations, cation–radicals, and dications have been uncovered in recent years; for example,

Cation

$\xrightarrow[-H^+]{-2e^-}$ CH_2^+ $\xrightarrow{HOAc}$ CH_2OAc

(*Ref. 41*)

Dication

$\xrightarrow{-2e^-}$ $\xrightarrow{2OAc^-}$ OAc AcO

(*Ref. 42*)

Cation Radical

C_6H_5 C_6H_5 $\xrightarrow{-e^-}$ C_6H_5 C_6H_5 $\xrightarrow{Pyr}$ Pyr^+ C_6H_5 C_6H_5

28

28 $\xrightarrow{-e^-}$ Pyr^+ C_6H_5 C_6H_5 $\xrightarrow{Pyr}$ Pyr^+ C_6H_5 Pyr^+ C_6H_5

(*Ref. 43*)

(Pyr = Pyridine)

It is interesting to note the effect of the 9,10 substituents on the reactivity of anthracene derivatives. The indicated oxidation of 9,10-dimethylanthracene was shown unequivocally (in contrast to the constant-current reduction of diphenylacetylene previously discussed) to involve a doubly charged intermediate, the anthracene dication.[42] Oxidation of 9,10-diphenylanthracene, on the other hand, was found to proceed to the *bis*-pyridine salt via an ECEC process.[43] These differences are presumably related at least in part to differences in both the redox properties of the corresponding radical cations and the nucleophilicity of trapping agents. (Parker has commented on this problem.)[42]

The preceding discussion will suggest that the variety of available trapping agents is limited only by the chemist's imagination. It must also be pointed out here, incidentally, that every electrolysis mixture already contains several potential traps for reactive intermediates. The solvent itself may, for example, sometimes become implicated in electrode reactions. While one usually strives to avoid this, there may be situations in which it is desirable. Furthermore, every electrolysis mixture contains a supporting electrolyte, and possible trace impurities, especially water.* For example, carbonium ions generated electrochemically during the oxidation of organic substrates may react either with the solvent or with traces of water in the solvent[44a]:

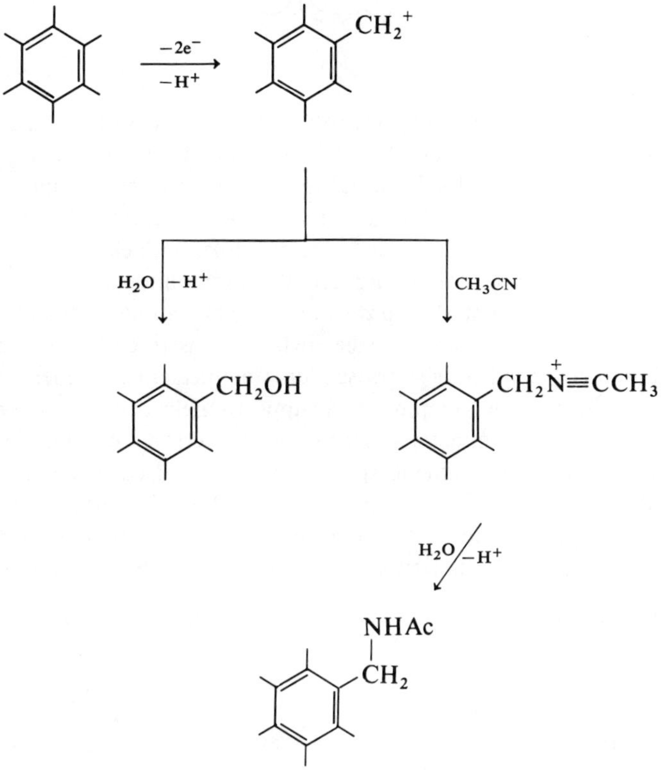

$\xrightarrow[-H^+]{-2e^-}$ CH_2^+

H_2O $-H^+$ CH_3CN

CH_2OH $CH_2\overset{+}{N}{\equiv}CCH_3$

H_2O $-H^+$

NHAc

CH_2

Proton abstraction (Hofmann elimination) by electrochemically generated carbanions on tetraalkylammonium ions used as supporting electrolyte would also fall into this category (p. 88). Since water, almost all organic solvents, and most anions of salts used as electrolytes are nucleophilic, reactions of electrochemically generated intermediates with the solvent or another constituent of the solution are particularly common during electrochemical oxidations, in which cationic

* Solvents dried by typical standard procedures often contain water at concentrations ≥ 10 m*M*, especially after even brief exposure to the atmosphere. Rigorous drying and manipulative procedures are generally required to reduce the water concentration below this level.[44] The nature of the problem will become clear when it is recalled that many electrochemical experiments are carried out with the concentration of electroactive material only 1–10 m*M*.

(electrophilic) species are generated. The anthracene radical action **29**, for example, is extremely shortlived under conditions where the corresponding radical anion **6** is stable.[44,45]

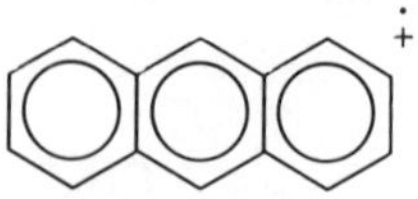

29

The nature of the supporting electrolyte may, on occasion, exert a dominant effect on the very course of an electrode reaction. In particular, a number of investigators have observed substantial differences in electrochemical behavior between lithium and tetraalkylammonium ions as the cation of the supporting electrolyte. Benzil[46] and azobenzene,[47] for example, both exhibit two 1-electron reduction waves in DMF containing tetraalkylammonium salts. Electrolysis of either substance at a potential atop the first wave affords solutions of the corresponding radical anion, whose existence can be demonstrated by esr spectroscopy. Cyclic voltammetry confirms a stepwise reduction mechanism under these conditions. When a lithium salt is employed as supporting electrolyte, however, polarography and cyclic voltammetry show a single 2-electron reduction step for each compound, and no paramagnetic species can be detected when electrolysis is carried out in the cavity of the esr spectrometer.[48,49] Furthermore, addition of lithium salts to a solution of the azobenzene radical anion **30** results in its instantaneous disproportionation into azobenzene and the dilithium salt **31**.[49]

$$2C_6H_5\dot{N}-\bar{N}C_6H_5 + 2Li^+ \longrightarrow C_6H_5N{=}NC_6H_5 + (C_6H_5)(Li)N{-}N(Li)(C_6H_5)$$

30 **31**

These results, as well as similar phenomena observed during the electrochemical reduction of quinones[50] and ferric acetylacetonate[51] in the presence of lithium ion, have been ascribed to the powerful coordinating ability of lithium. Thus, although disproportionation equilibria normally lie well on the side of the radical anion, that observed here for the azobenzene radical anion in the presence of lithium ion is apparently driven to the right by formation of two strong covalent bonds to lithium in **31**. Philp, Layloff, and Adams, suggested[48] an ECE mechanism for reduction of benzil in the presence of lithium ion:

$$C_6H_5COCOC_6H_5 \xrightarrow{e^-} [C_6H_5COCOC_6H_5]^{\dot{-}} \xrightarrow{Li^+}$$

$$C_6H_5CO\dot{C}(OLi)C_6H_5 \xrightarrow[Li^+]{e^-} C_6H_5C(OLi){=}C(OLi)C_6H_5$$

Disproportionation of the benzil radical anion, as with azobenzene, could also account for the observed result. These results imply that one may include reduction in an aprotic solvent in the presence of lithium or any strongly coordinating ion among the methods (Table 4.4) for effecting overall 2-electron reduction of unsaturated substrates. Conversely, lithium salts are to be avoided when one wishes to effect 1-electron reduction in aprotic solvents.*

4.2.3 Electrochemically Generated Reagents

Somewhat akin to the interception of electrochemically generated organic intermediates just discussed is the procedure of *in situ* electrochemical generation of a reagent, usually inorganic, for reaction with the organic substrate of interest. This has been discussed earlier (Section 4.1.3), but deserves mention in the present context and because it may not even be clear in some reactions which component of the mixture is actually the electroactive one. For example, electrochemical oxidation of aromatic hydrocarbons in the presence of anions such as cyanide, methoxide, and hydroxide, results in ring substitution[52]:

$$\mathrm{ArH} \xrightarrow[+\mathrm{X}^-]{-2e^-} \mathrm{ArX} + \mathrm{H}^+ \qquad (\mathrm{X} = \text{a nucleophilic anion})$$

The initial electron-transfer step in such reactions has been the object of considerable dispute. One could imagine initial oxidation of either the aromatic species or the anion, that is,

$$\mathrm{ArH} \xrightarrow{-e^-} \mathrm{ArH}\dot{+} \xrightarrow{\mathrm{X}^-} \mathrm{Ar}\dot{\mathrm{H}}\mathrm{X} \xrightarrow{-e^-} \mathrm{ArHX}^+ \xrightarrow{-\mathrm{H}^+} \mathrm{ArX}$$

or

$$\mathrm{X}^- \xrightarrow{[\mathrm{O}]} \mathrm{X}\cdot,\ \mathrm{X}_2,\ \text{or}\ \mathrm{X}^+ \xrightarrow{\mathrm{ArH}} \text{products}$$

It is now clear (Chapter 8) that most such reactions proceed by the first pathway, but the second may be followed in a few instances.

The electrochemical reduction of isolated acetylenic linkages at a spongy nickel[53] electrode in protic solvents should be mentioned under the present category.

$$\mathrm{RC{\equiv}CR} + 2e^- + 2\mathrm{H}^+ \longrightarrow \textit{cis-}\mathrm{RCH{=}CHR}$$

* Evans's observation (p. 255) that yields of "hydrodimer" from the electrochemical reduction of α,β-unsaturated ketones in dimethyl sulfoxide are much higher in the presence of lithium ion represents an exception to this rule. Apparently, coordination of the radical anion to lithium lowers the electron density on the radical anion. Consequently, not only is dimerization faster, but also the neutral radical resulting from coordination of the radical anion to lithium is a weaker nucleophile and is less able to initiate anionic polymerization of the starting material than is the radical anion.

Although once thought to involve direct reduction of the organic substrate, it is now clear that it actually proceeds via reduction by electrochemically generated hydrogen, with the electrode serving as catalyst. This conclusion is supported by the fact that no reaction is observed at metal cathodes (e.g., mercury), which are poor hydrogenation catalysts, and by the observed stereochemistry of reduction. Electrochemically generated hydrogen is undoubtedly responsible for a great deal of other chemistry that has been observed in the past at platinum cathodes in protic media.

Although a number of common inorganic reagents (Cl_2, Br_2, I_2, Fe^{+2}, H^+, OH^-) may be readily generated electrochemically *in situ*, there would appear to be relatively little synthetic advantage associated with this procedure as opposed to addition from an external source (except possibly where it might be desired to add the reagent slowly at extremely low concentrations. By controlling the current it would be possible, for example, to generate such reagents at steady-state concentrations as low as $10^{-6} M$.*).

Probably the most interesting electrochemically generated reagent from a synthetic point of view is the solvated electron. This is, in a sense, the simplest possible reducing agent, although it is at the same time very powerful and hence unselective. One may take advantage of the moderate stability of the solvated electron in solvents such as ammonia, low-boiling amines, and hexamethylphosphoramide (HMPA) to carry out some useful chemistry with it, with aid of a simple constant-current source (or at controlled potential if desired). Benkeser and coworkers have effected reduction of a variety of structural groups, including the aromatic ring, acetylenes, carbonyl compounds, imines, and so on, through generation of solvated electrons by electrolysis of a solution of lithium chloride at a platinum cathode in a low-boiling amine.[55] The procedure requires only catalytic quantities of lithium:

$$Li^+ + e^- \longrightarrow Li^0$$

$$Li^0 \xrightarrow{RNH_2} Li^+ + e^-(\text{solv})$$

In a useful modification, Benkeser and Kaiser[55a, b] placed a semipermeable partition between the anode and cathode during electrolysis of aromatic hydrocarbons by this technique. This partition prevents protons generated by oxidation of the solvent at the anode from protonating carbanions formed in the reduction. As a

* Studies of the *rates* of reaction of electrochemically generated reagents offer, on the other hand, the possibility of genuinely valuable application in physical organic kinetic studies. For example, rates of bromination of alkenes lying in the range 1–$10^6\ M^{-1}\ sec^{-1}$ have been measured electrochemically by Dubois.[54a] It has been estimated that rate constants over the range 10^{-4}–$10^6\ M^{-1}\ sec^{-1}$ are accessible using the technique of constant-current coulometry.[54b]

consequence, therefore, of the highly basic conditions in the cathode compartment of this divided cell, initially formed products are isomerized into conjugation and reduced further to the tetrahydro stage:

$$C_6H_6 + 2e^- + 2H^+ \longrightarrow \text{1,4-cyclohexadiene} \xrightarrow{\text{base-catalyzed isomerization}} \text{1,3-cyclohexadiene} \longrightarrow \text{cyclohexene}$$

When the anode and cathode compartments were not separated, the mixture did not become highly basic, and (since solvated electrons do not reduce isolated olefinic linkages) the initially formed unconjugated dihydroaromatics could be isolated. The reaction mixture in the undivided cell consists of a proton donor (protons formed at the anode) and a reducing agent (the solvated electron), and therefore contains the essential elements of the well-known Birch reduction.[56] In view of its simplicity, this may be the experimental method of choice for effecting the Birch reduction of many compounds, especially when large-scale preparations, which would otherwise require the possibly hazardous use of large amounts of lithium, are contemplated.

Arapakos, Scott, and Huber reported the somewhat surprising observation that many aromatic and aliphatic nitriles are reduced to hydrocarbons by solvated electrons generated via electrolysis of lithium chloride in HMPA[57]:

n-octyl nitrile ⟶ *n*-heptane (70%)

benzonitrile ⟶ benzene (100%)

phenylacetonitrile ⟶ toluene (90%)

The reduction of benzonitrile, in which the aromatic ring survives intact, indicates that carbon–carbon bond cleavage is actually the preferred pathway over Birch reduction of the aromatic ring. (The present author has been unable to find in the literature reports of submission of benzonitrile to the usual conditions of the Birch reduction.) Arapakos et al. suggested the following mechanism for the reduction:

$$RCN + e^-(solv) \longrightarrow [RCN]^{\dot{-}} \longrightarrow R\cdot + CN^-$$

$$R\cdot + e^-(solv) \longrightarrow R^- \longrightarrow RH$$

It was also reported that reduction of tridecanenitrile (**32**) is reduced to a mixture of 35% dodecane (**33**) and 65% tridecylamine (**34**) when the nitrile is added to a

$$\underset{\mathbf{32}}{n\text{-}C_{12}H_{25}CN} \longrightarrow \underset{\mathbf{33}}{n\text{-}C_{12}H_{26}} + \underset{\mathbf{34}}{n\text{-}C_{13}H_{27}NH_2}$$

solution of solvated electrons formed by prior dissolution of lithium in ethylamine. Cleavage of nitriles to hydrocarbons is, on the other hand, the preferred path over reduction to amines when solvated electrons are generated by constant-current electrolysis of a solution of lithium chloride and the nitrile in ethylamine (*cf.* the above-noted reduction of $n\text{-}C_7H_{15}CN$). The reason for this difference is not clear. It may be that the higher bulk concentration of solvated electrons in solutions of lithium in ethylamine favors further reduction of the initial radical anion to a dianion; that is,

$$RCN + e^-(solv) \rightleftharpoons RCN^{\dot{-}} + e^-(solv) \rightleftharpoons$$

$$RCN^{-2} \xrightarrow{2H^+} RCH{=}NH \xrightarrow[2H^+]{2e^-} RCH_2NH_2$$

The electrochemical method is obviously the preferred method for cleavage of nitriles.*

4.3 SURFACE PHENOMENA

Theory of the structure of the double layer in aqueous solution is fairly well advanced at present. The nature of surface phenomena, that is, effects due to adsorption and double-layer structure, is still in a relatively primitive stage for organic solutes and solvents, however. Therefore, these phenomena are at present largely empirical, and often not amenable to manipulation in a predictable way. This does not mean that they may be simply ignored, though, for electron transfer is, after all, a heterogeneous process, taking place at the highly structured electrode–solution interface. For this reason, and because many organic compounds are adsorbed on electrodes to a greater or lesser degree, one ought to expect a variety of organic electrochemical surface phenomena. There are described below a number of ways in which adsorption and double-layer effects have been observed to affect the course of electrode reactions. No attempt has been made to present a comprehensive list; rather, the discussion includes a representative

* The electrochemical cleavage of several aryl cyanides, such as 4-cyanopyridine,[58] 4-cyanobenzoic acid,[59] and *m*-cyanobenzonitrile,[59] under alkaline conditions, may also involve intermediate radical anions; these reactions do not appear, however, to involve solvated electrons, but rather direct electron transfer from the electrode to the organic substrate.[60]

sampling of the varieties of possible surface behavior, at the same time representing, it must be admitted, the author's subjective evaluation of those phenomena that are likely to be of the greatest interest in synthetic electrochemistry. It is to be hoped that this promising area will become more amenable to systematic application as more data on these phenomena become available.

4.3.1 Field Effects

It will be recalled (p. 57) that a molecule located in the double layer is immersed in a strong electrical field, which may reach strengths of 10^6–10^7 V/cm or higher, particularly at potentials well removed from the pzc. At such high field strengths, molecules possessing a permanent dipole moment will be highly oriented at the electrode surface by the field. In addition, if a molecule happens to be polarizable as well as polar, its dipole moment can be considerably enhanced, sometimes by a factor of two or more,[61] by the field. Effects associated with the ability of the field both to orient and to polarize substrates are undoubtedly very common, although they have not often been explicitly recognized. At potentials removed from the pzc most organic molecules are surely highly oriented, and often polarized, in the electrical field near the electrode surface. A number of electrochemical phenomena have been traced to these *field effects.*

Stilbenediols

Vincenz-Chodkowska and Grabowski reported in 1964 that benzil may be reduced to a mixture of *cis*- and *trans*-stilbenediols, **35** and **36**, respectively, in neutral or acidic aqueous ethanol[62]:

$$C_6H_5COCOC_6H_5 \xrightarrow{2e^-} \underset{\mathbf{35}}{(C_6H_5)(HO)C{=}C(C_6H_5)(OH)} + \underset{\mathbf{36}}{(C_6H_5)(HO)C{=}C(OH)(C_6H_5)}$$

It could be shown that **35** and **36** isomerize only slowly (several hours for complete conversion) to benzoin, the ultimate electrolysis product. It was demonstrated, furthermore, that changes in experimental parameters that would be expected to affect the intensity of the field at the electrode surface also affected the ratio of **35** to **36** very markedly. The parameters investigated included electrode potential, ionic strength, temperature, and the nature of the supporting electrolyte, and the ratio of **35** to **36** could be varied from ca. 1 : 50 to ca. 2 : 1, depending on conditions. Specifically, the *cis*-isomer **35** was the predominant product under conditions of high field strength. It was assumed that all reductions initially formed the *trans*-diol **36**, and the observed product ratios were related to the field effect on the equilibrium ratios and the rates of interconversion of **35** and **36**. It was suggested that polarization of the diols in the field should reduce the double-bond character of the central bond, and hence increase the interconversion rate.[62] While the results of Vincenz-Chodkowska and Grabowski, which have been confirmed and extended by Stapelfeldt and Perone,[63] almost certainly do indicate a field effect on the ratio of **35** to **36**, it is possible that, rather than facilitating the interconversion of **35** and **36**, the field actually affects the relative proportions of rotamers **37** and **38** in the starting material:

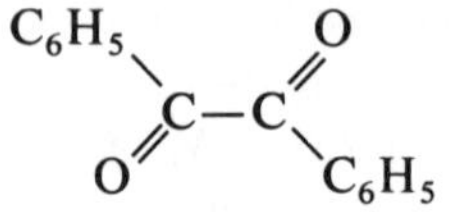

37 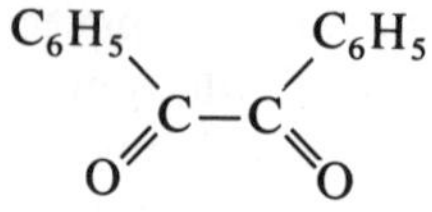 **38**

A combination of field effects on both starting material and product (and even intermediates in the reduction) would appear most reasonable as the origin of these stereochemical results. This notion has also recently been accepted by Grabowski et al.[64]

Aromatic Dianions

In both this and the previous chapter, it has been noted that reduction of aromatic hydrocarbons in media of low proton availability (aprotic or alkaline solvents) proceeds in stepwise fashion first to a radical anion and then to a dianion. Using modified Hückel molecular orbital calculations, Hoijtink predicted a striking field dependence of the electron density in the dianion of several aromatic hydrocarbons.[65] His calculations showed, for example, that the positions of highest electron density in the anthracene dianion **7**,

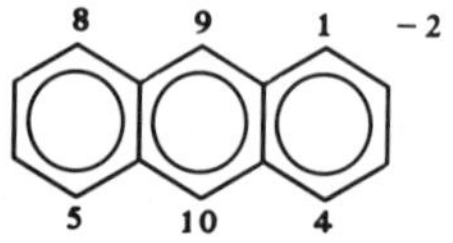

7

are carbons 9 and 10 when the molecule is in a field of strength $\leq 6 \times 10^7$ V/cm, but that carbons 1, 4, 5, and 8 bear most of the negative charge at field strengths $\geq 7 \times 10^7$ V/cm. Assuming that protonation occurs at the positions of highest electron density, Hoijtink suggested the following scheme for reduction at different field strengths:

7 → field strength $\leq 6 \times 10^7$ V/cm → **10**

7 → field strength $\geq 7 \times 10^7$ V/cm → [dihydroanthracene] → base-catalyzed isomerization → [dihydroanthracene] → $2H^+$, $2e^-$ → **39**

Similar field-dependent behavior was predicted for pyrene, benzanthracene, triphenylene, and terphenyl, but electron density in the naphthalene dianion was predicted to be highest at the α positions irrespective of field strength. Hoijtink then examined data from the literature concerning the course of reduction of these hydrocarbons under different conditions, such as electrochemical, catalytic, and dissolving-metal reductions. He suggested that field effects could account, *inter alia*, for the different products obtained from the electrochemical and sodium–ethanol reductions of anthracene, which produce respectively 9,10-dihydroanthracene (**10**), and a mixture of **10** and 1,2,3,4-tetrahydroanthracene (**39**). Inasmuch as sodium–alcohol reductions may actually involve an ECE pathway rather than dianions,[1] this comparison may not be apt, but Hoijtink has also suggested that protonation of radical anions should be field dependent in the same way as protonation of dianions.[65] There is great need for an extensive test of both the validity and the generality of these predictions, not only because of their theoretical interest, but because they are capable in principle of correlating a great deal of data in the literature concerning the position of protonation of species generated electrochemically and otherwise. It would appear that these predictions could best be tested in aprotic media by systematic investigation of changes in the same parameters (electrolyte, temperature, ionic strength, potential, etc.) used by Vincenz-Chodkowska and Grabowski in their studies of field effects on reduction of benzil to the stilbenediols. If Hoijtink's calculations are correct, changes in the position of highest electron density in aromatic dianions should take place over a rather narrow range of field strengths, and hence the position of protonation could be affected by relatively minor changes in experimental conditions.[65]

Carbonyl Compounds

The carbonyl group is polar and polarizable, and therefore could undoubtedly be affected by the field. One should, in particular, expect increased contribution to the resonance hybrid from the dipolar form **40b** when a carbonyl-containing

$$\underset{\textbf{40a}}{R_2C{=}O} \longleftrightarrow \underset{\textbf{40b}}{R_2\overset{+}{C}{-}O^-}$$

compound is immersed in the electrical field at the electrode surface during reduction. Carbonyl compounds should, therefore, be considerably more basic at the electrode surface than in bulk solution. A number of lines of evidence support this conjecture. Elving and Leone found, for example, that half-wave potentials for several alkyl aryl ketones shifted 59 mV for each unit change in pH, indicating that an electron and a proton must be added simultaneously to the carbonyl group in the transition state for reduction.[66] This behavior was attributed to polarization of the carbonyl group by the field. Grabowski and coworkers have also interpreted anomalous results obtained during an investigation of the polarographic reduction of substituted benzaldehydes as being due to field-enhanced basicity of the carbonyl group.[67] Finally, at the preparative level, it may be noted that the isolation of dimeric products (pinacols) from the electrochemical reduction of aromatic ketones and aldehydes under conditions where these carbonyl compounds are not appreciably protonated in bulk solution is further compelling evidence for field-induced carbonyl polarization (see Chapter 6).

Other Phenomena

The electrical effects of the field have also been observed during the electrochemical reduction of 2-iodopyridine, in which it was deduced that reduction must be preceded by polarization of the carbon–iodine bond[68]:

$$\text{2-iodopyridine} \xrightarrow{\text{field}} \text{(polarized: }\underline{N}^-\text{, }=I^+\text{)} \longrightarrow \text{products}$$

Supporting evidence for this postulate comes from adsorption studies, which indicate a much enhanced polarizability for 2-chloropyridine at the electrode surface, relative to pyridine.[69] Potential-dependent *ortho–para* ratios observed during electrochemical bromination of phenols have also been suggested as being due to field effects.[70]

4.3.2 Orientation of Substrates at the Electrode Surface

As a consequence of a number of effects, including chemical adsorption, physical adsorption, and/or electrostatic repulsion or attraction, organic molecules are often oriented in a very specific geometry at the electrode surface. A number of electrochemical phenomena, of which the following are representative, may be traced directly to orientational effects.

Erythrosin

Board and coworkers recently reported on the electrochemical reduction of erythrosin (**41**)[71]:

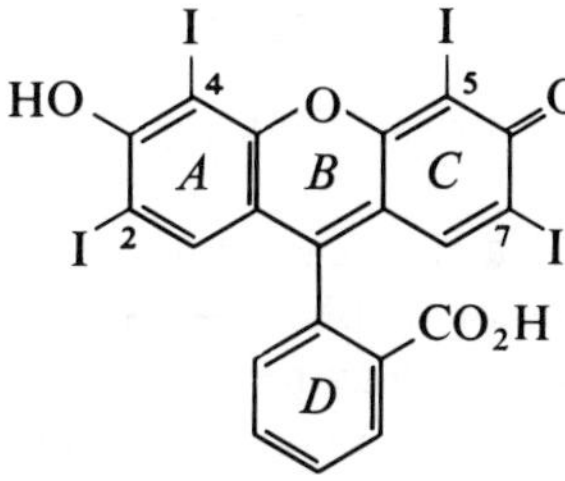

41

A polarogram shows three waves, at ca. −0.55, −0.80, and −1.2 V (*vs* S.C.E.), of relative heights, 1 : 2 : 2, respectively. The first wave is known to correspond to the 2-electron reduction of the quinoid system of ring *C* to the corresponding phenol. Because of the relative wave heights, the next two waves must involve 4 electrons each. The electrochemical reduction of aryl iodides ($ArI \rightarrow ArH$) consumes 2 electrons, and hence, since the molecule contains no other reducible functions, one is led to the surprising conclusion that two of the iodine atoms of **41** are easier to reduce than the other two. There is no apparent chemical reason for this difference, but Board and coworkers were able to show that it arises from the particular orientation assumed by **41** at the electrode surface. Capacitance measurements provided a value of Γ_{max}, the surface coverage at saturation (p. 116), corresponding to 1.43×10^{-10} mole/cm^2, or, from Avogadro's number, 117 Å^2/molecule. If ring *D* were coplanar with rings *A*, *B*, and *C*, and the molecule were to lie flat on the electrode, it would take up at least 140 Å^2/molecule. It is clear from scale models, however, that ring *D* must be perpendicular to the plane of the other three rings. If the molecule is adsorbed on the electrode in such a manner that the iodine atoms at 4 and 5 are nearest the mercury, and the carbon–iodine bonds to these atoms are *perpendicular* to the electrode surface, measurements from scale models indicate that each erythrosin molecule will take up 117 Å^2, exactly the value obtained experimentally. Given this assumed orientation of erythrosin at the electrode surface, one would then expect that the iodine atoms at carbons 4 and 5 should be more easily reduced than those at 2 and 7, since the former two are held closest to the electrode and hence should experience a more negative potential. Indeed, the major product from electrolysis at −1.0 V, that is, on the plateau of the first 4-electron wave, was 2,7-diiodofluoroscein (**42**)*:

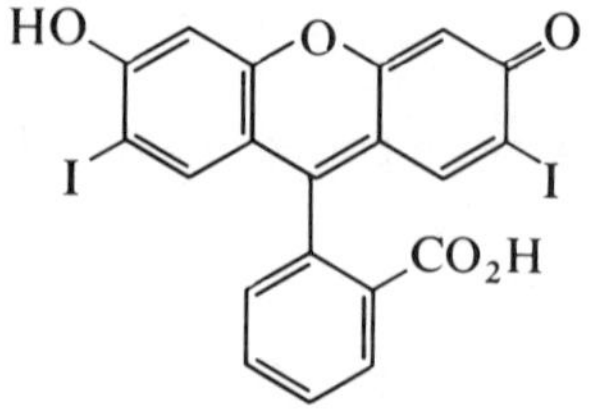

42

These experiments constitute a clear case in which the geometry assumed by an adsorbed species affects the course of its electrochemical reduction. They also show how measurements of Γ_{max} may be used to interpret the course of an apparently anomalous electrode reduction. It is possible to speculate that there may be systems in which a strongly adsorbed moiety in a large molecule holds another group near the electrode in optimum geometry for reduction, thus acting as a template for the electrode process.

Specific Orientation and Electron Transmission by Aromatic Substrates

Most neutral molecules are most strongly adsorbed on mercury at the electrocapillary maximum (ecm), where the electrode is uncharged, and are progressively desorbed at potentials increasingly anodic or cathodic of the ecm. This behavior can be clearly discerned by electrocapillary measurements, such as, for example, curve *B* in Figure 4.8 (curve *A* represents the solvent with no added adsorbent). A small but noticeable difference in the degree of adsorption of certain substrates is usually observed at potentials anodic of, and even slightly cathodic of, the ecm. In this region, aromatic compounds are generally found to be rather more strongly adsorbed than their aliphatic counterparts, as evidenced by a greater depression of the mercury–solution interfacial tension. (Compare curves *B* and *C* in Figure 4.8, which represent aliphatic and aromatic adsorbents, respectively.) This phenomenon has been observed, for example, for pyridine, several quinoline, benzene, and naphthalene derivatives, and coumarin, and it has generally been interpreted as being due to a tendency of the pi system of the aromatic species to lie flat on the electrode surface.[69, 72–75] A more quantitative test of the latter hypothesis involves the use of surface tension or capacitance data to obtain Γ_{max}, from which the actual area taken up by the molecule at the

* Reduction of the quinoid system of ring *C* must also occur at −1.0 V, in addition to removal of the two iodine atoms. The phenolic system in these compounds is easily oxidized, however, and the dihydro derivative of **42** presumably suffers air oxidation to the quinoid form during workup.

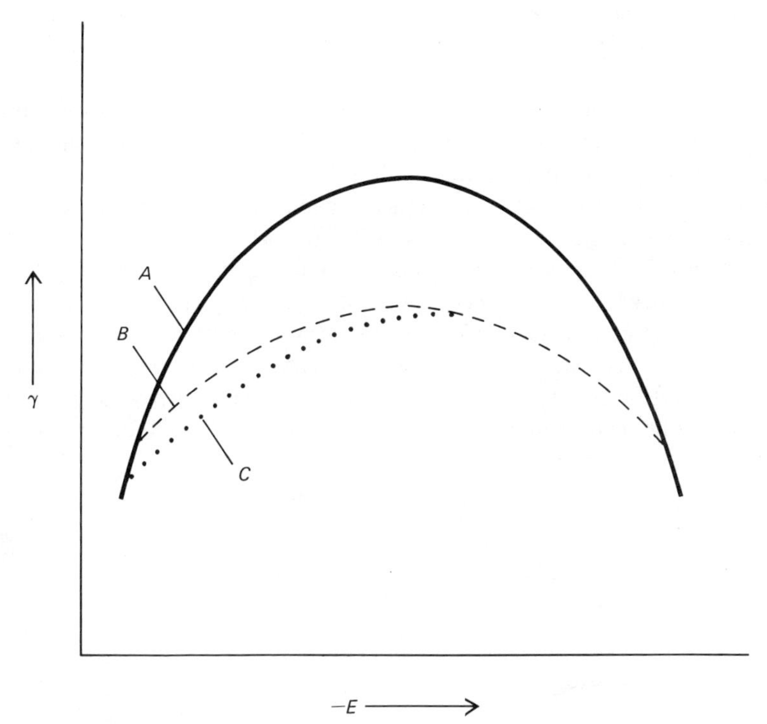

Figure 4.8 Representative electrocapillary curves. Curve A: plot of surface tension *vs* potential in the absence of adsorption. Curve B: addition of an aliphatic adsorbent to the solution. Curve C: addition of an aromatic adsorbent to the solution.

electrode surface, and hence its orientation, may be determined. When this is done, it does support the hypothesis that aromatic compounds are oriented flat on the electrode surface. Bockris and coworkers, for example, compared the adsorption characteristics of a number of molecules of general formula RX, where X was any of a variety of functional groups (amino, carboxyl, sulfhydryl, etc.) and R was either *n*-butyl, phenyl, or naphthyl.[74] In all cases studied, measurement of the area taken up by an individual molecule showed that the phenyl and naphthyl compounds lay flat (parallel to) the electrode at the ecm, while the *n*-butyl derivatives were found to be adsorbed perpendicular to the electrode. The real significance of this observation lies in the fact that aromatic compounds, when adsorbed parallel to the electrode surface in this manner, apparently *do not inhibit electron transfer* to other components of the solution.[69,72] The aromatic ring must act in such instances as an electron bridge to the electroactive substance, apparently via a mechanism involving its pi system. Aliphatic substances, however, do often *inhibit* electron transfer when adsorbed at the electrode surface.[76] This, coupled with the fact that an adsorbed film may affect two given electrode processes to differing extents, provides a potentially useful method for discriminating between otherwise equally electroactive functional groups.

A third mode of orientational behavior has been discerned with aromatic compounds containing a permanent dipole in the plane of the ring, such as pyridine, quinoline, coumarin, and aniline.[69,72,73,75] Taking coumarin as an example,[75] it is found that this substance occupies an area of 56 Å^2/molecule at potentials close to the ecm (0 V *vs* N.H.E.) in 1-*M* sodium sulfate, but that it occupies only 27 Å^2/molecule at more negative potentials (−0.2 to −1.1 V *vs* N.H.E.) (Figure 4.9).* These areas (56 Å^2 and 27 Å^2) happen to correspond closely to the respective areas calculated for a coumarin molecule adsorbed parallel and perpendicular to the electrode surface. Reorientation from parallel to perpendicular adsorption geometry takes place over a relatively narrow range of potentials, as may be seen from Figure 4.9 (this is also true for the other substances

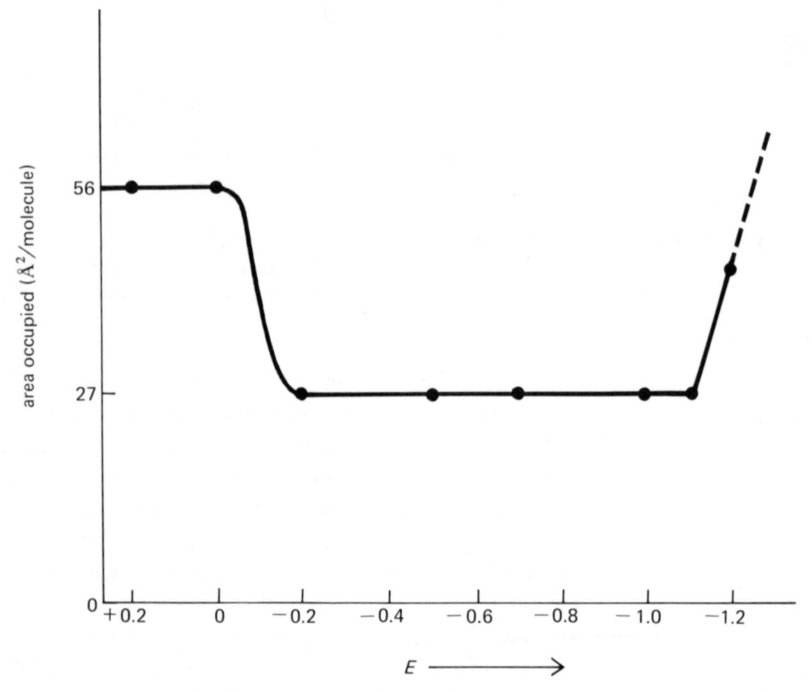

Figure 4.9 Area occupied by a single coumarin molecule at the mercury-solution interface as a function of potential. Solution: 5-m*M* coumarin in 1-*M* sodium sulfate. (L. K. Partridge, A. C. Tansley, and A. S. Porter, *Electrochim. Acta*, **11**, 520 [1966]. Reprinted by permission of Pergamon Press.)

* At potentials more cathodic than −1.1 V, coumarin begins to desorb from the electrode.[75]

mentioned above). It has been suggested by several authors that this reorientation is a field effect: as the electrode is made increasingly negative, a potential is eventually reached where the negative charge on the electrode is great enough to induce an orientation of the dipole moment of the adsorbed species perpendicular to the electrode surface.[75] The particular significance of this phenomenon for present purposes lies in the fact that electron transfer through the aromatic species is not possible when it is adsorbed perpendicular to the electrode surface. Partridge has shown, for example, that the reduction of nickel ion is unaffected by coumarin in the parallel geometry, but is completely inhibited under conditions where coumarin is adsorbed perpendicular to the electrode.[75] Partridge, Tansley, and Porter have supplied another dramatic example of the effect of the geometry of adsorbed coumarin on the course of an electrode process.[77] The polarographic behavior of cadmium in the presence of increasing amounts of coumarin is shown in Figure 4.10. Observe that in the absence of coumarin the half-wave potential for cadmium is ca. −0.6 V (*vs* N.H.E.), but as the coumarin concentration increases this wave is progressively diminished, and a new wave appears at −1.2 V. At the highest coumarin concentration used, the latter wave represents essentially the entire reduction process.* Furthermore, at intermediate

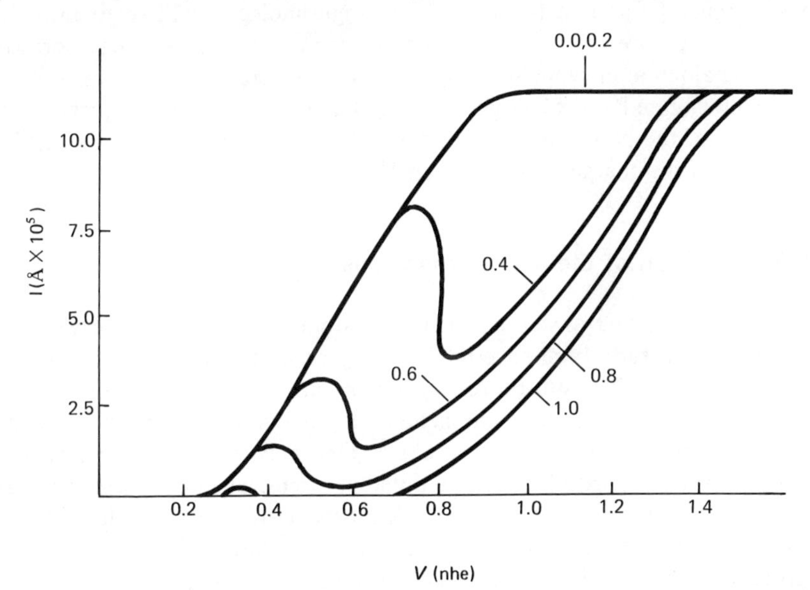

Figure 4.10 Polarographic reduction of cadmium(II) in the presence of increasing amounts of added coumarin at pH 4.6 (20°C). The number shown against each curve is the fractional saturation with coumarin. (L. K. Partridge, A. C. Tansley, and A. S. Porter, *Electrochim. Acta*, **14**, 226 [1969]. Reprinted by permission of Pergamon Press.)

* Coumarin itself is not reduced over this range of potentials.

coumarin concentrations (fractional coverages of 0.4–0.9) cadmium reduction begins normally, but then abruptly decreases. The potential where this decrease occurs is dependent on the concentration of coumarin. Partridge et al. were able to show from surface-tension measurements that the abrupt decrease in the first polarographic wave occurs at the point where coumarin reorients from parallel to perpendicular geometry.* The behavior observed in this system has been interpreted as follows. The reduction of cadmium begins normally, unaffected by the parallel-adsorbed coumarin. At the potential where coumarin changes orientation, however, electron transfer to cadmium is inhibited, and remains so until the potential becomes sufficiently negative that desorption of coumarin finally permits reduction to proceed.

These experiments indicate again the importance of measurements of Γ in studies of the effects of added adsorbents. The implications of these effects of coumarin on inorganic electrode processes will be apparent. Most organic substrates are at least slightly polar and/or polarizable, and hence are very likely oriented in a very special way at the electrode surface, this orientation may be potential dependent, and the electron-bridging capability of the adsorbate may therefore also depend on potential. If the substrate is adsorbed on mercury, measurement of the area taken up by a single molecule will often provide a guide to the nature of that adsorption. Conway and Barradas have drawn a possible analogy to enzyme–substrate interactions here: substrates may also be specifically oriented at the charged surface of an enzyme as a consequence of the interaction of the permanent dipole moment of the substrate with the field associated with the charge on the enzyme.[69]

4.3.3 Inhibition of Electrode Reactions

Inhibition of electron transfer to nickel and cadmium by perpendicularly oriented coumarin has already been encountered. In general, the inhibition of electrode processes by adsorbed substances, although rather frequently observed, is a complex and often puzzling phenomenon. Heyrovsky and Kuta have described some of the forms inhibition may take.[78] One may observe, *inter alia*, cases where two processes are inhibited to a different extent, where all electron-transfer processes are inhibited, or where the adsorbent does not interfere at all.** Examples of the latter two cases are, respectively, adsorption of long-chain hydrocarbons on the electrode, which is thought to form a film impervious to electron transfer,[76] and adsorption of aromatic compounds flat on the electrode

* This potential is dependent on the bulk concentration of coumarin.[76]

** Not pertinent here are changes in $k_{f,h}$, the electron-transfer rate constant, as a result of a shift in the potential of the OHP due to the introduction of an adsorbed ionic species into the double layer.

surface, where electron transmission apparently can occur through the pi system.[72,75] Inhibition of one of two possible electrode reactions is probably of the most potential synthetic interest, but it is the least understood phenomenon. At least some of such cases appear to arise not from any direct interference with any electrochemical step, but through inhibition of chemical reactions following electron transfer. For example, dimerization, disproportionation, or protonation of an intermediate may be inhibited by a slowing of the diffusion rate of the intermediate by the film. Anomalous effects observed during electrochemical reduction of aromatic nitro compounds in the presence of surface-active agents are thought to be of this type.[79]

Meites and coworkers have described a useful and striking example of inhibition during the electrochemical reduction of *p*-chlorophenylpropiolic acid methyl ester (**43**).[80] The polarographic behavior of this substance,

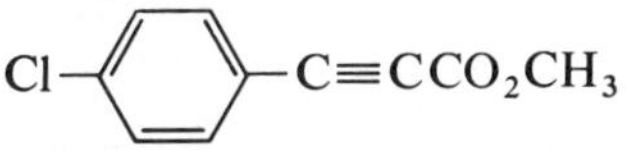

$$Cl-C_6H_4-C{\equiv}CCO_2CH_3$$

43

$$\mathbf{43} \xrightarrow{-1.20\ V} Cl-C_6H_4-CH{=}CHCO_2CH_3$$

44

$$\xrightarrow{-1.60\ V} Cl-C_6H_4-CH_2CH_2CO_2CH_3$$

45

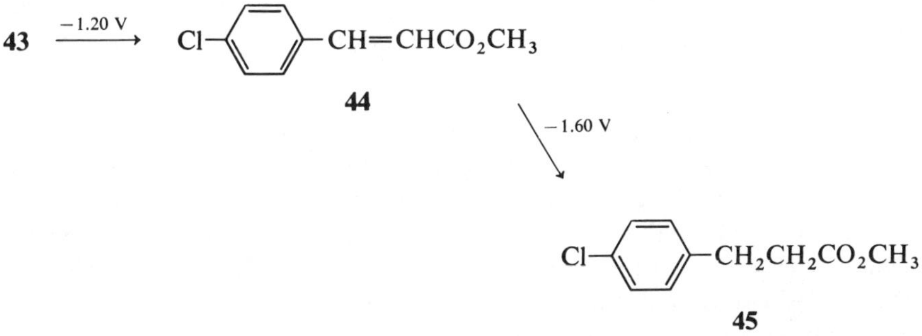

in 25% ethanol using tetraalkylammonium salts ($R_4N^+X^-$) as supporting electrolyte was found to depend on the nature of R (Figure 4.11). When tetramethylammonium ion is the cation of the supporting electrolyte, two 2-electron waves are observed. As the size of R increases, the first wave diminishes and the second grows, until when R is butyl or larger a single 4-electron wave is found. The two 2-electron waves correspond to stepwise reduction of **43** to methyl *p*-chlorophenylcinnamate (**44**) and thence to dihydrocinnamate (**45**). It is clear then that when R is a long-chain group, the electrode process consists of a direct 4-electron reduction to **45**. Tetraalkylammonium ions are known to be strongly adsorbed on mercury, particularly when R is butyl or larger[81]; the observed electrochemical behavior, therefore, is undoubtedly an adsorption effect. Meites suggested that

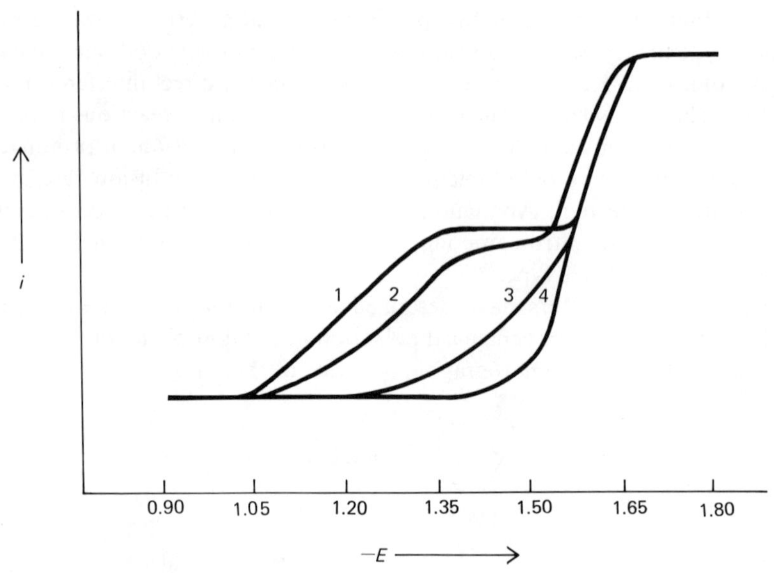

Figure 4.11 Polarograms of methyl *p*-chlorophenylpropiolate in 0.1-*F* (a) tetramethylammonium chloride, (b) tetraethylammonium bromide, (c) tetra-*n*-propylammonium iodide, and (d) tetra-*n*-butylammonium iodide. (S. R. Missan, E. I. Becker, and L. Meites, *J. Amer. Chem. Soc.*, **83**, 59 [1961].)

it arises from a field effect on the orientation of **43** at the electrode surface, but alternative explanations appear to be possible. Whatever its origin, the synthetic applications of this discovery are apparent: one may reduce **43** (and presumably related compounds) either to the alkene or to the saturated compound by proper choice of electrolyte. In a complex situation, one might wish to examine the effects of several cations, such as, Me_4N^+, Bu_4N^+, and Li^+ (p. 148) on polarographic behavior.

$$\mathbf{45} \xleftarrow[25\%\ \mathrm{EtOH}]{Bu_4N^+I^-,\ 4e^-} \mathbf{43} \xrightarrow[25\%\ \mathrm{EtOH}]{Me_4N^+Cl,\ 2e^-} \mathbf{44}$$

4.3.4 Acceleration of Electrode Processes

Among the myriad adsorption effects that have been recognized, it has been found that, unlike the preceding examples of inhibition, adsorption can sometimes *facilitate* reactions that otherwise might not be possible. One example of this has already been met in the case of erythrosin, where adsorption results in two of the four iodine atoms being substantially easier to reduce than the other two.[71] This case represents a rather general situation: when the electroactive species is itself the adsorbent, certain of its electrochemical and chemical properties may be enhanced by any special orientation it may have at the electrode or because it is present in higher concentrations at the electrode surface than in bulk solution. (By contrast, inhibition is most often observed when a substance is electrolyzed in the presence of a surface-active agent.) Ross has pointed out a case where adsorption apparently has a major effect on a reaction by accelerating one of two possible processes.[82] The well-known *Kolbe reaction* involves oxidation of carboxylate anions, often in water or methanol. When carried out at platinum anodes, the oxidation of RCO_2^- generally affords dimeric products, RR, strongly suggestive of radical intermediates (Chapter 8). However, when the electrode is composed of carbon and other conditions are the same, products indicative of carbonium ions (alcohols, ethers, rearranged materials) are the principal products (the so-called Hofer–Moest reaction). This difference has been explained by Ross as possibly due to the presence on the carbon surface of numerous para-

$$R\cdot + CO_2 \xleftarrow[e^-]{\text{platinum anode}} RCO_2^- \xrightarrow[-2e^-]{\text{carbon anode}} R^+ + CO_2$$

magnetic centers that "strongly attach any radicals formed, prevent their desorption, and thus promote further oxidation to the carbonium ion."[82] Although this argument appears basically reasonable, it cannot represent the entire difference between the two reactions, for there is compelling evidence that dimer formation during oxidation of carboxylates at platinum also arises from adsorbed intermediates (Chapter 8).[83] The *nature* of the adsorption process on the two electrode materials must also be involved; this is supported by the fact that the ratio of Kolbe to Hofer–Moest products at carbon depends on the history of the particular carbon used as anode.[84]

Coupling of an adsorbed radical (*cf.* the Kolbe reaction) is often more efficient at the electrode surface than in bulk solution. The previously mentioned reduction of acetone in acid solution in the presence of acrylonitrile (p. 145), which results in efficient carbon–carbon bond formation, has been shown to proceed via *adsorbed* 2-hydroxy-2-propyl radicals.[40a] The oxidation of aromatic amines also involves adsorbed intermediates (Chapter 8).

References

[1]A. J. Fry and R. G. Reed, *J. Amer. Chem. Soc.*, **91**, 6448 (1969).
[2]A. J. Fry, M. A. Mitnick, and R. G. Reed, *J. Org. Chem.*, **35**, 1232 (1970).
[3a]J. D. Anderson, M. M. Baizer, and E. J. Prill, *J. Org. Chem.*, **30**, 1645 (1965); [b]M. M. Baizer and J. D. Anderson, *J. Org. Chem.*, **30**, 3138 (1965).
[4]K. Slotta and R. Kethur, *Chem. Ber.*, **71**, 335 (1938).
[5]O. Manousek, P. Zuman, and O. Exner, *Coll. Czech. Chem. Commun.*, **33**, 3979 (1968).
[6]W. J. Koehl, Jr., *J. Org. Chem.*, **32**, 614 (1967).
[7]M. R. Rifi, *J. Amer. Chem. Soc.*, **89**, 4442 (1967).
[8]J. G. Traynham and J. S. Dehn, *J. Amer. Chem. Soc.*, **89**, 2139, 2799 (1967).

[9]H. C. Brown, *Hydroboration*, Menlo Park, Calif., Benjamin, 1962.
[10]H. Schecter, D. E. Ley, and L. Zeldin, *J. Amer. Chem. Soc.*, **74**, 3664 (1952).
[11a]W. J. Seagers and P. J. Elving, *J. Amer. Chem. Soc.*, **72**, 5183 (1950); [b]S. Deswarte and J. Armand, *Compt. Rend.*, **258**, 3865 (1964).
[12a]J. Volke and A. M. Kardos, *Coll. Czech. Chem. Commun.*, **33**, 2560 (1968); [b]V. D. Bezuglyi, L. V. Kononenko, A. F. Korunova, V. N. Dmitrieva, and B. L. Timan, *J. Gen. Chem. U.S.S.R.*, **39**, 1647 (1969).
[13]A. Hickling, *Trans. Faraday Soc.*, **38**, 27 (1942).
[14a]F. Haber, *Z. Elektrochem.*, **4**, 510 (1897); [b]F. Haber, *Z. Physik. Chem.*, **32**, 193 (1900).
[15]S. Swann, Jr., in A. Weissberger, ed., *Technique of Organic Chemistry*, 2nd ed., New York, Wiley, 1956, vol. 2, pp. 385–523.
[16a]T. Kitagawa, T. P. Layloff, and R. N. Adams, *Anal. Chem.*, **35**, 1086 (1963); [b]J. G. Lawless and M. D. Hawley, *J. Electroanal. Chem.*, **21**, 365 (1969); [c]D. E. Bartak, W. C. Danen, and M. D. Hawley, *J. Org. Chem.*, **35**, 1206 (1970).
[17]L. L. Miller and E. Riekena, *J. Org. Chem.*, **34**, 3359 (1969).
[18]H. McConnell, *J. Chem. Phys.*, **24**, 764 (1956); [b]S. Weissman, *J. Chem. Phys.*, **25**, 890 (1956); [c]M. Bersohn, *J. Chem. Phys.*, **24**, 1066 (1956).
[19]J. W. Sease and R. Filas, private communication.
[20a]J. G. Lawless, D. E. Bartak, and M. D. Hawley, *J. Amer. Chem. Soc.*, **91**, 7121 (1969); [b]M. Mohammad, J. Hajdu, and E. M. Kosower, *J. Amer. Chem. Soc.*, **93**, 1792 (1971).
[21a]J. W. Sease, S. Nickol, and F. Burton, private communication; [b]J.-M. Saveant, *J. Electroanal. Chem.*, **29**, 87 (1971).
[22] M. E. Peover, in A. J. Bard, ed., *Electroanalytical Chemistry*, New York, Marcel Dekker, 1967, vol. 2, pp. 1–51.
[23a]C. L. Perrin, *Progr. Phys. Org. Chem.*, **3**, 165 (1965); [b]P. Zuman, *Progr. Phys. Org. Chem.*, **5**, 81 (1967).
[24]P. H. Given and M. E. Peover, *J. Chem. Soc.*, 385 (1960).
[25]J. R. Jezorek, and H. B. Mark, Jr., *J. Phys. Chem.*, **74**, 1627 (1970).
[26a]J. L. Sadler and A. J. Bard, *J. Amer. Chem. Soc.*, **90**, 1979 (1968); [b]R. F. Michielli and P. J. Elving, *J. Amer. Chem. Soc.*, **90**, 1989 (1968).
[27]A. J. Fry and W. E. Britton, unpublished experiments.
[28]G. J. Hoijtink, J. van Schooten, E. de Boer, and W. I. Aalbersberg, *Rec. Trav. Chim.*, **73**, 355 (1954).
[29]G. J. Hoijtink and J. van Schooten, *Rec. Trav. Chim.*, **71**, 1089 (1952).
[30a]I. Bergman, *Trans. Faraday Soc.*, **50**, 829 (1954); [b]I. Bergman, *Trans. Faraday Soc.*, **52**, 690 (1956).
[31]H. Lund, *Acta Chem. Scand.*, **13**, 249 (1959).
[32a]R. Pasternak, *Helv. Chim. Acta*, **31**, 753 (1948); [b]P. J. Elving and J. T. Leone, *J. Amer. Chem. Soc.*, **80**, 1021 (1958).
[33]R. J. Gillespie, *Accounts Chem. Res.*, **1**, 202 (1968).
[34]S. Wawzonek and D. Wearring, *J. Amer. Chem. Soc.*, **81**, 2067 (1959).
[35a]R. M. Kopchik and J. A. Kampmeier, *J. Amer. Chem. Soc.*, **90**, 6733 (1968); [b]O. Simamura, in E. L. Eliel and N. L. Allinger, eds., *Topics in Stereochemistry*, New York, Wiley, 1969, vol. 4, pp. 1–38; [c]L. A. Singer and J. Chen, *Tetrahedron Lett.*, 4849 (1969).
[36a]H. O. House and V. Kramar, *J. Org. Chem.*, **28**, 3362 (1963); [b]H. O. House and B. M. Trost, *J. Org. Chem.*, **30**, 1341 (1965).
[37]P. H. Given and M. E. Peover, *J. Chem. Soc.*, 465 (1960).
[38]T. J. Curphey, C. W. Amelotti, T. P. Layloff, R. L. McCartney, and J. H. Williams, *J. Amer. Chem. Soc.*, **91**, 2817 (1969).

[39]K. Sugino and T. Nonaka, *J. Electrochem. Soc.*, **112**, 1241 (1965).
[40a]O. R. Brown and K. Lister, *Disc. Faraday Soc.*, **45**, 106 (1968); [b]M. M. Baizer and J. L. Chruma, *J. Electrochem. Soc.*, **118**, 450 (1971).
[41]V. D. Parker and R. N. Adams, *Tetrahedron Lett.*, 1721 (1969).
[42]V. D. Parker, *Chem. Commun.*, 848 (1969).
[43]G. Manning, V. D. Parker, and R. N. Adams, *J. Amer. Chem. Soc.*, **91**, 4584 (1969).
[44a]K. S. V. Santhanam and A. J. Bard, *J. Amer. Chem. Soc.*, **88**, 2669 (1966); [b]J. L. Mills, R. Nelson, S. G. Shore, and L. B. Anderson, *Anal. Chem.*, **43**, 157 (1971).
[45]S. A. Cruser and A. J. Bard, *J. Amer. Chem. Soc.*, **91**, 267 (1969).
[46]R. H. Philp, Jr., R. L. Flurry, and R. A. Day, Jr., *J. Electrochem. Soc.*, **111**, 328 (1964).
[47]J. L. Sadler and A. J. Bard, *J. Amer. Chem. Soc.*, **90**, 1979 (1968).
[48]R. H. Philp, Jr., T. Layloff, and R. N. Adams, *J. Electrochem. Soc.*, **111**, 1189 (1964).
[49]G. H. Aylward, J. L. Garnett, and J. H. Sharp, *Chem. Commun.*, 137 (1966).
[50a]M. E. Peover and J. D. Davis, *J. Electroanal. Chem.*, **6**, 46 (1963); [b]M. E. Peover and J. D. Davis, in G. J. Hills, ed., *Polarography—1964*, New York, Wiley, 1966, vol. 2, p. 1003.
[51]R. W. Murray and L. K. Hiller, Jr., *Anal. Chem.*, **39**, 1221 (1967).
[52]L. Eberson and K. Nyberg, *J. Amer. Chem. Soc.*, **88**, 1686 (1966).
[53]K. N. Campbell and E. E. Young, *J. Amer. Chem. Soc.*, **65**, 965 (1943).
[54a]J. E. Dubois and M. Ropars, *J. Chim. Phys.*, **65**, 2000, 2009 (1968); [b]J. Janata and H. B. Mark, Jr., in A. J. Bard, ed., *Electroanalytical Chemistry*, New York, Marcel Dekker, 1969, vol. 3, pp. 1–55.
[55a]R. A. Benkeser and E. M. Kaiser, *J. Amer. Chem. Soc.*, **85**, 2858 (1963); [b]R. A. Benkeser, E. M. Kaiser, and R. F. Lambert, *J. Amer. Chem. Soc.*, **86**, 5272 (1964); [c]R. A. Benkeser and S. J. Mels, *J. Org. Chem.*, **34**, 3970 (1969); [d]R. A. Benkeser and S. J. Mels, *J. Org. Chem.*, **35**, 261 (1970).
[56]A. J. Birch and H. Smith, *Quart. Rev.*, **12**, 17 (1958).
[57a]P. G. Arapakos, M. K. Scott, and F. E. Huber, Jr., *J. Amer. Chem. Soc.*, **91**, 2059 (1969); [b]P. G. Arapakos and M. K. Scott, *Tetrahedron Lett.*, 1975 (1968).
[58]J. Volke and A. M. Kardos, *Coll. Czech. Chem. Commun.*, **33**, 2560 (1968).
[59]O. Manousek, P. Zuman, and O. Exner, *Coll. Czech. Chem. Commun.*, **33**, 3979 (1968).
[60]P. H. Rieger, I. Bernal, W. H. Reinmuth, and G. K. Fraenkel, *J. Amer. Chem. Soc.*, **85**, 683 (1963).
[61]H. S. Harned and B. B. Owen, *Physical Chemistry of Electrolytic Solutions*, 3rd ed., New York, Van Nostrand Reinhold, 1958, p. 138ff.
[62]A. Vincenz-Chodkowska and Z. R. Grabowski, *Electrochim. Acta*, **9**, 789 (1964).
[63]H. E. Stapelfeldt and S. P. Perone, *Anal. Chem.*, **41**, 623 (1969).
[64]Z. R. Grabowski, B. Czochralska, A. Vincenz-Chodkowska, and M. S. Balasiewicz, *Disc. Faraday Soc.*, **45**, 145 (1968).
[65]G. J. Hoijtink, in P. Delahay and C. W. Tobias, eds., *Advances in Electrochemistry and Electrochemical Engineering*, New York, Wiley, 1970, vol. 7, pp. 221–281.
[66]P. J. Elving and J. T. Leone, *J. Amer. Chem. Soc.*, **80**, 1021 (1958).
[67]Z. R. Grabowski and E. T. Bartel, *Roczn. Chem.* **34**, 611 (1960); see *Chem. Abstr.*, **55**, 11135 (1961).
[68]R. F. Evilia and A. J. Diefenderfer, *J. Electroanal. Chem.*, **22**, 407 (1969).
[69]B. E. Conway and R. G. Barradas, *Electrochim. Acta*, **5**, 319 (1961).
[70]R. Landsberg, H. Lohse, and U. Lohse, *J. Prakt. Chem.*, [*4*], **12**, 253 (1961).
[71]P. W. Board, D. Britz, and R. V. Holland, *Electrochim. Acta*, **13**, 1575, 1633 (1968).
[72]B. B. Damaskin, *Electrochim. Acta*, **9**, 231 (1964).
[73]S. Bordi and G. Papeschi, *J. Electroanal. Chem.*, **20**, 297 (1969).
[74]E. Blomgren, J. O'M. Bockris, and C. Jesch, *J. Phys. Chem.*, **65**, 2000 (1961).

[75]L. K. Partridge, A. C. Tansley, and A. S. Porter, *Electrochim. Acta*, **11**, 517 (1966).
[76]E. B. Weronski, *Trans. Faraday Soc.*, **58**, 2217 (1962).
[77]L. K. Partridge, A. C. Tansley, and A. S. Porter, *Electrochim. Acta*, **14**, 223 (1969).
[78]J. Heyrovsky and J. Kuta, *Principles of Polarography*, New York, Academic, 1966.
[79]L. Holleck, B. Kastening, and R. D. Williams, *Z. Elektrochem.*, **66**, 396 (1962).
[80]S. R. Missan, E. I. Becker, and L. Meites, *J. Amer. Chem. Soc.*, **83**, 58 (1961).
[81a]G. Sutra, *J. Chim. Phys.*, **43**, 189, 279 (1946); [b]G. Darmois and E. Darmois, *Compt. Rend.*, **238C**, 971 (1954).
[82a]S. D. Ross and M. Finkelstein, *J. Org. Chem.*, **34**, 2923 (1969); [b]S. D. Ross, *Trans. N.Y. Acad. Sci.*, **30**, 901 (1968).
[83]A. K. Vijh and B. E. Conway, *Chem. Rev.*, **67**, 623 (1967).
[84]D. L. Muck and E. R. Wilson, *J. Electrochem. Soc.*, **117**, 1358 (1970).

5 CLEAVAGE OF SINGLE BONDS

The emphasis in this chapter will be on methods for reductive electrochemical cleavage of single bonds. Methods will be described for cleavage both by direct electron transfer to the substrate and by reaction with electrochemically generated reagents, the latter being a new and potentially powerful technique for reduction of organic compounds.

It should be pointed out that there are in this area, as in much of organic electrochemistry, a large number of unanswered mechanistic questions. In many cases the products of electrolysis are known empirically, but the proper experimentation (polarography, cyclic voltammetry, controlled-potential electrolysis, etc.) necessary to an understanding of electrode mechanism is lacking. Nevertheless, enough is often known to allow one to decide whether a contemplated reaction is feasible electrochemically. For example, it is known that the ease of reductive cleavage tends to be related, for a series of heteroatoms X, to the carbon–X bond strength, so that among alkyl halides one finds the order of decreasing ease of reduction: $I > Br > Cl$. Furthermore, bonds activated by adjacent unsaturation, such as those that are benzylic, allylic, or *alpha* to a carbonyl, nitro, or cyano group, are generally considerably easier to cleave than alkyl, vinyl, or phenyl bonds.

Relatively little is known about the stereochemical course of most electrochemical reductions, including those discussed in this chapter. This is unfortunate, since such information is not only extremely helpful in determining the mechanism of an organic electrode process, but also it is precisely the kind of information required by the chemist who proposes to use the reaction synthetically. There is much interest at present in electroorganic stereochemistry, however, and it may be anticipated that the stereochemical course of many organic electrode processes will be known in substantial detail in the near future. Because of its mechanistic and synthetic importance, stereochemical information will be included, where available, for all reactions discussed. In addition to the usual stereochemical factors familiar to the organic chemist, it is likely that surface effects on stereochemistry will be found to be important in some instances, since intermediates are both generated and react close to the massive electrode surface.

What follows is *not* an exhaustive compilation of all known examples of a given reaction. Instead, an attempt has been made to present a sampling of sufficient variety to illustrate the scope of the reaction. It is hoped that these examples, together with the accompanying mechanistic information, will be useful as a guide to the potential utility of a given reaction and to the development of new chemistry based on it. Mechanistic speculation by the author is generally so identified where it appears.

5.1 DIRECT ELECTROCHEMICAL CLEAVAGE

In this section there will be discussed reactions in which bond breakage occurs in the species undergoing electron transfer at the electrode, as distinguished from the electrocatalytic processes of Section 5.2. However, even in the present section, it will be seen that there is a variety of reduction mechanisms. There are processes, for example, in which bond breaking has already begun in the transition state for electron transfer, and others in which bond breaking occurs after the electron-transfer step, as in reactions occurring via amalgam intermediates.

5.1.1 Alkyl Halides

General Features

Although electrochemical dehalogenations had been known for a number of years, it was not until 1949 that there appeared, in a classic paper by von Stackelberg and Stracke, a systematic ordering of the relative ease of reduction of a wide variety of alkyl halides and polyhalides.[1] von Stackelberg and Stracke investigated a large number of compounds, principally by polarography but also in a few instances by controlled-potential electrolysis, and were able to formulate a number of useful generalizations. Modified somewhat from the form first stated by von Stackelberg and Stracke, these are as follows.

1. Allylic and benzylic halides are much easier to reduce than the corresponding halides; for example, allyl bromide has $E_{1/2} = -1.29$ V (*vs* S.C.E.) in 75% dioxane containing tetraethylammonium bromide, while under the same conditions *n*-butyl bromide has $E_{1/2} = -2.27$ V. Vinyl halides are somewhat more difficult to reduce than saturated halides (half-wave potentials of -2.47 and -2.09 V for vinyl and ethyl bromide, respectively). Half-wave potentials of aryl halides do not differ systematically from those of saturated halides:

Compound	E (*vs* S.C.E.), V
Iodobenzene	−1.62
Ethyl iodide	−1.67
Bromobenzene	−2.32
n-Butyl bromide	−2.27

The double bond in unsaturated halides is not electrochemically reducible.

2. Ease of reduction decreases in the order I > Br > Cl > F.* Indeed, except under special circumstance, such as R = benzyl or allyl, alkyl chlorides and fluorides are not electrochemically reducible at all.

3. Geminal and vicinal polyhalides are easier to reduce than simple halides. For example, one observes the following orders of decreasing ease of reduction:

$$CCl_4 > CHCl_3 > CH_2Cl_2 > CH_3Cl$$

$$BrCH_2CH_2Br > CH_3CHBr_2 > CH_3CH_2Br$$

4. An increase in the length of the aliphatic chain attached to a carbon atom reduces the ease of reduction:

methyl bromide > *n*-butyl bromide > *n*-octyl bromide

This is probably an adsorption effect.

5. Half-wave potentials (with a few exceptions to be noted below) are pH independent. This indicates that proton transfer cannot occur either before or in the transition state for reduction.

von Stackelberg and Stracke isolated naphthalene and ethane from controlled-potential electrolysis of 1-bromonaphthalene and ethyl bromide, respectively. From this evidence, it was suggested that reduction involves 2 electrons/molecule of halide, and the following reduction mechanism was proposed:

$$RX + e^- \longrightarrow R\cdot + X^-$$

$$R\cdot + e^- \longrightarrow R^- \xrightarrow{H^+} RH$$

It was suggested that the rate-determining step in the reduction involves transfer of a single electron to the substrate, with bond breakage and formation of a free radical.** It was assumed that the second electron transfer to produce a carbanion

* Fluorine compounds were not investigated by von Stackelberg and Stracke, but they have since been shown to fall into this sequence, e.g., benzyl chloride and benzyl fluoride have half-wave potentials of −2.1 and −2.7 V (S.C.E.), respectively, in DMF containing tetraethylammonium bromide.[2a]

** Some investigators have suggested[3] that this first step is composite, consisting of initial electron transfer to generate a radical anion, which subsequently decomposes:

$$RX + e^- \longrightarrow [RX]^{\dot{-}}$$

$$[RX]^{\dot{-}} \longrightarrow R\cdot + X^-$$

It is not clear, however, whether $[RX]^{\dot{-}}$ should be considered a true intermediate or merely a transition state. If an intermediate, it is extremely shortlived: cyclic voltammetry on a wide variety of alkyl halides gives no evidence of reversibility even at the highest scan rates employed.[2] Studies of the reduction of alkyl halides by hydrated electrons, which do involve intermediates of structure $[RX]^{\dot{-}}$, in fact argue against such intermediates in the electrochemical process. For example, rates of reduction of alkyl chlorides by hydrated electrons are unrelated to carbon–halogen bond strength, in striking contrast to electrochemical reduction.[4a] The lifetime of such intermediates is very short; e.g., the species $[CH_3I]^{\dot{-}}$ has a lifetime of less than 10^{-7} sec in water,[4b, c] suggesting that the question of whether $[RX]^{\dot{-}}$ is an intermediate or a transition state in the electrochemical reduction of RX will be extremely difficult to establish with any certainty.

is extremely fast. This is tantamount to suggesting that the standard potential for reduction of the radical is anodic of the potential required for the first electron-transfer step, so that the radical is reduced almost as rapidly as it is formed. This assumption appears to be reasonable, although, with one or two exceptions to be met later, formal potentials for reduction of simple free radicals are largely unknown. The von Stackelberg and Stracke mechanism has proved basically sound in the light of many subsequent investigations. The intermediacy of radicals, corroborating the stepwise nature of the reduction, has been established, for example, through isolation in some reductions of dimeric products (RR), organometallic compounds from reduction at mercury[2a,b] or lead[2c] cathodes, and rearranged hydrocarbons (*cf.* reduction of **1** and **2**), from partial

$$\underset{\mathbf{1}}{CH_2{=}CH(CH_2)_3CH_2I} \xrightarrow{e^-} CH_2{=}CH(CH_2)_3CH_2\cdot \longrightarrow \text{(cyclopentyl)}{-}CH_2\cdot$$

$$\longrightarrow \text{(cyclopentyl)}CH_2CH_2\text{(cyclopentyl)} + \text{(cyclopentyl)}{-}CH_3 \qquad (\textit{Ref. 2b})$$

$$\underset{\mathbf{2}}{C_6H_5C(CH_3)_2CH_2Cl} \xrightarrow{e^-} C_6H_5C(CH_3)_2CH_2\cdot \longrightarrow (CH_3)_2\dot{C}CH_2C_6H_5$$

$$\xrightarrow[H^+]{e^-} (CH_3)_2CHCH_2C_6H_5 \qquad (\textit{Ref. 5})$$

loss of configuration during reduction of stereoisomeric vinyl iodides,[6] and also from polarographic data.[11,16] Likewise, the formation of carbanions in the second step can be demonstrated either coulometrically, by trapping the carbanion with carbon dioxide,[7] or by isolation of Hofmann elimination products when reduction is carried out in aprotic solvents containing quaternary ammonium halides as electrolyte[2,3a]:

$$RBr \xrightarrow{e^-} R\cdot \xrightarrow{e^-} R^- \xrightarrow{Et_4N^+} RH + CH_2{=}CH_2 + Et_3N$$

A certain amount of the alkene derived from the alkyl halide is also obtained, particularly at high alkyl halide concentrations, where the probabilities both of Hofmann elimination upon the alkyl halide by the carbanion and disproportionation of the radical are enhanced. The product distribution from reduction of *n*-hexyl bromide in DMF containing tetraethylammonium bromide is typical: hexane (70%), 1-hexene (12%), dihexylmercury (2%), and *n*-hexanol (7%), the latter formed by hydroxide ion displacement upon starting material.[2a] (Hydroxide ion is formed from the small amounts of water always present in even the best samples of DMF.)[2a] To the extent that radical-derived products and destruction of starting material by the intermediate carbanion are involved, the coulometric *n* value is, of course, lower than the value of 2 predicted by the von Stackelberg and Stracke mechanism (it is 1.46 in the reduction of *n*-hexyl bromide above). For synthetic purposes, it is advisable to add a good proton donor, such a phenol, to the mixture to protonate the carbanion as formed, thus averting carbanion-promoted elimination upon the starting material.

Although the von Stackelberg–Stracke mechanism provides an adequate overall picture of the reduction mechanism, a number of subsidiary mechanistic points are not yet clear. The orientation of the C—X bond with respect to the electrode surface during reduction is, for example, a matter of some dispute. Some investigators have suggested a backside approach by the electron-rich electrode, that is, a transition-state geometry akin to that in the S_N2 displacement.[8] A very real difficulty with this suggestion lies in the fact that there is no correlation at all between S_N2 reactivity and half-wave potential. Neopentyl[2a] and bridgehead norbornyl, adamantanyl, bicyclooctyl, and triptycyl bromides,[9,10] for example, are all reduced at potentials not very different from those required for reduction of *n*-butyl or *n*-octyl bromide. Furthermore, allyl chloride and allyl tosylate are reduced at the *same* potential, despite the considerably higher S_N2 reactivity of the latter.[11] The ready reducibility of bridgehead bicyclic halides suggests instead that in the transition state for reduction the carbon–halogen bond is oriented either perpendicularly to the electrode with the halogen atom nearest the electrode surface, or parallel to the electrode surface. This point has not been settled as yet. As evidence for the perpendicular orientation, however, one might cite the data on the electrochemical reduction of erythrosin (Section 4.3.2) and the fact that axial halides in rigid chair cyclohexanes or *exo*-halides in the norbornyl system are reduced more easily than their corresponding equatorial or *endo*-isomers.[8b,12] In both of the latter cases, the halogen reduced more easily is more sterically accessible. An indication of the magnitude of the steric preference for *exo*-attack in norbornyl halides comes from studies of reduction of 2,2-dichloronorborane labeled with ^{36}Cl in the *exo*-position. It is found that the product *endo*-norbornyl chloride contains only 7% of the label.[13]

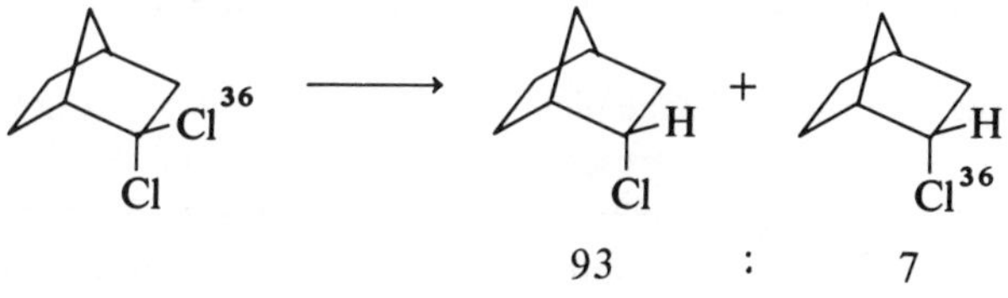

Cl^{36}, Cl ⟶ H, Cl + H, Cl^{36}

93 : 7

The perpendicular geometry, it must be admitted, is counter to expectation based on the direction of the dipole moment of the carbon–halogen bond (but see reference 14*a* and 14*b*) but is reasonable based on the importance of London attractive forces generally in organic halogen compounds,[14c] and it is also consistent with the strong specific adsorption of halide ions onto mercury at negative potentials.[15] It is likely that surface-tension measurements could answer this question of orientation by providing the value of Γ_{max} and hence the area covered by a single molecule.

The only light shed thus far on the assumption that radical intermediates have reduction potentials considerably anodic of the potentials necessary to break the carbon–halogen bond comes from work by Hoffmann, Hodgson, Maricle, and Jura,[16] and later by Petrovich and Baizer.[11] The first investigators observed two 1-electron polarographic waves at −1.59 and −2.49 V (*vs* S.C.E.) for *t*-butyl iodide in 1,2-dimethoxyethane containing tetrabutylammonium perchlorate (TBAP). A single 2-electron wave at −2.39 V was observed for *t*-butyl bromide. The two waves of *t*-butyl iodide were interpreted as arising from initial reduction of *t*-butyl iodide to *t*-butyl radical at −1.59 V and subsequent reduction of the radical to *t*-butyl carbanion at −2.49 V. The bromide is reduced at the same potential as the second wave of the iodide, so the radical would indeed be reduced immediately upon formation this case. If the polarographic waves have been correctly assigned, the standard potential for reduction of a *t*-butyl radical is ca. −2.2 to 2.4 V *vs* S.C.E. Unfortunately, controlled-potential electrolyses were not carried out to confirm the polarographic assignments. The *t*-butyl iodide case is probably the most favorable instance one could ever expect for polarographic resolution of the two distinct steps in the von Stackelberg–Stracke mechanism, because (1) iodides are the easiest halides to reduce, and (2) tertiary radicals should be harder to reduce than either primary or secondary radicals, as a consequence of the relative energies of primary, secondary, and tertiary radicals and the corresponding carbanions (Figure 5.1). The assignments of Hoffmann et al. therefore appear reasonable, but in view of the paucity of data on reduction potentials of free radicals, confirmation of the polarographic assignments made by these workers is badly needed, as is confirmation of a similar result obtained by Petrovich and Baizer, who observed two 1-electron waves for allyl bromide in DMSO and ascribed the second wave (at −1.56 *vs* S.C.E.) to reduction of allyl radical to allyl anion.[11] In both of these cases, one would expect radical-derived products from controlled-potential reduction at the first wave, and carbanion-derived products from reduction at the second wave.

Data on the overall stereochemistry of electrochemical reduction of alkyl halides is relatively sparse at present. Webb, Mann, and Walborsky found predominant retention during electrochemical reduction of optically active 2,2-diphenyl-1-methyl-cyclopropyl bromide,[3a] and other groups have observed predominant retention in the reduction of 1,1-dihalocyclopropanes.[17] These results are consistent with the known low rates of isomerization of cyclopropyl carbanions,[18] but may not for that very reason be representative of the stereochemistry of reduction of halides of less special structure. The same point may

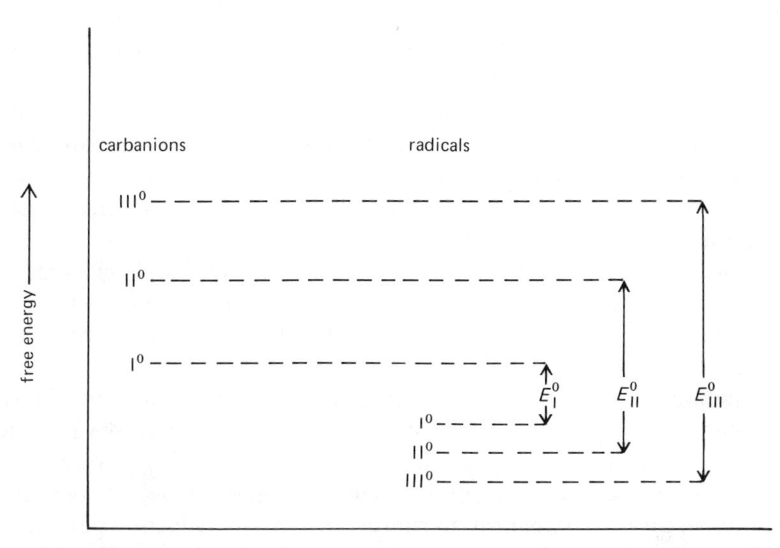

Figure 5.1 Relative ease of reduction of primary, secondary, and tertiary radicals to the corresponding carbanions.

be made with respect to the high degree of stereochemical retention, noted above, associated with the electrochemical reduction of radioactively labeled 2,2-dichloronorbornane.[13] In the absence of adsorption effects, one would not expect preservation of stereochemical integrity during electrochemical reduction of an acylic alkyl halide, since alkyl carbanions generally interconvert rapidly. Yet this is precisely the tenor of a report by Czochralska, who found that optically active 2-chloro-2-phenylpropionic acid is reduced with predominant *inversion* of configuration.[19] This highly surprising result, as with the predominant inversion also observed in reduction of cyclopropyl halides containing adjacent carboxyl and carboalkoxyl groups,[20] may very well be a surface phenomenon, particularly in view of the known strong adsorption of carboxyl groups on mercury.[21] For this reason, it would be interesting to know the stereochemical course of reduction of an optically active halide lacking adsorbable functional groups that might otherwise obscure fundamental stereochemistry.

Aryl and Vinyl Halides

Inspection of Table 4.1 reveals that vinyl and aryl halides are normally reduced at potentials that are not too cathodic, if at all, of those required to reduce saturated alkyl halides. This, of course, differs considerably from the behavior of aryl and vinyl halides under either S_N1 or S_N2 displacement conditions and provides evidence against either of the latter as models for the electrochemical reduction. The relative ease of electrochemical reducibility of these halides is of considerable synthetic interest because of the contrast with their inertness under many other reaction conditions. Since the nature of the halogen, that is, whether it is iodine, bromine, or chlorine, is usually more important in determining its ease of reducibility than its structural type, one can discriminate between two or more reducible functional groups in a manner not possible with other reagents. A number of examples of synthetically valuable selective reduction that illustrate this point have already been met (Section 4.1.1).

Of further synthetic interest, side reactions such as dimer and dialkylmercury formation are normally negligible with aryl and vinyl halides, and hence nearly quantitative yields of the corresponding hydrocarbon may usually be expected upon reduction.

As indicated in Chapter 4, it has been suggested that reduction of certain unsaturated halides involves initial addition of the electron to the pi system, followed by cleavage of the sigma bond to halogen. It is likely, however, that only in a rather extended pi system is the lowest unfilled molecular orbital low enough to permit this mechanism to subvert the normal reduction path. Simple vinyl and phenyl halides are probably reduced via direct reduction of the carbon–halogen bond.* However, reduction of nitrophenyl and nitrobenzyl,[23] polyphenylethylenic,[24] and naphthyl[25a] halides, and halobenzophenones[25b] undoubtedly does proceed via initial electron transfer to the pi system. (It must be noted here, however, that Miller has suggested that even reductions of simple vinyl halides may involve this indirect path.[24])

Activated Systems

Although in simple alkyl, aryl, and vinyl systems the structure of group R in RX does not affect reduction potential nearly as much as does the identity of the halogen, it is found that halides containing α-unsaturation, such as α-halocarbonyl compounds or benzylic or allylic halides, are conspicuously easier to reduce than the corresponding saturated halides. Benzyl chlorides, for example, are almost as easy to reduce as iodobenzenes or iodoalkanes, and much more easily reduced than alkyl or aryl bromides (Table 4.1). The general ease of reduction of these halides is undoubtedly related to formation of an especially stable radical in the initial reduction step, and suggests that bond breaking is sufficiently advanced in the reduction transition state for the stability of the incipient radical to become important in lowering the activation energy for reduction.

* It has been estimated that benzene is reduced to its radical anion at potentials ca. 1.0 V cathodic of naphthalene.[22] Such potentials cannot be reached electrochemically in organic solvents because of prior reduction of the electrolyte.

Yields of radical-derived products, that is, dimers and dialkylmercury compounds, are exceptionally high from benzylic and allylic halides. Grimshaw and Ramsey,[26] for example, observed a single polarographic wave for, and isolated only dibenzyls and dibenzylmercury compounds from, the reduction of benzylic bromides in ethanolic solvents; only traces of the corresponding toluenes were observed. (The study was complicated, naturally, by simultaneous solvolysis of the halides under these conditions.) These authors suggested a reduction mechanism similar to that proposed for allyl bromide by Petrovich and Baizer,[11] that is, that benzyl radical should be harder to reduce than benzyl bromide, and further, that in protic solvents, potentials sufficiently negative to effect this second reduction step cannot be reached before the solvent decomposition potential is reached. The ratio of dibenzylmercury compound to dibenzyl from the reduction of 3,4-dichlorobenzyl bromide was found to depend on the potential at which reduction is carried out.[26] This unusual phenomenon is almost certainly a double-layer effect, and was so recognized by Grimshaw and Ramsey.

α-Halocarbonyl compounds, unlike the allylic and benzylic systems, are generally reduced polarographically and preparatively to the corresponding hydrocarbon in a clean 2-electron step. This indicates that the reduction potentials of acylalkyl radicals, $RCO\dot{C}\langle$, are anodic of the reduction potential of the carbon–halogen bond; this ease of reduction is consistent with the stabliity of the enolate anion formed in the second electron transfer, as well as with the electron-withdrawing power of the acyl group. The latter also facilitates the first electron transfer, as can be seen from the fact that phenacyl bromide is easier to reduce than ethyl bromide by 1.9 V.[1,27]

Reduction of α-halocarbonyl compounds, except for the fluorides which will be mentioned later, is generally found to be pH independent, indicating that proton transfer occurs subsequent to electron transfer, and hence that the enolate anion must be a discrete intermediate.[28] This suggests that electrochemical reduction of α-haloketones in the absence of proton donors should provide a method of preparation of solutions of stable enolate anions, whose chemistry could then be studied independently. Attempts have already been made, in fact, to alkylate electrochemically generated enolates, but yields were low because of Hofmann elimination by the enolate on the tetraalkylammonium ions used as supporting electrolyte.[29] It is likely that one might improve the yields in such studies by the use of metal ion salts as electrolytes, as well as by rigorous solvent drying procedures.

ω-Fluoroacetophenone (**3**) presents an interesting contrast to the usual generalization that reduction of alkyl halides is pH independent.[30] Although this *is* true for **3** in alkali, in acidic media increasing the acidity *facilitates* reduction. Since this pH dependence means that a proton must be involved in the transition state for reduction, a "push–pull" mechanism was suggested by Elving and Leone for reduction of **3** in acid. This anomalous pH dependence is most likely

$$\underset{\mathbf{3}}{C_6H_5COCH_2F} \xrightarrow[H^+]{2e^-} \left[\underset{\text{electrode}}{C_6H_5COCH_2 \cdots \overset{H^+}{\vdots}F} \right]^- \xrightarrow{H^+} \underset{\mathbf{4}}{C_6H_5COCH_3}$$

due to the marked tendency of fluoride to enter into hydrogen bonding. Stocker and Jenevein also found a case of unusual reduction behavior of α-ketofluorine.[31] They observed that the order of decreasing ease of reduction of the ω-polyfluoroacetophenones is

$$\underset{\mathbf{5}}{C_6H_5COCHF_2} > C_6H_5COCF_3 > \underset{\mathbf{3}}{C_6H_5COCH_2F}$$

since interruption of the controlled-potential electrolysis of **5** before completion provides, along with starting material, only **3** (27%) and **4** (37%). The reason for this order of reduction is unclear, but it is probably related to the generally high chemical stability of perfluoroalkyl groups such as that in **5**.

Geminal Polyhalides

Halogen substitution on carbon facilitates reduction of a second halogen bonded to the same carbon, and the effect is cumulative with increasing halogen substitution. This is nicely illustrated by the polarographic behavior of the polychloromethanes. Carbon tetrachloride exhibits, in aqueous DMF, three 2-electron waves at −0.25, −1.49, and −2.17 V (*vs* S.C.E.), respectively, these corresponding to successive stepwise reductions to chloroform, methylene chloride, and methyl chloride[32] (Figure 5.2). Methyl chloride is, of course,

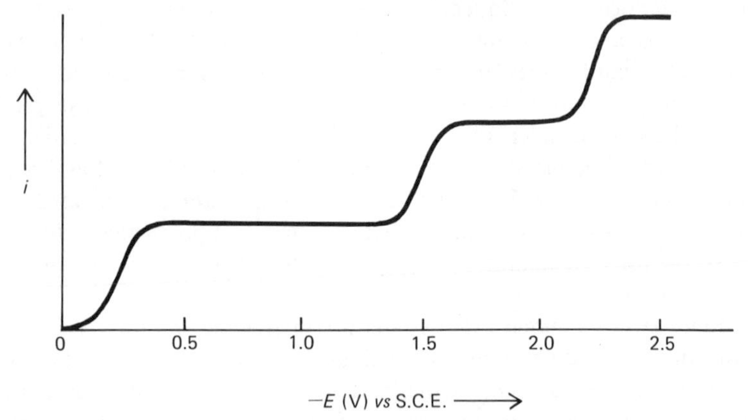

Figure 5.2 Polarographic behavior of carbon tetrachloride in aqueous dimethylformamide. (S. Wawzonek and R. C. Duty, *J. Electrochem. Soc.*, **108**, 1136 [1961].)

electrochemically inert, and no wave is observed for its reduction. In accordance with these assignments, chloroform exhibits two waves corresponding to the last two of carbon tetrachloride, and methylene chloride has only one wave at −2.14 V, for reduction to methyl chloride. The ability of halogen atoms to facilitate reduction is, like that of the carbonyl group, due both to their electronegativity and to their ability to stabilize the incipient radical in the transition state for the first step.[33] Halogen atoms also share with the carbonyl group the ability to stabilize carbanions, thus facilitating the second electron transfer and accounting for the absence of radical-type side reactions during reduction of geminal halides. The labilizing effect of halogen upon reduction is quite large, as can be seen from Figure 5.2 and Table 4.1. Since successive steps are generally separated by at least 0.5 V, it is a simple matter to effect selective removal of one or more geminal halogens at controlled potential, in yields close to quantitative. Taking advantage of the differences in reducibility of halogens, one may also selectively reduce mixed geminal polyhalides. The isomeric 7-bromo-7-chloronorcaranes are thus converted quantitatively to the corresponding chlorides upon reduction.[17a]

Wawzonek and Duty found that the electrochemical reduction of carbon tetrachloride follows a quite different course in aprotic media, however.[32] In anhydrous acetonitrile, carbon tetrachloride exhibits 2-electron waves at −0.25 and −0.64 V. The following scheme, involving dichlorocarbene as an intermediate, was proposed to account for the polarographic behavior under these aprotic conditions:

$$CCl_4 \xrightarrow[-0.25\ V]{2e^-} Cl^- + \bar{C}Cl_3$$

$$\bar{C}Cl_3 \longrightarrow Cl^- + :CCl_2$$

$$:CCl_2 \xrightarrow[-0.64\ V]{2e^-} CCl_2{}^{-2}$$

When reduction was carried out in the presence of tetramethylethylene, a small amount of the adduct **6** was formed, supporting their mechanistic interpretation

Cl Cl

6

of the polarographic behavior. (The low yield is unsurprising in view of the other reaction pathways, including reduction, available to the carbene.) Likewise, electrochemical reduction of 2,2-dichloronorbornane and the stereoisomeric 2-bromo-2-chloronorbornanes in dry DMF affords not only *endo*-norbornyl chloride (38%) (p. 173), but also nortricyclene (62%).[13] The latter product is also apparently derived from an intermediate carbene, which is known to insert intramolecularly to afford nortricyclene.[34] The scope of this carbene-forming reaction, and any possible synthetic use, is unknown at present. When found to occur but not wanted, it can be subverted by addition of certain proton donors to the reaction mixture.[13]

The electrochemical reduction of *gem*-dihalocyclopropanes has been investigated.[17] Typically, yields of cyclopropyl halides are close to quantitative, and the less stable *cis*-isomer **7** predominates in the product mixture by as much as 8 or 10 to 1 over the *trans*-isomer (**8**); this ratio may be increased through

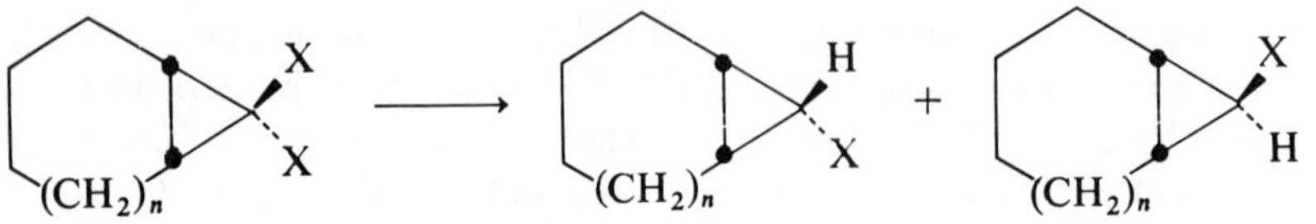

($n=1$ or 3; X = Cl or Br) **7** **8**

addition of proton donors. The same result can also be effected through the use of tributyltin hydride or methylmagnesium iodide as selective reductants,[35] but the electrochemical technique has some attraction for reduction of molecules containing functional groups liable to attack by either of the latter reagents, as shown by the following examples:

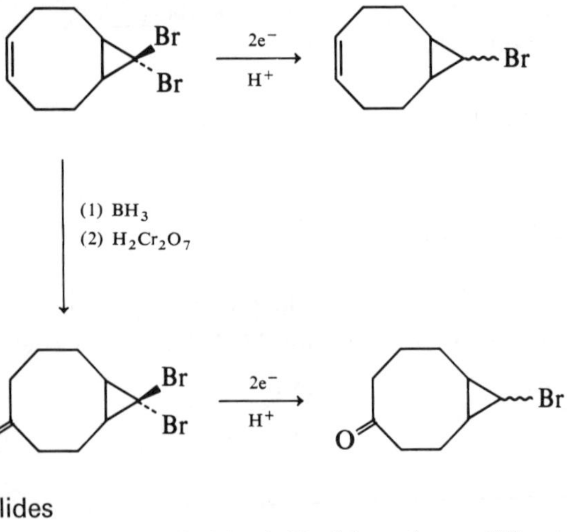

$2e^-$, H^+

(1) BH_3
(2) $H_2Cr_2O_7$

$2e^-$, H^+

(*Ref. 36*)

Vicinal Dihalides

The electrochemical reduction of vicinal dihalides takes a different course from that of other halides and dihalides. While reduction of the latter apparently involves stepwise formation of free radicals and carbanions, vicinal dihalides are generally reduced cleanly and concertedly to olefins. Although olefin formation could conceivably occur via β elimination from an intermediate carbanion (path *A*), the evidence clearly points toward a concerted reduction (path *B*) in

Path *A*

Path *B*

most cases. A strong argument for concerted reduction lies in the fact that the reduction products are olefins even in protic solvents, which should intercept carbanionic intermediates. The best evidence of all for the concerted nature of the reduction, though, comes from the observation that half-wave potentials for reduction of vicinal dihalides are considerably anodic of the half-wave potential of corresponding monohalides [*cf.* ethyl bromide, $E_{1/2} = -2.08$ V, and 1,2-dibromoethane, $E_{1/2} = -1.52$ V (*vs* S.C.E.)].[1,10] These differences are much too large to be accounted for on inductive grounds. Since alkyl halide reductions are electrochemically irreversible, half-wave potentials are indicative of relative rates of reduction, and the data therefore indicate that vicinal dihalides possess a low energy (i.e., concerted) reduction pathway not available to monohalides. This has been demonstrated in a striking manner by Zavada, Krupicka, and Sicher,[10] who found that the polarographic half-wave potentials for reduction of a series of rigid cyclic and bicyclic vicinal dibromides obey a Karplus-type relationship; that is, a plot of $E_{1/2}$ *vs* the dihedral angle φ between the two halogen atoms exhibits minima in $E_{1/2}$ at $\varphi = 0°$ and 180°, and a maximum at $\varphi = 90°$ (Figure 5.3). Reduction is therefore facilitated when the two atoms are coplanar, as expected for a concerted process. Typical data are as follows:

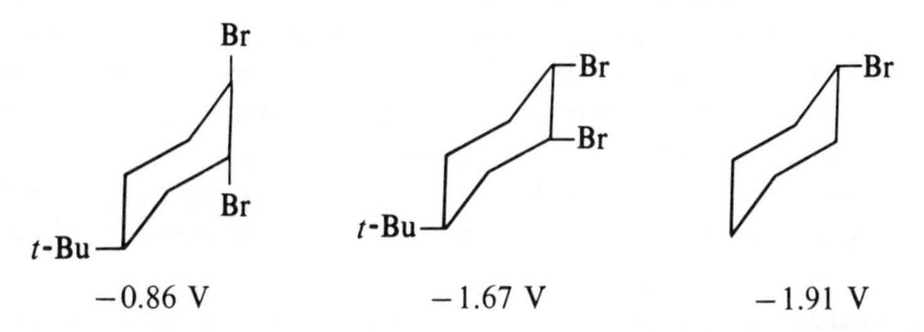

It was also found that *meso*-5,6-dibromodecane (−1.14 V) is reduced somewhat more easily than its *dl*-isomer (−1.24 V), a result in keeping with a preferred *anti*-coplanar transition-state geometry.* Finally, Koch and McKeon showed that rigid *trans*-diaxial dibromocyclohexanes are reduced cleanly to olefins, as expected for a concerted reduction mechanism.[37a]

Vicinal dihaloalkenes also undergo concerted reduction when the two halogens are disposed *trans* about the double bond. This is demonstrated by the work of Jura and Gaul, who found marked differences between the polarographic behavior of *trans*- and *cis*-2,3-dichloroacrylonitrile (**9**).[38] The *cis*-isomer (*cis*-**9**) exhibits three 2-electron polarographic waves, due to a straightforward reduction

$$\underset{cis\text{-}\mathbf{9}}{\text{Cl(H)C=C(Cl)CN}} \xrightarrow{2e^-} CH_2{=}C(Cl)CN \xrightarrow{2e^-} CH_2{=}CHCN \xrightarrow{2e^-} CH_3CH_2CN$$

$$\underset{trans\text{-}\mathbf{9}}{\text{Cl(H)C=C(CN)Cl}} \xrightarrow{2e^-} \underset{\mathbf{10}}{[HC{\equiv}CCN]} \xrightarrow{2e^-} CH_2{=}CHCN \xrightarrow{2e^-} CH_3CH_2CN$$

$$Cl_2CHCCl_2CN \xrightarrow{2e^-} trans\text{-}\mathbf{9} \xrightarrow{4e^-} CH_2{=}CHCN \xrightarrow{2e^-} CH_3CH_2CN$$

sequence producing successively 2-chloroacrylonitrile, acrylonitrile, and propionitrile. The *trans*-isomer, on the other hand, exhibits a 4-electron wave, followed by a 2-electron wave corresponding to reduction of acrylonitrile to propionitrile. The 4-electron wave was ascribed to a composite sequence involving the initial concerted 2-electron reduction of *trans*-**9** to propiolonitrile (**10**), followed by immediate 2-electron reduction of **10** to acrylonitrile. It was ascertained in a control experiment that **10** is more easily reduced than *trans*-**9**, so it would indeed be reduced immediately upon formation. These workers also observed that 2,2,3,3-tetrachloropropionitrile exhibits, in succession, a 2-electron wave, a 4-electron wave at the same potential as that of *trans*-dichloroacrylonitrile, and a 2-electron wave. This sequence was interpreted as being due

* One compound, *cis-exo*-2,3-dibromonorbornane, fell off the curve defined by nine other rigid dihalides. This compound is substantially harder to reduce than predicted from the curve ($E_{1/2}$ found: −1.53 V; predicted: ca. −1.21 V). If the correlation of $E_{1/2}$ with φ is valid, this may indicate distortion of the ring system in this compound, a phenomenon observed in other norbornanes.[37b]

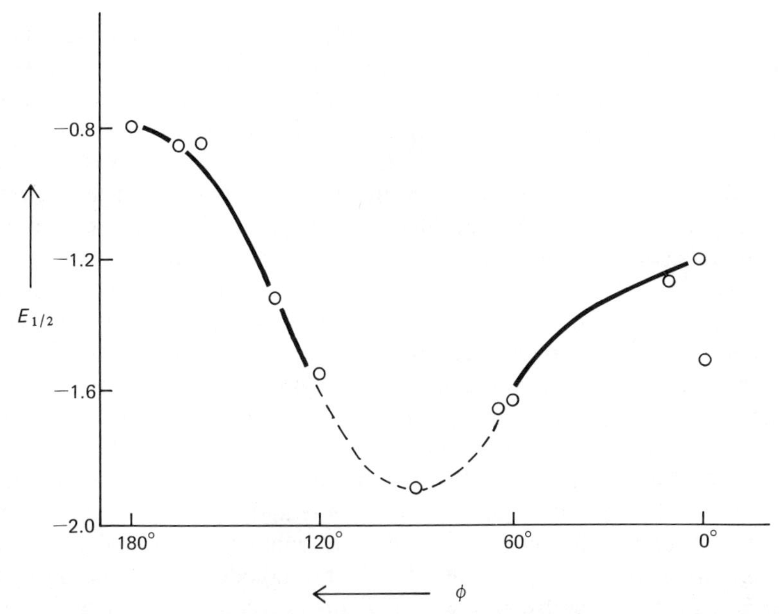

Figure 5.3 Plot of half-wave potentials of rigid vicinal dibromides *vs* estimated values of the dihedral angle, ϕ, between the carbon–bromine bonds. (J. Zavada, J. Krupicka, and J. Sicher, *Coll. Czech. Chem. Commun.*, **28**, 1670 [1963]. Reprinted by permission of Academia [publishing house of the Czechoslovak Academy of Sciences].)

to successive concerted dehalogenations. The concerted *trans*-dehalogenations observed by Jura and Gaul have their chemical precedent, of course, in numerous examples in the literature of acetylene formation by stereospecific *trans*-elimination across the olefinic bond.[39]

Other vicinal dehalogenations that may be of potential synthetic use and that illustrate the scope of the reaction include the following:

$$(C_6H_5)_2C{=}CX{-}CX{=}C(C_6H_5)_2 \xrightarrow{2e^-} (C_6H_5)_2C{=}C{=}C{=}C(C_6H_5)_2$$

(X = Cl or Br) *(Ref. 40)*

$$o\text{-}C_6H_4Br_2 \xrightarrow{2e^-} C_6H_4 \text{ (benzyne)} \xrightarrow[2H^+]{2e^-} C_6H_6$$

(Ref. 41)

Cl, Cl, Cl (trichlorinated spiro-lactone) $\xrightarrow{4e^-}$ $OCH_2CO_2^-$ *(Ref. 42)*

In the first example, Kemula and Kornacki demonstrated polarographically the clean electrochemical formation of the cumulene system; it is reduced immediately upon formation to 1,1,4,4-tetraphenyl-1,3-butadiene, which may be reduced further to 1,1,4,4-tetraphenyl-2-butene and 1,1,4,4-tetraphenylbutane at more negative potentials.[40] In the second instance, Wawzonek and Wagenknecht observed that 1,2-dibromobenzene exhibits a single 4-electron wave in aprotic solvents. As indicated, they suggested an initial concerted 2-electron reduction to form benzyne, which undergoes subsequent 2-electron reduction to benzene.[41] The transient benzyne was trapped as its Diels–Alder adduct (in low yield) by electrolysis in the presence of furan. Sease has ascribed anomalies in the electrochemical behavior of *o*-bromoiodobenzene and *o*-diiodobenzene to formation of benzyne as an intermediate.[25] Preparative-scale reduction of *o*-bromoiodobenzene in DMF containing lithium chloride as supporting electrolyte afforded chlorobenzene (13%), strong evidence for intermediate benzyne. Electrolysis with tetraethylammonium perchlorate as supporting electrolyte produced a quantity of benzene and bromobenzene amounting to only 9% of the theoretical quantity, and coulometry showed that only 2 electrons were consumed per molecule of *o*-bromoiodobenzene, indicating that benzyne can disappear by pathways that do not consume electrons. Polarographic and preparative-scale reductions of *o*-chloroiodobenzene proceeded normally to chlorobenzene, with no evidence for benzyne as an intermediate.[25] The third example above is of some interest because it demonstrates the concerted elimination of vicinal chlorine and oxygen[42]; it is also closely related to the quantitative electrochemical reduction of hexachlorocyclohexane to benzene.[43]

Vinylogous vicinal dehalogenations have been reported by Covitz.[44] 1,4-Bis-(bromomethyl)-benzene (**11**) is converted to [2.2]-paracyclophane (5–10%) and polyxylylene upon reduction; the *ortho*-dibromide **12** affords only the *ortho*-xylylene polymer. These eliminations are also concerted, since the half-wave potentials of **11** and **12** are, respectively, 0.52 and 0.71 V anodic of that of the *meta*-isomer, and the products of electrolysis are unchanged in the

CH_2Br / CH_2Br **11** $\xrightarrow{2e^-}$ CH_2 / CH_2 $\longrightarrow$ + polymer

CH_2Br / CH_2Br **12** $\xrightarrow{2e^-}$ CH_2 / CH_2 $\longrightarrow$ polymer

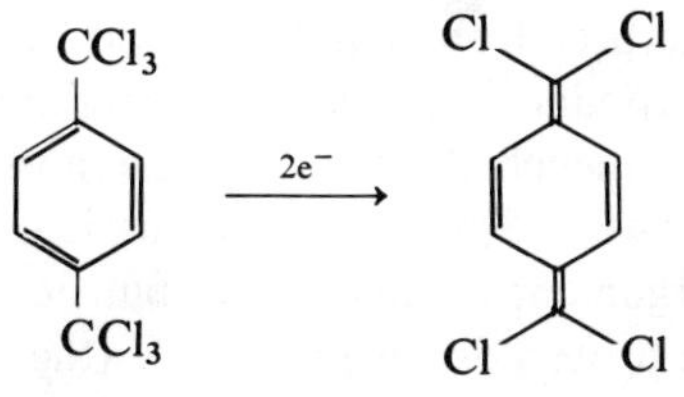

CCl_3 $\xrightarrow{2e^-}$ Cl Cl

CCl_3 Cl Cl

13

presence of aqueous hydrochloric acid, which would protonate carbanionic intermediates in a stepwise mechanism. In a related report from the same laboratory, it was shown that tetrachloro-*p*-xylylene (**13**) can be isolated in good yields by electrolysis of hexachloro-*p*-xylene at $-10°C$.[45]

Nonvicinal Dihalides

In most cases, halogens separated by more than two saturated carbon atoms are reduced independently of one another. Reduction of 1,5-dibromopentane, for example, proceeds cleanly to pentane in DMF.[46a] There are, however, some reports of ring formation during reduction of nonvicinal halides. Rifi has reported electrochemical synthesis of strained compounds by reduction of nonvicinal dihalides.[46] In general, these reductions do not appear to be concerted.

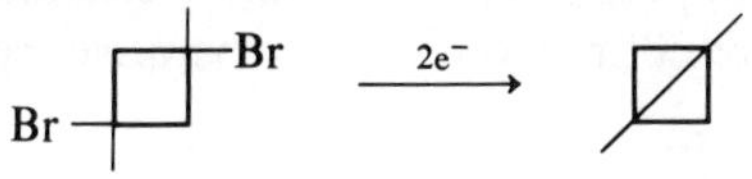

Br Br $\xrightarrow{2e^-}$

14

Cl Cl Cl Cl $\xrightarrow{2e^-}$ Cl Cl

Br Br $\xrightarrow{2e^-}$

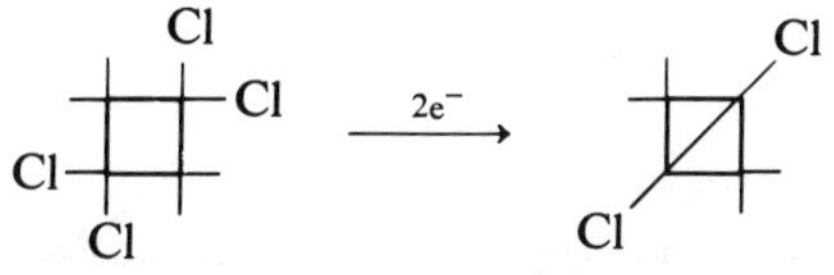

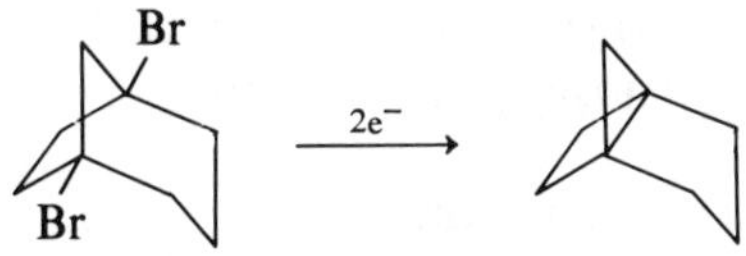

Evidence for stepwise reduction comes both from isolation of side products indicative of carbanion intermediates and from the observation that reduction of either *cis*- or *trans*-**14** affords the bicyclobutane. The pathway suggested for these reductions by Rifi involves an intramolecular S_N2 displacement:

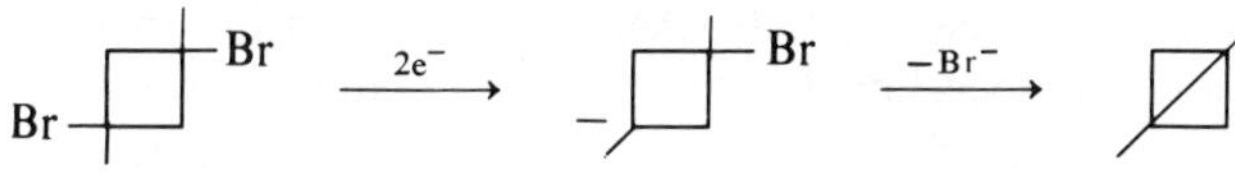

Br Br $\xrightarrow{2e^-}$ $-$ Br $\xrightarrow{-Br^-}$

14

Rifi has favored the interpretation that formation of cyclopropane upon reduction of 1,3-dibromopropane[46a] proceeds in *concerted* fashion, since cyclopropane formation is not quenched by electrolysis in the presence of a ten-fold excess of water, and because 1,3-dibromopropane is easier to reduce than the 1,4-dibromide.[46c] The argument is reasonable, but not necessarily compelling, since water is not a good proton donor in DMF[47] (the electrolysis solvent) and since the small differences in reduction potential may be due instead to inductive effects. If the reduction of 1,3-dibromopropane to cyclopropane is actually concerted, the problem arises as to why **14** is reduced in stepwise fashion. Perhaps neither isomer of **14** can assume the correct transition-state geometry for concerted reduction. Another unanswered question in this area is whether cyclobutanes can be prepared by electrochemical reduction of 1,4-dihalides. Rifi has reported isolation of cyclobutane (25%) and *n*-butane (75%) from reduction of 1,4-dibromobutane in DMF,[46a] while under apparently identical conditions Diefenderfer and Dougherty found only *n*-butyl bromide or *n*-butane, depending on electrolysis potential.[48]

A possible case of concerted elimination of two distant halogen atoms has been reported by Zavada, Krupicka, and Sicher,[10] who found that the isomeric 1,6-dibromocyclodecanes are somewhat easier to reduce than cyclodecyl bromide and that the dibromides appear to consume only 2 electrons upon reduction.[10] Since the compounds were not reduced on a preparative scale (presumably they

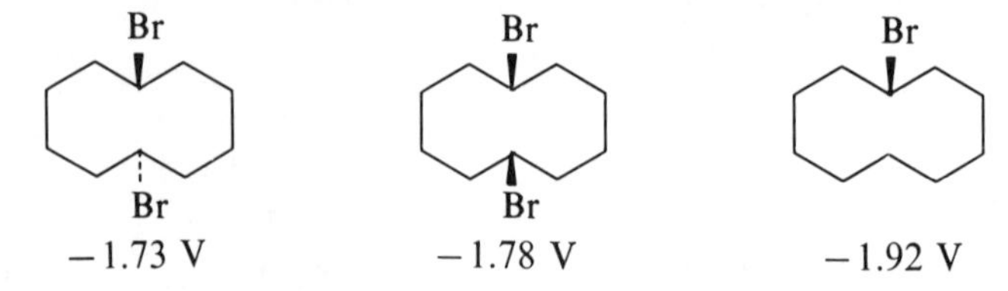

should afford decalins), the polarographic evidence must at present be considered interesting and suggestive, but only conjectural. A conceivable alternative pathway would involve normal reduction of one halogen to a carbanion, which then displaces the other halogen in S_N2 fashion.

Synthetic Scope and Limitations

A number of comments have been made in this and earlier chapters with respect to the synthetic merits of electrochemical dehalogenation, particularly with respect to its selectivity. These remarks need not be repeated here. It may be useful, however, to recapitulate the electrochemical characteristics of specific structural types, with some indication of preferred conditions for their reduction.

The controlled-potential electrochemical reduction of vinyl, aryl, geminal polyhalides and α-halocarbonyl compounds generally proceeds in yields close to quantitative, particularly under conditions where the intermediate carbanions are rapidly protonated. Typical conditions might include, for example, dipolar aprotic (DMF, CH_3CN, or DMSO) or protic (alcohol or aqueous glyme) solvent, containing at least an equivalent of phenol or acetic acid as proton donor. Protic solvents have the obvious advantage of low electrical resistance (especially important in large-scale electrolysis), but the disadvantage that one cannot reach as negative potentials as in aprotic solvents.

Vicinal dihalides are normally converted cleanly to olefins upon reduction. In synthetic schemes where an olefinic group is temporarily blocked by conversion to the *vic*-dibromide, there may be an advantage to electrochemical dehalogenation over conventional methods in certain cases, since electrolysis may be effected in aprotic media if desired.

Nonvicinal dihalides sometimes undergo conversion to cyclic products. Either protic or aprotic media might be used, with the latter naturally preferred where a stepwise nonconcerted reduction is suspected. The scope of this ring-forming reaction is not at all clear at present. Rifi has reported that reduction of 1-bromo-3-bromomethylcyclobutane affords bicyclo-[1.1.1]-pentane in hexamethylphosphoramide but not in DMF.[46c]

Saturated alkyl halides are reduced to the corresponding hydrocarbons in good, but not quantitative, yields. Side reactions, which can amount to 20% or more of the product, consist of formation of dialkylmercury compounds, olefins (formed both by disproportionation of the intermediate radical and by E2 elimination upon the starting material by the intermediate carbanion), dimers (usually formed in quite low yields), and occasionally, rearranged products.[2] One can improve the situation by electrolysis in the presence of a proton donor, but since radical-derived products cannot be eliminated, it is normally not possible to obtain hydrocarbons in yields greater than 80–85%. Reduction of saturated alkyl halides is probably better effected with reagents such as lithium aluminum hydride or chromous ion.[49]

The synthetic situation is more complicated with allylic and benzylic halides. It appears that the chlorides and fluorides are dehalogenated in good yields (*cf.* the isolation of toluene in 75% yield from electrolysis of benzyl chloride,[7] and the single 2-electron polarographic wave observed for allyl chloride).[11] The preferred conditions for this conversion would include electrolysis in a dipolar aprotic solvent (to avoid spontaneous solvolysis) containing an equivalent of phenol. The corresponding bromides have been investigated relatively little, particularly in aprotic solvents. Radical-derived products appear to be more important, however, with the bromides. In fact, it will be recalled that in ethanol, benzylic bromides afford only dibenzyls and dibenzylmercury compounds.[26] Benzylic and allylic iodides apparently react spontaneously with mercury to form mercury compounds.[7, 50]

5.1.2 Ammonium Salts

Electrochemical cleavage of the carbon–nitrogen single bond can generally be effected only when nitrogen bears a positive charge, so that it can leave as a neutral entity. We are concerned here, therefore, only with the electrochemistry of ammonium salts. This may be illustrated, for example, by the polarographic data on α-aminoacetophenones, whose reduction to acetophenone is pH dependent, becoming easier in media of increasing acidity.[51] Mechanisms of reductive cleavage of ammonium salts are not understood nearly as well as those of alkyl halides. The principal reason for this lies in an almost complete lack, with a few exceptions, of voltammetric data (polarography or cyclic voltammetry) on such compounds, complicated further by the fact that investigators have usually employed constant-current electrolysis, so that it is often unclear what species is actually being reduced. It appears, these difficulties notwithstanding, that there are at least two distinct mechanistic paths by which ammonium ions may be cleaved. These pathways include (1) electrochemical reduction with bond breaking in the transition state for electron transfer, and (2) reduction via intermediate quarternary ammonium amalgams.

Electron Transfer with Simultaneous Bond Breakage

The electrochemical cleavage of quarternary nitrogen compounds can generally be effected at platinum, lead, aluminum, or mercury cathodes, in either protic or aprotic solvents, when the carbon–nitrogen bond is activated by an allyl, benzyl, α-carbonyl, and so on, group:

$$C_6H_5COCH_2N^+R_3 \xrightarrow[Hg]{2e^-} C_6H_5COCH_3 \qquad (Ref.\ 51)$$

$$4\text{-}(CH_2N^+H_2R)\text{pyridine} \xrightarrow[Hg]{2e^-} 4\text{-}CH_3\text{-pyridine} \qquad (Ref.\ 52)$$

(R = H or C_6H_5)

$$C_6H_5CH{=}CHCH_2N^+Et_3 \xrightarrow[Al]{2e^-} \underset{80\%}{C_6H_5CH{=}CHCH_3} + \underset{20\%}{C_6H_5CH_2CH{=}CH_2} \qquad (Ref.\ 53)$$

$$\mathbf{15} \xrightarrow[Pb]{4e^-,\ H^+} \mathbf{16} \qquad (Ref.\ 54b)$$

It appears that these reductions proceed by a mechanism similar to that proposed for activated halides, that is, electron transfer to the ammonium ion with bond breaking in the transition state and consequent stabilization of the transition state by delocalization of the incipient radical:

$$R{-}N^+R_3 + e^- \longrightarrow [R \cdots NR_3]^{\cdot} \longrightarrow NR_3 + R\cdot$$

$$R\cdot \xrightarrow[\text{fast}]{e^-} R^- \longrightarrow RH$$

If free radicals are indeed intermediates in the reduction, one would expect, as with alkyl halides, subsequent reduction to carbanions, dimerization, or hydrogen abstraction from the solvent. The latter two pathways do not appear to be of importance in protic media, but are observed to some extent in aprotic solvents:

$$\underset{\text{optically active}}{C_6H_5CH(CH_3)N^+(CH_3)_3} \xrightarrow[\text{DMF, Al}]{e^-} \underset{\text{optically inactive}}{[C_6H_5CH(CH_3){-}]_2} \qquad (\textit{Ref. 55})$$

$$\underset{\mathbf{17}}{C_6H_5CH(C_6H_5)N^+(CH_3)_2} \begin{cases} \xrightarrow[H_2O]{e^-,\ Hg} C_6H_5CH_3 \\ \xrightarrow[CH_3CN]{e^-,\ Hg} C_6H_5CH_3 + C_6H_5CH_2CH_2C_6H_5 \end{cases} \qquad (\textit{Ref. 56})$$

Bard and Mayell demonstrated the intermediacy of the benzyl radical in reduction of **17** in DMF by (1) obtaining an esr signal from the electrolysis mixture, (2) coulometry, which showed that $n < 2$ under conditions where toluene is the major product, and (3) observation that the ratio of bibenzyl to toluene is increased at high concentrations of starting material.[56] Bard and Mayell also showed, by a treatment based on the polarographic equation for a totally irreversible process (Eq. 2.50), that the rate-determining reduction step involves a single electron.

The interesting reductive cleavage of cyclic α-aminoketones (e.g., **15** → **16**) was investigated by Leonard and Swann and their coworkers.[54] Yields ranged from 42 to 73% in the several cases investigated, but no attempt was made to maximize yields, and these could undoubtedly be improved upon. The concomitant reduction of the carbonyl group can be avoided by carrying out the reduction at controlled potential.[51] Acyclic α-aminoketones may, of course, be cleaved with equal facility.

Reduction via Quarternary Ammonium Amalgams

Lead and mercury both have very high overvoltages for hydrogen discharge, and hence very negative potentials can be reached with either metal as cathode material. Electrochemical reactions involving simple electron transfer, such as reduction of aromatic hydrocarbons or the activated ammonium salts discussed above, should for this reason be quite similar at lead and mercury. This is usually true. It has been found, however, that certain other reductions *not* possible at lead cathodes *can* be carried out at mercury, such as cleavage of aryltrialkylammonium salts:

$$\underset{\mathbf{18}}{ArN^+R_3} \begin{cases} \xrightarrow[\text{Pb}]{e^-,\ H_2O,\ 25^\circ C} \text{N.R.} \\ \xrightarrow[\text{Hg}]{e^-,\ H_2O,\ 25^\circ C} ArH + R_3N \end{cases} \qquad (\textit{Ref. 57})$$

Such behavior is strongly suggestive of *involvement of the cathode itself in the reduction mechanism.* The nature of this involvement in the present case is almost certainly via the corresponding *quarternary ammonium amalgam*:

$$R_4N^+ + e^- \xrightarrow{\text{Hg}} R_4N{-}Hg_n \qquad n \simeq 12\text{–}13 \qquad (\textit{Ref. 59})$$

A number of investigations have confirmed and elaborated upon the first report of preparation of quarternary ammonium amalgams by McCoy and Moore[58] over 50 years ago.* As one might expect, quarternary ammonium amalgams have much the same properties as do the alkali metal amalgams: they react with water (*vigorously*) to form hydrogen, the quarternary ammonium hydroxide, and mercury; they have crystalline or semicrystalline structures; and, of particular interest, they are excellent reducing agents.[58,59] The latter point will be discussed separately later in this chapter. Unlike alkali metal amalgams, however, quarternary ammonium amalgams, when isolated in a pure state, free of solvent, are thermally unstable at relatively low temperatures. This instability arises from a decomposition pathway not available to the metal amalgams:

$$R_4N{-}Hg_n \xrightarrow{\Delta} n\text{Hg} + R_3N + R\cdot$$

Methyl radicals have been identified by the Paneth metallic-mirror method as decomposition products of tetramethylammonium amalgam,[60] and several investigators have also identified the tertiary amine fragment from the decompositions. It is noteworthy that quarternary ammonium amalgams have only been isolated under conditions where hydrogen evolution is slow or impossible, such as when the solvent was absolute ethanol or DMF. In water, electrolysis of tetraalkylammonium salts usually results simply in evolution of hydrogen, although the formation of the amalgam as a reactive intermediate is evidenced in some instances by liberation of a colloidal gray suspension of amalgam and mercury into solution upon passage of current, followed by reaction of the amalgam with water to form hydrogen and hydroxide ion.[57,58] The suspension apparently arises because the surface tension of the mercury is decreased sufficiently by adsorption of the amalgam that the mercury becomes colloidal.

The cleavage of aryltrialkylammonium ions such as **18** at a mercury cathode probably involves the corresponding amalgam as an intermediate, since direct reduction of the aryl–nitrogen bond is not feasible at lead cathodes, at

* The synthesis of ammonium amalgam itself antedates the report of McCoy and Moore by more than a century: Seebeck, *Ann. Chim.*, **66**, 191 (1808).

which similar potentials can be reached. If this postulate is true, it follows that of the two pathways open to an amalgam in water, that is, thermal decomposition or reaction with water to liberate hydrogen, arylalkylammonium amalgams follow the first path preferentially* and tetraalkylammonium amalgams proceed by the second. Reductive cleavage via the amalgam would be visualized as

$$Ar\overset{+}{N}R_3 \xrightarrow[Hg]{e^-} ArNR_3{-}Hg_n \xrightarrow[-nHg]{\Delta} NR_3 + Ar\cdot \xrightarrow[H^+]{e^-} ArH$$

We have seen thus far two types of cleavage: those of activated ammonium salts, such as **17**, which proceed at mercury, platinum, or lead cathodes, and those of arylalkylammonium salts, such as **18**, which proceed only at mercury and are postulated here to involve amalgam intermediates. Is there any reason to suspect amalgams in the reduction of activated ammonium salts at mercury? No. When presented with two modes of reaction, a compound prefers the lower-energy pathway. Since activated salts are cleaved even at potentials accessible on platinum, direct cleavage must be the lower-energy path for these species. Only in arylalkylammonium salts must direct cleavage be so high in energy that the amalgam route becomes the easier.

Synthetic Scope and Limitations

The cleavage of activated ammonium salts to the corresponding amine and hydrocarbon generally proceeds in good to high yields in protic solvents. α-Aminoketones generally suffer not only cleavage of the carbon–nitrogen bond but also simultaneous reduction of the carbonyl group. The latter can be avoided if desired through electrolysis at controlled potential. Lead has generally been used as the cathode material, but mercury should presumably serve equally well.

The order of ease of reduction is allyl $\simeq$ benzyl $\simeq$ α-carbonyl $\gg$ aryl $\gg$ alkyl. This order has been put to good synthetic advantage by Horner, who has developed a procedure for synthesis of a variety of mixed tertiary amines and quarternary ammonium salts not readily synthesized by other means[57]:

$$R_1NH_2 + 2C_6H_5CH_2Cl \longrightarrow R_1N(CH_2C_6H_5)_2 \xrightarrow{R_2Br} R_1R_2N^+(CH_2C_6H_5)_2Br^-$$

$$\xrightarrow[20°C,\ H_2O]{2e^-} R_1R_2NCH_2C_6H_5 \xrightarrow{R_3X} R_1R_2R_3N^+CH_2C_6H_5Br^-$$

$$\xrightarrow[20°C,\ H_2O]{2e^-} R_1R_2R_3N \xrightarrow{R_4Br} R_1R_2R_3R_4N^+Br^-$$

* Horner has suggested, on the other hand, that these reductions, and in fact all reductive cleavages of ammonium salts at mercury, involve not amalgams, but cathode complexes, $[R_4N]^{-2}[Hg]$, whose exact structure and mode of reaction is unclear.[57]

Certain reductions of quarternary ammonium ions similar to those above have been carried out by Wrobel and coworkers.[61] They used platinum electrodes in the apparatus of Benkeser and Kaiser,[62] and effected reduction (80–90% yields) in moist liquid ammonia containing sodium iodide. They have shown that the reductions actually proceed by electrochemically generated sodium, that is, by solvated electrons. Here too, as in reductions at mercury, N-benzyl and N-aryl bonds are cleaved in preference to N-alkyl bonds:

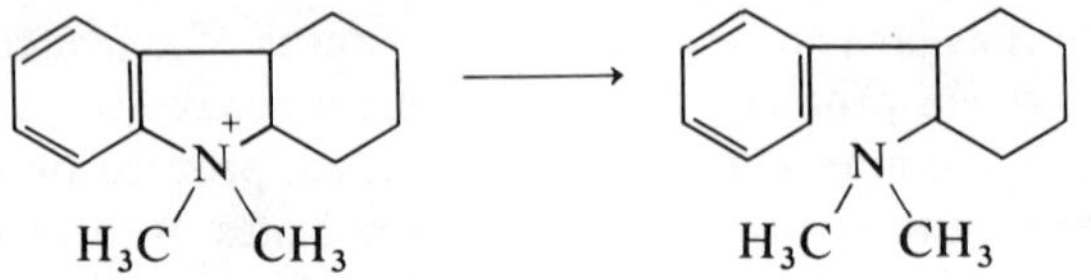

(*Ref. 61a*)

(*Ref. 61b*)

These cleavages may also be effected using sodium in ammonia, and should be in small-scale work. There would, on the other hand, be some advantage in using the electrochemical method for large-scale preparations, where large amounts of sodium would otherwise be required.

5.1.3 Phosphonium Salts

Quarternary phosphonium salts may be cleaved into the corresponding hydrocarbon and tertiary phosphine in aqueous solution at platinum, lead, or

$$R_3P^+R^1X^- \xrightarrow[H_2O]{2e^-} R_3P + R^1H$$

mercury cathodes.[63] At platinum, with its lower hydrogen overvoltage, only activated carbon–phosphorus bonds are cleaved, but at lead and mercury* cathodes, aryl and alkyl groups are cleaved. The order of increasing ease of reductive cleavage of substituents appears to be

$$\text{alkyl(I° or II°)} \simeq C_6H_5 < \text{alkyl(III°)} \simeq C_6H_5CH_2$$

One of the most interesting and useful features of the electrochemical cleavage of the phosphonium salts is the fact that it occurs with retention of stereochemical configuration.[65] It thus opens a quite convenient synthetic route to optically active tertiary phosphines and quarternary phosphonium salts, such as

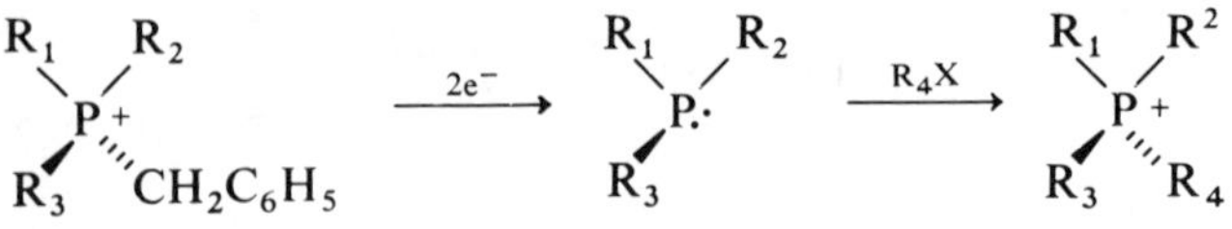

$$R_1R_2R_3P^+CH_2C_6H_5 \xrightarrow{2e^-} R_1R_2R_3P: \xrightarrow{R_4X} R_1R^2R_3P^+R_4$$

* Quarternary phosphonium amalgams can be prepared, by electrolysis in aprotic solvents at low temperature.[64]

Horner has suggested[63] that these reductions proceed by pentacovalent hydrides, R_4PH, which subsequently decompose to R_3P and RH, but the reaction might equally well involve direct electron transfer to the phosphonium salt with bond breaking in the transition state. Indeed, the subsequent discovery by Horner that reduction of optically active phosphonium salts by lithium aluminum hydride (a reaction that probably does involve pentacovalent phosphorus hydrides) affords phosphines that are mostly racemized,[66] suggests that such intermediates are *not* involved in the electrochemical reduction.

Synthetic Scope and Limitations

The electrochemical cleavage of phosphonium salts proceeds particularly cleanly when the leaving group is activated, such as benzylic or allylic. Unlike the situation with amines, phenyl and alkyl substituents are cleaved with comparable facility. Horner's valuable synthetic route to optically active phosphines and phosphonium salts has already been mentioned. More generally, Horner has pointed out that an extremely wide variety of mixed phosphines and phosphonium salts can be synthesized by alternate alkylation and cleavage of benzylic phosphines[63]:

$$P(CH_2C_6H_5)_3 \xrightarrow{R_1X} R_1P^+(CH_2C_6H_5)_3X^- \xrightarrow[H_2O,\ 20°C]{2e^-} R_1P(CH_2C_6H_5)_2$$

$$\xrightarrow{R_2X} R_1R_2P^+(CH_2C_6H_5)_2 \xrightarrow[H_2O,\ 20°C]{2e^-} R_1R_2RCH_2C_6H_5$$

$$\xrightarrow{R_3X} R_1R_2R_3P^+CH_2C_6H_5X^- \xrightarrow[H_2O,\ 20°C]{2e^-} R_1R_2R_3P$$

$$\xrightarrow{R_4X} R_1R_2R_3R_4P^+X^-$$

As enumerated by Horner, the advantageous features of this electrochemical route include simplicity of operation and workup, good yields (80–90%), and high purity of the product phosphines.

Bestmann, Graf, and Hartung have developed a general procedure for synthesis of α-substituted β-keto esters, based on the electrochemical cleavage of phosphonium salts and utilizing ylides of type **19**.[67]

$$RCH_2COCl + \underset{\textbf{19}}{\underset{P(C_6H_5)_3}{\overset{\|}{R^1CCO_2Et}}} \longrightarrow RCH_2CO\overset{R^1}{\underset{^+P(C_6H_5)_3}{C}}CO_2Et$$

$$\xrightarrow[H_2O,\ 25°C]{2e^-,\ Hg} RCH_2COCH(R^1)CO_2Et$$

It should probably be pointed out here that, unlike phosphonium salts, phosphine oxides (even those containing benzylic carbon–phosphorus bonds) are apparently quite stable against electrochemical cleavage. Horner and

Neumann have reported that reduction of phenyldibenzylphosphine oxide, $C_6H_5(C_6H_5CH_2)_2PO$, affords cyclohexadienyldibenzylphosphine oxide, $C_6H_7(C_6H_5CH_2)_2PO$ (presumably a mixture of double-bond isomers) in 93% yield; benzene (5%) (but *no* toluene) was also formed.[68]

5.1.4 The Carbon–Oxygen Bond

Reductive electrochemical cleavage of the carbon–oxygen bond has not been investigated extensively. Reductions are generally pH dependent, proceeding more efficiently under acidic conditions, where oxygen is protonated and the species removed is thus neutral.[27] Carbon–oxygen bonds *alpha* to carbonyl or cyano groups, and also the cinnamyl moiety, may be cleaved electrochemically, but benzylic (and of course phenyl and alkyl) groups are *not* removed (*cf.* the stability of benzyl phenyl ether toward reduction[27]).

The cleavage studied most is that of α-ketols and their derivatives. Lund examined a series of phenacyl derivatives, $C_6H_5COCH_2OR$, where R is acetyl, benzoyl, phenyl, or hydrogen.[27] Half-wave potentials (*vs* S.C.E.) at pH 1 were −0.75, − 0.75, −0.83, and −0.85 V, respectively, for the four derivatives. There appears to be some correlation between half-wave potential and the leaving group ability of the substituent OR, but differences are small. Lund showed that it is the phenacyl group that is removed in each case, by isolation of acetophenone from controlled-potential electrolyses. Nonaromatic α-ketols and ketol acetates, on the other hand, cannot normally be selectively reduced to the corresponding ketone, since the half-wave potential for reduction of the carbonyl group is too close to that of the α-substituent. Such compounds are therefore reduced further to the corresponding saturated alcohol[69]:

$$\underset{\underset{O}{\|}}{RC}\overset{\overset{OR^{11}}{|}}{C}HR^1 \xrightarrow{4e^-} \underset{\underset{OH}{|}}{RCH}CH_2R^1 \qquad (R^{11} = \text{H or Ac})$$

Interestingly, the latter reduction appears to be stereospecific, because when it was applied to a number of substrates in the steroid series only the equatorial alcohol could be isolated.[69] Yields of alcohol from the steroidal ketol or ketol acetate were generally high (90–100%):

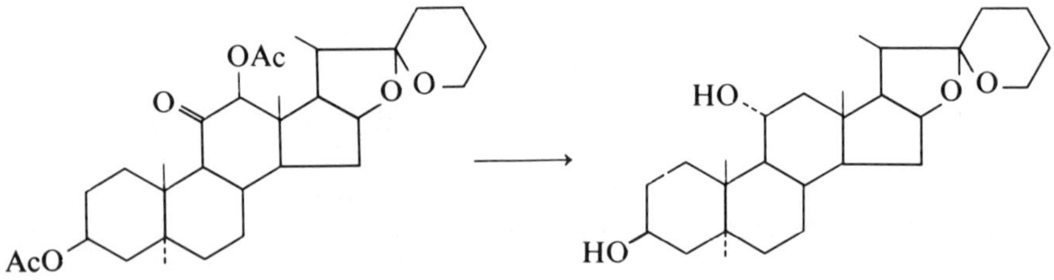

A number of mandelonitriles and atrolactic acid derivatives have been reduced in good yield under neutral conditions.[70, 71]

$$\text{R-}C_6H_4\text{-}CH(OR^1)CN \longrightarrow \text{R-}C_6H_4\text{-}CH_2CN \qquad (\textit{Ref. 70})$$

$(R = H \text{ or } OCH_3;\ R^1 = COCH_3, COC_6H_5, CO_2Et, H)$

$$C_6H_5C(OCOC_6H_5)(CH_3)CO_2R \longrightarrow C_6H_5CH(CH_3)CO_2R \qquad (\textit{Ref. 71})$$

$(R = H \text{ or } CH_3)$

The latter reductions have been shown to proceed with loss of stereochemical integrity,[71] in contrast to the reported stereochemical course of reduction (inversion) upon reduction of 2-chloro-2-phenylpropionic acid.[19]

Synthetic Scope and Limitations

Carbon–oxygen bonds may be cleaved in protic media when adjacent to one or more powerfully activating groups such as the carbonyl or cyano functions. Bonds doubly activated may be cleaved without reduction of the activating group, such as reduction of mandelonitriles, atrolactic acid, and phenacyl derivatives, and reduction of pseudoesters to phthalides:

$$\text{3-R-3-}OR^1\text{-phthalide} \longrightarrow \text{3-R-3-H-phthalide} \qquad (\textit{Ref. 72})$$

$(R = C_6H_5 \text{ or } H;\ R^1 = CH_3, C_2H_5, C_6H_5, COCH_3)$

Reduction of aliphatic ketols and ketol acetates results in both cleavage of the carbon–oxygen bond and reduction of the carbonyl group to the alcohol.[69] Yields of the alcohol (90–100%) are better than can be obtained by reduction of the ketol with sodium–ammonia or other conventional reducing agents. (Removal of the α-hydroxyl substituent of the ketol without reduction of the carbonyl is best carried out by a nonelectrochemical method, such as hydrogen iodide in acetic acid.[73])

Mairanovskii and coworkers have shown that the O-benzyl, O-cinnamyl, and O-trityl derivatives of alcohols and carboxylic acids (and presumably also phenols) may be cleaved electrochemically under neutral conditions (95%

aqueous DMF), thus permitting use of these derivatives as blocking groups in synthetic procedures.[74] Anderson, Baizer, and Petrovich observed the unusual reductive cleavage of a phenyl vinyl ether[74e]:

$$C_6H_5OCH{=}CHCO_2Et \longrightarrow CH_2{=}CHCO_2Et$$

This reaction is probably initiated by addition of an electron to the pi system, followed by breakage of the carbon–oxygen sigma bond.

5.1.5 Other Single–Bond Cleavages

The Carbon–Sulfur Bond

What little is known about the cleavage of the carbon–sulfur bond is in complete accord with expectation based on analogies to the other reactions already discussed. Thus, the only bonds generally cleaved are those in which sulfur bears a positive charge, and adjacent activating groups (carbonyl, benzyl, etc.) facilitate reduction.[27,75] The strong adsorption of sulfur compounds on

$$C_6H_5COCH_2\overset{+}{S}R_2 \xrightarrow{2e^-} C_6H_5COCH_3 + R_2S \qquad (Ref.\ 75)$$

mercury results, in the case of triphenylsulfonium ion, in mercury actually participating *chemically* in the initial electron-transfer step; final products are diphenylmercury and diphenyl sulfide.[76] (Triphenylsulfonium amalgam may be an intermediate in this reduction.) Similar behavior is observed with trialkylsulfonium salts, which afford complex mixtures of products upon reduction.[77b] The electrochemical cleavage of the carbon–sulfur bond in α-thiophenylcinnamonitrile is unusual, because it represents removal of a neutral sulfur atom.[77b] The reaction may well be initiated by addition of an electron to the pi system, followed by bond breakage. The 2-electron polarographic wave observed with some arylalkyl sulfides may be due to a similar phenomenon.[77c]

The Carbon–Carbon Bond

Despite its strength, several examples of cleavage of the carbon–carbon single bond have been reported. In every case the bond cleaved is attached to a pi system, and it is probable that all proceed via initial electron transfer to the pi system, followed by breakage of the sigma bond. Stradins, Tutane, Neilands, and Vanags observed a 2-electron polarographic reduction wave at ca. -0.6 V (S.C.E.) for several indanedione dehydrodimers (**20**) in DMF, and identified the indanedione enolate **21** spectroscopically after controlled-potential electrolysis of **20**.[78] Predissociation of **20** into radicals before electron transfer was excluded, since esr signals could not be observed from **20** except where Ar was *p*-dimethylaminophenyl, and then only at temperatures greater than 60°C. Interestingly, enolate **21** was shown by these workers to undergo further 2-electron reduction to a trianion at -1.67 V, a reduction similar to several reported by Bauld and coworkers.[79]

$$\mathbf{20} \xrightarrow[-0.6\ \mathrm{V}]{2e^-} 2\ \mathbf{21}$$

$$\mathbf{21} \xrightarrow[-1.67\ \mathrm{V}]{2e^-} [\ \ ]^{-3}$$

The other examples of cleavage of carbon–carbon single bonds all involve reduction of cyano compounds to the corresponding hydrocarbons and cyanide ion. Such cleavages are observed upon reduction of a number of substituted aryl nitriles[80–82] in neutral, alkaline, and aprotic media and proceed via intermediate radical anions, as has been demonstrated by esr spectroscopy.[83] In acid solution, the corresponding benzylamine is formed via reduction of the conjugate acid of the nitrile:

$$\mathrm{ArH} \xleftarrow[\substack{-\mathrm{CN}^- \\ +\mathrm{H}^+}]{2e^-} \mathrm{ArCN} \xrightarrow[4\mathrm{H}^+]{4e^-} \mathrm{ArCH_2NH_2}$$

Volke and Kardos also observed this mechanistic dichotomy (cleavage of cyanide ion in base and reduction to the amine in acid) with cyanopyridines, and suggested that the orientation of the substrate at the electrode surface plays an important role in the reduction.[82] The polarogram of the cyanopyridines exhibits a pronounced dip in neutral solution (Figure 5.4). At the maximum (ca. -0.9 V), the major product (80%) from reduction of 4-cyanopyridines is the benzylamine, while at potentials corresponding to the bottom of the dip (and beyond), the principal product is pyridine (90%). It was suggested that this

$$\text{pyridine} \xleftarrow[-1.4\ \mathrm{V}]{2e^-} \text{4-CN-pyridine} \xrightarrow[-0.9\ \mathrm{V}]{4e^-} \text{4-}\mathrm{CH_2NH_2}\text{-pyridine}$$

remarkable behavior is associated with a change of surface orientation of the substrate with potential, as observed for other aromatic substrates (Section 4.3.2). It was further proposed that, at potentials where the amine is formed, the cyano group is nearest the electrode surface, while at more negative potentials

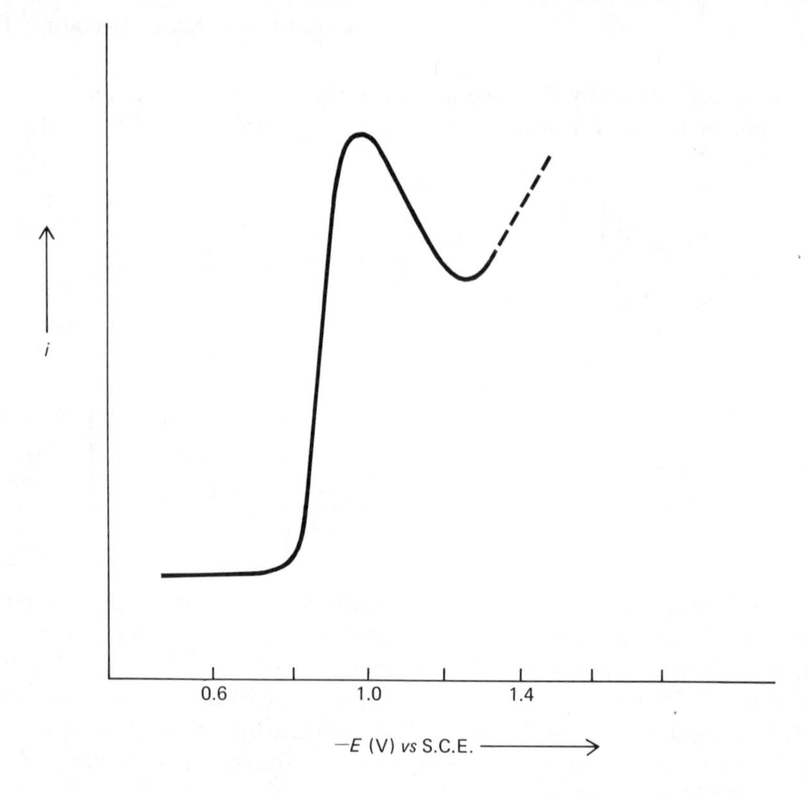

Figure 5.4 Polarogram of 4-cyanopyridine at pH 7.4 (phosphate buffer). (J. Volke and A. M. Kardos, *Coll. Czech. Chem. Commun.*, **33**, 2563 [1968]. Reprinted by permission of Academia [publishing house of the Czechoslovak Academy of Sciences].)

dipolar repulsion should orient the molecule perpendicular to the electrode surface, with the cyano group farthest from the surface. The dip is postulated to occur at the potential where a change of orientation between the two geometries takes place. As was seen in Chapter 4, such changes in orientation often do occur rather abruptly with changes in potential.

The cleavage of aliphatic nitriles to hydrocarbons and cyanide ion by electrochemically generated solvated electrons has already been discussed (Section 4.2.3).

5.2 ELECTROCATALYTIC REDUCTIONS

All of the reactions discussed thus far in this chapter involve more or less direct reduction, in the sense that bond breakage occurs in the same species that undergoes the initial electron-transfer step. There is another possible mode of electrochemical reduction, far less well studied in organic systems. This, the so-called "catalytic" process (Section 3.1.1), involves electrochemical generation of a species that may itself then effect chemical reduction of a second component of the solution. There is reason to believe that future research in this area will uncover much new, useful, and otherwise inaccessible synthetic electrochemistry. At the present time, however, only two organic electrocatalytic reagents have been investigated in any detail: the naphthalene radical anion and tetramethylammonium amalgam. Since tetraalkylphosphonium and trialkylsulfonium amalgams,[64] as well as a number of other aromatic radical anions, are excellent reducing agents, the scope of available organic electrocatalytic processes promises to expand considerably.

5.2.1 Naphthalenide Ion

The reduction of alkyl chlorides to the corresponding hydrocarbons by a catalytic process involving electrochemically generated naphthalene radical anion has already been mentioned (Sections 3.4 and 4.1.3). Alkyl chlorides are the only organic species that have been thus investigated,* but there is no doubt that this process may be extended to a wide variety of other substrates, the only requirements being that the substrate be electroinactive but reduced by naphthalenide ion.

One may legitimately raise a question concerning the relative merits of using the preformed *vs* the electrochemically generated naphthalene radical anion, or indeed any reagent generated *in situ* electrochemically. In small-scale operations, it is not generally likely that the electrochemical generation of reagent will have sufficient advantage over the use of preformed reagent to warrant the necessary extra investment in time and equipment, except where *in situ* generation of reagent at extremely low steady-state concentration levels is desired. In large-scale preparations, however, electrochemical generation has the obvious advantage of requiring only catalytic amounts of electroactive material (here naphthalene), both considerably simplifying isolation of the product and eliminating the problems involved in preparation and manipulation of large amounts of a highly reactive species such as the naphthalene radical anion. Large-scale synthesis by the electrochemical route has the further advantage of safety in the present case, since the necessity of handling large amounts of sodium is avoided and the rate of generation of the radical anion can easily be controlled through adjustment of the electrolysis current. The same holds true for reductions involving electrochemically generated sodium or lithium in ammonia, lithium in methylamine, and so on.

* In the inorganic domain, van Tamelen and his coworkers have effected catalytic fixation of molecular nitrogen to ammonia and hydrazine using electrogenerated naphthalene radical anion to reduce titanium(IV) to the active (II) state. This process is doubly catalytic, since electrons are transferred sequentially from the electrode to naphthalene, thence to titanium(IV), and finally to nitrogen.[84]

5.2.2 Tetramethylammonium Amalgam

Tetraalkylammonium amalgams lie far above hydrogen in the electromotive series, and hence are excellent reducing agents. It has been demonstrated by Myatt and Todd[85] that the reduction potentials of several tetraalkylammonium amalgams are at least -2.7 V (*vs* S.C.E.) since they all reduce biphenyl ($E^0 = -2.7$ V) and naphthalene ($E^0 = -2.5$ V) to the corresponding radical anions. [By way of comparison, sodium amalgam ($E^0 = -2.1$ V) will not effect these reductions.] There is further evidence that reduction potentials become more negative as one proceeds through alkyl groups of increasing size, from tetramethylammonium to tetrabutylammonium.[59]

Littlehailes and Woodhall reported that the amalgams reduce alkyl halides, activated olefins, and aromatic nitro compounds, but few details were given.[59] Their report described isolation and reaction of the pure amalgams. Horner and his coworkers, on the other hand, have simplified the use of amalgams by reacting them *in situ*, that is, in the presence of electron acceptors, so that an electrocatalytic cycle involving the quarternary ammonium ion and amalgam is set up. Horner and Neumann, for example, electrolyzed ethanolic solutions of tetramethylammonium chloride containing a wide variety of aryl amides [either $ArCONR_2$ or RCON(Ar)R], aryl sulfonamides, and aryl esters.[86] Carbon–nitrogen, sulfur–nitrogen, and carbon–oxygen bonds, respectively, were cleaved by the intermediate tetramethylammonium amalgam in good to high yields (Table 5.1).

Table 5.1 Reduction of amides, sulfonamides, and esters by electrochemically generated tetramethylammonium amalgam[a]

Substrate	Products (%)
N-Benzoylpiperidine	Piperidine (78) + benzyl alcohol (76)
N-Benzoylaniline	Aniline (94) + benzyl alcohol (53)
N-Acetylaniline	Aniline (94)
N-Benzyl-N-acetylaniline	N-Benzylaniline (95)
N-Acetylpiperidine	N.R.
N-Benzylphenylacetamide	N.R.
N-*p*-Toluenesulfonyl-N-benzylaniline	N-Benzylaniline (98)
N-Methanesulfonylaniline	N.R.
Methyl benzoate	Benzyl alcohol (91)
Methyl phenylacetate	N.R.

Source: L. Horner and H. Neumann, *Chem. Ber.*, 98, 3462 (1965). (Reprinted by permission of Verlag Chemie.)

[a] Mercury cathode, methanol, tetramethylammonium chloride, 5°C.

It will be noted that N-benzyl bonds are not cleaved under these neutral conditions. Horner pointed out that the reductions illustrated in Table 5.1 represent a convenient, mild, procedure for removing N-terminal benzoyl and

N-sulfonyl protecting groups that should therefore be of particular value in peptide synthesis. Simple peptide bonds are not touched (*cf.* the attempted reductions of N-acetylpiperidine and N-benzylphenylacetamide), and the blocking group is removed without racemization. A further experimental advantage derives from the fact that the peptides, when protected with the benzoyl or tosyl function, are quite soluble in alcohol while the free peptides are not, so that the latter usually precipitate from solution in high purity as the cleavage proceeds. Furthermore, the unwanted cleavage fragments (toluenesulfinate from N-tosyl peptides and benzyl alcohol from N-benzoyl peptides) are also soluble in alcohol and do not contaminate the product. *Neither the formyl nor the benzyloxycarbonyl protecting groups are removed* by tetramethylammonium amalgam, thus providing for the possibility of selective cleavage in complex peptide synthesis. Also, the S-benzyl group is removed considerably more slowly than either the benzoyl or tosyl groups, permitting either selective cleavages (*cf.* the conversion of **22** to **23**) or, if desired, complete reduction of a doubly protected peptide such as **22**. Horner and Neumann examined the reductive cleavage of a number of peptides and oligopeptides, of which a few examples are illustrative:

Ts—L(−)—Tyr	L(−)—Tyr	96%
Ts—Gly—Gly	Gly—Gly	91%
$TsNHCH(CH_2SCH_2C_6H_5)CO_2H$ (**22**)	$H_2NCH(CH_2SCH_2C_6H_5)CO_2H$ (**23**)	86%
Bz—Gly—Gly	Gly—Gly	76%
Bz—Gly—DL—Phe	Gly—DL—Phe	63%

Horner and Neumann have also found that sulfones, RSO_2R^1, are readily cleaved by tetramethylammonium amalgam at room temperature when at least one of the R groups is aryl, allyl, or benzyl.[87] This constitutes a considerably milder method for cleavage of sulfones than other methods commonly employed. Sodium, potassium, and ammonium amalgams are all ineffective in cleaving sulfones, emphasizing the reducing power of the tetraalkylammonium amalgams. It can be seen (Table 5.2) that sterically hindered sulfones are cleaved readily, although steric hindrance does affect the direction of cleavage (*cf.* the reduction of mesityl phenyl sulfone).

In a related reaction, Horner and Singer effected electrochemical cleavage of alkyl *p*-toluenesulfonates[88]:

$$\text{ROTs} \xrightarrow[\text{Me}_4\text{N}^+\text{Cl}^-,\ \text{EtOH}]{2e^-,\ \text{Hg},\ 8\text{–}10°\text{C}} \text{ROH}$$

Unlike the preceding reactions, the cleavage of alkyl tosylates is apparently a direct electrochemical cleavage, since a polarographic wave can be observed for tosylates before solvent (or electrolyte) decomposition begins. (Mann has investigated the mechanism of reduction in dipolar aprotic solvents, where the

Table 5.2 Cleavage of sulfones by electrochemically generated tetramethylammonium amalgam[a]

Substrate	Products (%)
Diphenylsulfone	Benzenesulfinate (87) + benzene (62)
Dimesitylsulfone	Mesitylenesulfinate (74) + mesitylene (93)
Mesitylphenylsulfone	Benzenesulfinate (80) + mesitylene (88)
Methylphenylsulfone	Benzenesulfinate (85) + methane (84) ethane (1.1)
Phenylbenzylsulfone	Benzenesulfinate (90) + toluene (82)
Methylbenzylsulfone	Methanesulfinate (86) + toluene (73)
Di-*n*-propylsulfone	N.R.
Di-β-phenethylsulfone	N.R.

Source: L. Horner and H. Neumann, *Chem. Ber.*, 98, 1715 (1965). (Reprinted by permission of Verlag Chemie.)

[a] Mercury cathode, methanol, tetramethylammonium chloride, 5°C.

reduction is more complex.)[89] Of special synthetic interest, the alcohols from reduction of optically active tosylates were formed not only in high yields (75–95%) but with retention of configuration (<1% racemization).[88] The tosyl group may therefore be removed electrochemically under mild conditions and in considerably cleaner stereochemical fashion than with other reducing agents such as lithium aluminum hydride, Raney nickel, or sodium in liquid ammonia.

5.2.3 Synthetic Scope

The electrochemical cleavage of N-benzoyl and N-tosyl amides by tetramethylammonium amalgam would appear to merit further attention as a possible general technique in peptide synthesis, in view of its simplicity of operation and workup, good yields, selectivity, and mild conditions. Further, as Horner has shown, the cleavage of sulfones by the amalgam provides a convenient route to many sulfinic acids that could be synthesized only with considerable difficulty by other routes.

At the present time, the only substrates that have been allowed to react with electrochemically generated naphthalene radical anion are alkyl and aryl chlorides, which are reduced cleanly to the corresponding hydrocarbons. Because of the similar reduction potentials of naphthalene and tetramethylammonium ion, however, it is very likely that the naphthalene radical anion and tetramethylammonium amalgam will eventually be found to reduce much the same kinds of substrates. (In fact, sulfonamides are cleaved by the naphthalenide ion as well as by amalgams).[90] It appears that the two reagents may actually complement one another quite nicely, the naphthalenide ion being used for electrocatalytic reactions in aprotic media and the amalgams being used in protic solvents. Furthermore, the evidence indicating that tetrabutylammonium ion is reduced at more negative potentials than is tetramethylammonium suggests that tetrabutylammonium amalgam will be found to be a sufficiently more potent reducing agent than tetramethylammonium amalgam that the way will be opened to the carrying out of reductions impossible with the latter.

Finally, it should be pointed out here that many constant-current reductions in the literature that have been thought to involve direct electrochemical reduction may actually proceed by an electrocatalytic pathway involving tetra-alkylammonium amalgams. These two modes of reduction can presumably be distinguished by the proper voltammetric experimentation.

References

[1]M. von Stackelberg and W. Stracke, *Z. Elektrochem.*, **53**, 118 (1949).
[2a]J. W. Sease and R. C. Reed, private communication; [b]J. W. Sease and W. L. Morgan, private communication; [c]M. Fleischmann, D. Pletcher, and C. J. Vance, *J. Electroanal. Chem.*, **29**, 325 (1971).
[3a]J. L. Webb, C. K. Mann, and H. M. Walborsky, *J. Amer. Chem. Soc.*, **92**, 2042 (1970); [b]L. W. Marple, L. E. I. Hummelstedt, and L. B. Rogers, *J. Electrochem. Soc.*, **107**, 437 (1960).
[4a]M. Anbar and E. J. Hart, *J. Phys. Chem.*, **71**, 3700 (1967); [b]J. K. Thomas, *J. Phys. Chem.*, **71**, 1919 (1967); [c]M. Anbar, *Advan. Phys. Org. Chem.*, **7**, 115 (1969).
[5a]L. Eberson, *Acta Chem. Scand.*, **22**, 3045 (1968); [b]E. Grovenstein, Jr., and Y.-M. Cheng, *Chem. Commun.*, 101 (1970).
[6]A. J. Fry and M. A. Mitnick, *J. Amer. Chem. Soc.*, **91**, 6207 (1969).
[7]S. Wawzonek, R. C. Duty, and J. H. Wagenknecht, *J. Electrochem. Soc.*, **111**, 74 (1964).
[8a]P. J. Elving, *Record Chem. Progr.*, **14**, 99 (1953); [b]F. L. Lambert, A. H. Albert, and J. P. Hardy, *J. Amer. Chem. Soc.*, **86**, 3155 (1964).
[9]J. W. Sease, P. Chang, and J. L. Groth, *J. Amer. Chem. Soc.*, **86**, 3154 (1964).
[10]J. Zavada, J. Krupicka, and J. Sicher, *Coll. Czech. Chem. Commun.*, **28**, 1664 (1963).
[11]J. P. Petrovich and M. M. Baizer, *Electrochim. Acta*, **12**, 1249 (1967).
[12a]A. M. Wilson and N. L. Allinger, *J. Amer. Chem. Soc.*, **83**, 1999 (1961); [b]P. Kabasakalian and J. McGlotten, *Anal. Chem.*, **34**, 1440 (1962).
[13a]A. J. Fry, M. A. Mitnick, and R. G. Reed, *Extended Abstracts, 21st Meeting*, Comite Internationale de Thermodynamique et de Cinetique Electrochimiques, Prague, Czechoslovakia, September, 1970, p. 403; [b]A. J. Fry and R. G. Reed, *J. Amer. Chem. Soc.*, **93**, 553 (1971); A. J. Fry and R. G. Reed, *J. Amer. Chem. Soc.*, in press.
[14a]S. Trasatti and G. Olivieri, *J. Electroanal. Chem.*, **27**, App. 7 (1970); [b]R. A. Caldwell and S. Hacobian, *Austr. J. Chem.*, **21**, 1403 (1968); [c]E. L. Eliel, N. L. Allinger, S. J. Angyal, and G. A. Morrison, *Conformational Analysis*, New York, Wiley, 1965, p. 17.
[15a]E. B. Weronski, *Electrochim. Acta*, **14**, 259 (1968); [b]V. J. Puglisi, G. L. Clapper, and D. H. Evans, *Anal. Chem.*, **41**, 279 (1969).
[16]A. K. Hoffmann, W. G. Hodgson, D. L. Maricle, and W. H. Jura, *J. Amer. Chem. Soc.*, **86**, 631 (1964).
[17a]R. E. Erickson, R. Annino, M. D. Scanlon, and G. Zon, *J. Amer. Chem. Soc.*, **91**, 1767 (1969); [b]A. J. Fry and R. H. Moore, *J. Org. Chem.*, **33**, 1283 (1968).
[18]H. M. Walborsky, F. J. Impastato, and A. E. Young, *J. Amer. Chem. Soc.*, **86**, 3283 (1964).

[19]B. Czochralska, *Chem. Phys. Lett.*, **1**, 239 (1967).
[20]R. Annino, R. E. Erickson, J. Michalovic, and B. McKay, *J. Amer. Chem. Soc.*, **88**, 4424 (1966).
[21a]G. A. Korchinskii, *Ukr. Khim. Zh.*, **29**, 1031 (1963); see *Chem. Abstr.*, **60**, 7681g (1964); [b]H. B. Herman, Ph.D. dissertation, Syracuse University, 1964.
[22]H. W. Sternberg, R. E. Markby, I. Wender, and D. M. Mohilner, *J. Electrochem. Soc.*, **113**, 1060 (1966).
[23a]J. G. Lawless and M. D. Hawley, *J. Electroanal. Chem.*, **21**, 365 (1969); [b]M. Mohammad, J. Hajdu, and E. M. Kosower, *J. Amer. Chem. Soc.*, **93**, 1792 (1971).
[24]L. L. Miller and E. Riekena, *J. Org. Chem.*, **34**, 3359 (1969).
[25a]J. W. Sease and R. Filas, private communication; [b]J.-M. Saveant, *J. Electroanal. Chem.*, **29**, 87 (1971); [c]L. Nadjo and J.-M. Saveant, *J. Electroanal. Chem.*, **30**, 41 (1971).
[26]J. Grimshaw and J. S. Ramsey, *J. Chem. Soc.* (B), 60 (1968).
[27]H. Lund, *Acta Chem. Scand.*, **14**, 1927 (1960).
[28]P. J. Elving and R. E. Van Atta, *Anal. Chem.*, **27**, 1908 (1955).
[29]C. S. McDowell, Ph.D. dissertation, Indiana University, 1967.
[30]P. J. Elving and J. T. Leone, *J. Amer. Chem. Soc.*, **79**, 1546 (1957).
[31]J. H. Stocker and R. M. Jenevein, *Chem. Commun.*, 934 (1968).
[32]S. Wawzonek and R. C. Duty, *J. Electrochem. Soc.*, **108**, 1135 (1961).
[33]E.g., A. N. Nesmeyanov, R. Kh. Freidlina, V. N. Kost, and M. Ya. Khorlina, *Tetrahedron*, **16**, 94 (1961).
[34]W. G. Dauben and F. G. Willey, *J. Amer. Chem. Soc.*, **84**, 1497 (1962).
[35a]D. Seyferth, H. Yamazoki, and D. L. Alleston, *J. Org. Chem.*, **28**, 703 (1963); [b]D. Seyferth and B. Prokai, *J. Org. Chem.*, **31**, 1702 (1966).
[36]A. J. Fry, unpublished experiments.
[37a]H. R. Koch and M. G. McKeon, *J. Electroanal. Chem.*, **30**, 331 (1971); [b]C. Altona and M. Sundaralingam, *J. Amer. Chem. Soc.*, **92**, 1995 (1970).
[38]W. H. Jura and R. J. Gaul, *J. Amer. Chem. Soc.*, **80**, 5402 (1958).
[39]G. Kobrich, *Angew. Chem. Int. Ed.*, **4**, 49 (1965).
[40]W. Kemula and J. Kornacki, *Roczniki Chem.*, **36**, 1835, 1857 (1962); see *Chem. Abstr.*, **59**, 218 (1963).
[41]S. Wawzonek and J. H. Wagenknecht, *J. Electrochem. Soc.*, **110**, 420 (1963).
[42]H. Lund, *Acta Chem. Scand.*, **17**, 2139 (1963).
[43]K. Schwabe and H. Frind, *Z. Physik. Chem.* (*Leipzig*), **196**, 342 (1951).
[44]F. Covitz, *J. Amer. Chem. Soc.*, **89**, 5403 (1967).
[45]H. Gilch, *J. Polymer Sci.*, (A-1), **4**, 1351 (1966).
[46a]M. R. Rifi, *J. Amer. Chem. Soc.*, **89**, 4442 (1967); [b]M. R. Rifi, *Extended Abstracts, 21st Meeting*, Comite Internationale de Thermodynamique et de Cinetique Electrochimiques, Prague, Czechoslovakia, September 1970, p. 452; [c]M. R. Rifi, *Tetrahedron Lett.*, 1043 (1969).
[47]J. R. Jezorek and H. B. Mark, Jr., *J. Phys. Chem.*, **74**, 1627 (1970).
[48]J. A. Dougherty and A. J. Diefenderfer, *J. Electroanal. Chem.*, **21**, 531 (1969).
[49]J. K. Kochi and J. W. Powers, *J. Amer. Chem. Soc.*, **92**, 137 (1970).
[50]N. S. Hush and K. B. Oldham, *J. Electroanal. Chem.*, **6**, 34 (1963).
[51a]P. Zuman and V. Horak, *Coll. Czech. Chem. Commun.*, **26**, 176 (1961); [b]Ya. I. Stradins, I. K. Tutane, A. K. Aren, and G. Ya. Vanags, *J. Gen. Chem. U.S.S.R.*, **35**, 1333 (1965).
[52a]J. Volke, R. Kubicek, and F. Santavy, *J. Gen. Chem. U.S.S.R.*, **25**, 871 (1960); [b]H. Lund, *Acta Chem. Scand.*, **17**, 2325 (1963).
[53]M. Finkelstein, R. C. Petersen, and S. D. Ross, *J. Amer. Chem. Soc.*, **81**, 2361 (1959).

[54a]N. J. Leonard, S. Swann, Jr., and E. H. Mottus, *J. Amer. Chem. Soc.*, **74**, 6251 (1952); [b]N. J. Leonard, S. Swann, Jr., and J. Figueras, Jr., *J. Amer. Chem. Soc.*, **74**, 4620 (1952); [c]N. J. Leonard, S. Swann, Jr., and G. Fuller, *J. Amer. Chem. Soc.*, **76**, 3193 (1954).
[55a]S. D. Ross, M. Finkelstein, and R. C. Petersen, *J. Amer. Chem. Soc.*, **82**, 1582 (1960); [b]S. D. Ross, M. Finkelstein, and R. C. Petersen, *J. Amer. Chem. Soc.*, **92**, 6003 (1970).
[56]J. S. Mayell and A. J. Bard, *J. Amer. Chem. Soc.*, **85**, 421 (1963).
[57]L. Horner and A. Mentrup, *Ann.*, **646**, 49 (1961).
[58]H. N. McCoy and W. C. Moore, *J. Amer. Chem. Soc.*, **33**, 273 (1911).
[59a]J. D. Littlehailes and B. J. Woodhall, *Chem. Commun.*, 665 (1967); [b]J. D. Littlehailes and B. J. Woodhall, *Disc. Faraday Soc.*, **45**, 187 (1968).
[60]G. B. Porter, *J. Chem. Soc.*, 760 (1954).
[61a]J. T. Wrobel, K. M. Pazdro, and A. S. Bien, *Chem. Ind.*, 1760 (1966); [b]J. T. Wrobel and A. R. Krawczyk, *Chem. Ind.*, 656 (1969); [c]see also L. Horner and H. Roder, *Chem. Ber.*, **101**, 4179 (1968).
[62a]R. A. Benkeser and E. M. Kaiser, *J. Amer. Chem. Soc.*, **85**, 2858 (1963); [b]R. A. Benkeser, E. M. Kaiser, and R. F. Lambert, *J. Amer. Chem. Soc.*, **86**, 5272 (1964).
[63]L. Horner and A. Mentrup, *Ann.*, **646**, 65 (1961).
[64]W. R. T. Cottrell and R. A. N. Morris, *Chem. Commun.*, 409 (1968).
[65]L. Horner, H. Winkler, A. Rapp, A. Mentrup, H. Hoffmann, and P. Beck, *Tetrahedron Lett.*, 161 (1961).
[66]L. Horner and M. Ernst, *Chem. Ber.*, **103**, 318 (1970).
[67a]H. J. Bestmann, G. Graf, and H. Hartung, *Angew. Chem. Int. Ed.*, **4**, 596 (1965); [b]H. J. Bestmann, G. Graf, H. Hartung, S. Kolewa, and E. Vilsmaier, *Chem. Ber.*, **103**, 2794 (1970).
[68]L. Horner and H. Neumann, *Chem. Ber.*, **102**, 3953 (1969).
[69a]P. Kabasakalian and J. McGlotten, *Anal. Chem.*, **31**, 1091 (1959); [b]P. Kabasakalian, J. McGlotten, A. Basch, and M. D. Yudis, *J. Org. Chem.*, **26**, 1738 (1961).
[70]S. Wawzonek and J. D. Frederickson, *J. Electrochem. Soc.*, **106**, 325 (1959).
[71a]R. E. Erickson and C. M. Fischer, private communication; [b]R. E. Erickson and C. M. Fischer, *J. Org. Chem.*, **35**, 1604 (1970).
[72a]S. Wawzonek, H. A. Laitinen, and S. J. Kwiatkowski, *J. Amer. Chem. Soc.*, **66**, 827 (1944); [b]H. Lund, *Acta Chem. Scand.*, **14**, 359 (1960).
[73]W. Reusch and R. LeMahieu, *J. Amer. Chem. Soc.*, **86**, 3068 (1964).
[74a]S. Ya. Mel'nik, V. G. Mairanovskii, M. A. Miropol'skaya, and G. I. Samokhvalov, *J. Gen. Chem. U.S.S.R.*, **38**, 1445 (1968); [b]V. G. Mairanovskii, A. Ya. Veinberg, and G. I. Samokhvalov, *J. Gen. Chem. U.S.S.R.*, **38**, 642 (1968); [c]A. Ya. Veinberg, V. G. Mairanovskii, and G. I. Samokhvalov, *J. Gen. Chem. U.S.S.R.*, **38**, 643 (1968); [d]J. B. Hendrickson and C. Kandal, *Tetrahedron Lett.*, 343 (1970); [e]M. M. Baizer, *J. Org. Chem.*, **31**, 3890 (1966), Table II, footnote *d*.
[75a]P. Zuman and S.-Y. Tang, *Coll. Czech. Chem. Commun.*, **28**, 829 (1963); [b]J. M. Saveant, *Compt. Rend.*, **258**, 585 (1964).
[76]P. S. McKinney and S. Rosenthal, *J. Electroanal. Chem.*, **16**, 261 (1968).
[77a]J. H. Wagenknecht and M. M. Baizer, *J. Electrochem. Soc.*, **114**, 1095 (1967); [b]M. M. Baizer, *J. Org. Chem.*, **31**, 3847 (1966).
[78]J. Stradins, I. Tutane, O. Neilands, and G. Vanags, *Dokl. Akad. Nauk. S.S.S.R.*, **166**, 631 (1966); see *Chem. Abstr.*, **64**, 12524 (1966).
[79]N. L. Bauld and M. S. Brown, *J. Amer. Chem. Soc.*, **89**, 5413 (1967).
[80]H. Lund, *Acta Chem. Scand.*, **17**, 2325 (1963).
[81]O. Manousek, P. Zuman, and O. Exner, *Coll. Czech. Chem. Commun.*, **33**, 3979 (1968).

[82]J. Volke and A. M. Kardos, *Coll. Czech. Chem. Commun.*, **33**, 2560 (1968).
[83]P. H. Rieger, I. Bernal, W. H. Reinmuth, and G. K. Fraenkel, *J. Amer. Chem. Soc.*, **85**, 683 (1963).
[84a]E. E. van Tamelen and D. A. Seeley, *J. Amer. Chem. Soc.*, **91**, 5194 (1969); [b]E. E. van Tamelen, R. B. Fechter, and S. W. Schneller, *J. Amer. Chem. Soc.*, **91**, 7196 (1969); [c]E. E. van Tamelen, *Accounts Chem. Res.*, **3**, 361 (1970).
[85]J. Myatt and P. F. Todd, *Chem. Commun.*, 1033 (1967).
[86]L. Horner and H. Neumann, *Chem. Ber.*, **98**, 3462 (1965).
[87]L. Horner and H. Neumann, *Chem. Ber.*, **98**, 1715 (1965).
[88]L. Horner and R.-J. Singer, *Chem. Ber.*, **101**, 3329 (1968).
[89]P. Yousefzadeh and C. K. Mann, *J. Org. Chem.*, **33**, 2716 (1968).
[90]W. D. Closson, S. Ji, and S. Schulenberg, *J. Amer. Chem. Soc.*, **92**, 650 (1970).

6 REDUCTION OF MULTIPLE BONDS

The electrochemical behavior of functional groups containing multiple bonds, such as carbonyl or nitro compounds, tends to be more complex than the systems treated in the preceding chapter, since a greater diversity of reaction pathways is generally available. Mechanistic progress in this area has often been slow because of a lack in many cases of voltammetric data, as well as the fact that many experiments have been run under constant-current conditions, and with a sometimes bewildering variety of solvents, cathode materials, and electrolytes.* It will be clear to the reader by now that the author, like most others presently in the field, is a strong advocate of controlled-potential electrolysis, used in conjunction with voltammetric data, in organic electrosynthesis. Disregarding its selectivity, which even by itself would argue for the use of controlled-potential electrolysis, the particular merit of this approach in the present context is that it is less empirical than the constant-current technique. The way is opened, therefore, for prediction of the correct experimental conditions for preparative electrolysis, and also new chemistry, from voltammetric data, mechanistic reasoning, and analogy to known systems. For these reasons, what follows will not be an extensive survey of the electrochemistry of all unsaturated functional groups (indeed, such surveys already exist and can be quite helpful in scanning the literature when a new electrochemical reaction is being planned).[2,3] Instead, a few of the most important and best-studied functional groups have been selected for their diversity of reaction under a variety of experimental conditions, and the reasons for this diversity are discussed as fully as possible. With these concepts in hand, the chemist should be in a position to plan new chemistry and to understand the behavior of more complex systems such as heterocycles or polyfunctional molecules. There are really no other reasons than these for discussion, for example, of the reasons why a carbonyl compound is converted to an alcohol under a given set of conditions, for there now exist many convenient chemical reagents for that particular conversion that would be preferred over electrochemical methods. The *principles* that govern whether a pinacol or alcohol is formed upon reduction of the carbonyl compound are completely general, however, and can be applied to many other systems.

The division of topics between this chapter and the next (on conjugated systems) is rather arbitrary. Generally speaking, reactions discussed in this

* Stocker has also commented on the difficulty in organizing the massive literature in this area into a coherent unit.[1]

chapter include those in which a single functional group, such as the nitro group, is the electrophore, or electroactive unit, while the next chapter includes systems where an extended pi system is the electrophore. It will soon be clear that the initial electrochemical step in reduction of both kinds of compound is addition of an electron to the lowest unfilled orbital of the electrophore to form a radical. The radical thus produced will have a variety of reactive options at its disposal, depending on both its structure and environmental factors such as electrode potential, solvent, pH, and so on. Taking this point of view, many similarities in the behavior of otherwise unlike groups will be evident.

6.1 THE CARBONYL GROUP

6.1.1 Aldehydes and Ketones

Pinacol *vs* Alcohol

Diaryl ketones such as benzophenone exhibit polarographic behavior very similar to that observed for aromatic hydrocarbons (Chapters 3 and 4). In acid ($pH \leq 4$) benzophenone exhibits two diffusion-controlled 1-electron waves.[4] The first wave is due to reduction of the protonated carbonyl to a radical (**1**) and the second to further reduction of this radical to a carbanion:

$$(C_6H_5)_2C{=}O \xrightarrow{H^+} (C_6H_5)_2C{=}O^+H \xrightarrow[E_1]{e^-} \underset{\mathbf{1}}{(C_6H_5)_2\dot{C}OH}$$

$$\mathbf{1} \xrightarrow[E_2]{e^-} (C_6H_5)_2\bar{C}OH \xrightarrow{H^+} (C_6H_5)_2CHOH$$

In agreement with this hypothesis, benzophenone affords benzpinacolone (**3**) in quantitative yield upon controlled-potential electrolysis at the first wave at pH 1.3.[4]

$$(C_6H_5)_2C{=}O \xrightarrow[E_1]{e^-,\,H^+} \mathbf{1} \longrightarrow \underset{\mathbf{2}}{(C_6H_5)_2C(OH){-}C(OH)(C_6H_5)_2}$$

$$\mathbf{2} \xrightarrow[\text{pinacol rearr.}]{H^+} \underset{\mathbf{3}}{C_6H_5C({=}O)C(C_6H_5)_3} + H_2O$$

At pH 1.3, only a very small fraction of the benzophenone molecules in solution is protonated. Therefore, if the species diffusing to the electrode surface were the conjugate acid, the polarographic reduction wave would be much smaller than it actually is. Furthermore, if protonation were occurring in the diffusion layer near the mercury drop, the wave should exhibit the characteristics of a kinetic wave (Section 3.1.1), rather than of a wave controlled by diffusion. Elving has suggested that this apparently paradoxical behavior arises because the neutral ketone diffuses to the electrode surface (hence the diffusion control of wave height), where it is polarized and protonated in the electrical field at the electrode surface (pp. 57–156).[5] The pH dependence of the first wave further implicates a proton in the transition state for reduction. That the polarity and polarizability of the carbonyl group is an important determinant of its polarographic behavior is demonstrated by the highly polar ketones, tropone and 2,7-dimethyltropone, which are reduced as their conjugate acids even at pH 10.[6]

In neutral and weakly alkaline media ($4.9 \leq pH \leq 11.3$), benzophenone exhibits a single 2-electron wave.[4] In this pH range, the proton availability near the electrode surface is lower, and so the neutral ketone is the electroactive species; it is reduced to a radical anion (**4**), which then immediately abstracts a proton from the solvent. The resulting radical (**1**) is then reduced, completing the ECE process (Section 4.2.1) and resulting in the overall 2-electron wave. Pasternak isolated comparable amounts of benzhydrol and benzpinacol from

$$(C_6H_5)_2CO + e^- \longrightarrow \underset{\mathbf{4}}{(C_6H_5)_2\dot{C}{-}\bar{O}} \xrightarrow{H^+} \mathbf{1} \xrightarrow[H^+]{e^-} (C_6H_5)_2CHOH$$

reduction at pH 4.9; at pH 8.6 the benzhydrol was accompanied by only a trace of benzpinacol, as a result of the change of mechanism with pH.[4] The small amount of pinacol isolated at pH 8.6 probably represents a kinetic phenomenon: the rate of dimerization of **1** is apparently high enough in the preparative experiment to compete with the rate of electron transfer. If this postulate is true, one could eliminate the undesired pinacol either by electrolyzing at more negative potentials, where electron transfer is faster, or by reducing a more dilute solution of the ketone to minimize the (bimolecular) pinacol formation.

Benzophenone exhibits two 1-electron waves in aprotic or strongly alkaline media.[4,7] Under these conditions of low proton availability, the ketyl **4** is not protonated on the polarographic time scale, so a single 1-electron wave is observed, followed at a more negative potential by the wave for reduction to the dianion. In an aprotic solvent such as DMF, solutions of **4** are quite stable. Reaction of the ketyl *in situ* with carbon dioxide or ethyl iodide initiates an ECE sequence in each case, since the products are neutral radicals like **1** and so are readily reducible[8]:

$$(C_6H_5)_2\dot{C}{-}O^- + EtI \longrightarrow (C_6H_5)_2\dot{C}{-}OEt \xrightarrow[H^+]{e^-} (C_6H_5)_2CHOEt \qquad (Ref.\ 8)$$

$$(C_6H_5)_2\dot{C}{-}O^- + CO_2 \longrightarrow (C_6H_5)_2\dot{C}OCO_2^- \xrightarrow{e^-} (C_6H_5)_2\bar{C}{-}OCO_2^-$$

$$\xrightarrow{CO_2} (C_6H_5)_2C(CO_2^-)OCO_2^- \xrightarrow[2H^+]{-CO_2} (C_6H_5)_2C(CO_2H)OH \qquad (Refs.\ 3\ and\ 8)$$

or

$$(C_6H_5)_2\dot{C}{-}\bar{O} + CO_2 \longrightarrow (C_6H_5)_2C(CO_2^-){-}O\cdot \xrightarrow[2H^+]{e^-} (C_6H_5)_2C(CO_2H)OH$$

The instability of benzpinacol in strong alkali preludes controlled-potential reduction of benzophenone in this medium; presumably, benzpinacol would be the primary product.

Reduction of arylalkyl ketones and aryl aldehydes (ArCOR, R = alkyl or H) is, with modifications, similar to that of diaryl ketones. In acid, two 1-electron waves are observed,* corresponding to reduction of the protonated carbonyl compound and neutral radical, respectively.[4, 5] Reduction at the first wave in solutions of moderate acidity affords the pinacol, usually in very high yield.[1, 4, 9] (In stronger acid, the pinacol may suffer dehydration or rearrangement.) Reduction at a potential corresponding to the second polarographic wave at pH 4 ought to afford the alcohol instead of the pinacol. This postulate does not appear ever to have been tested explicitly, but data obtained by Evans and Woodbury in a study of the electrochemical reduction of dibenzoylmethane support the assumption.[10]

In the pH range 7.2–11.3, a single 2-electron wave is observed for arylalkyl ketones; hence, alcohols are the expected products of preparative-scale reduction. Elving and Leone did observe a coulometric n value of 1.9 upon electrolysis of acetophenone at pH 9, but did not attempt to identify the product, which was

* There is some confusion in the literature on this point; a number of authors have not observed the second wave. Both waves are cathodic of the corresponding pair for diaryl ketones, and the second wave is sufficiently cathodic that it is often obscured by the solvent discharge, particularly at pH < 3.

presumably the alcohol.[5] It appears that some pinacol is formed even though polarographic indications are of a 2-electron reduction.[4] Like the reduction of diaryl ketones in neutral media discussed above, this is an example of the situation described in Section 3.4, where kinetic complications cause preparative-scale and polarographic results to differ. Apparently, here again dimerization of the intermediate radicals can compete noticeably with electron transfer in the more concentrated solutions used for preparative reduction. Pinacol formation could presumably be minimized by the same expedients suggested above for diaryl ketones; in fact, it has been observed that, under conditions where a mixture of pinacol and alcohol are formed, the yield of the latter can be increased by diluting the solution or by electrolysis at more negative potentials.[11]

In alkaline (pH > 12) and aprotic media, two 1-electron waves are observed for ArCOR. Unlike diaryl ketones, reduction to the pinacol can be carried out successfully in strong alkali, since these pinacols are not base sensitive. The pinacol is also the product when reduction is carried out in an aprotic solvent such as DMF.[8] Apparently, alkylaryl ketyls are less stable than diaryl ketyls because of decreased resonance delocalization.

Evans has effected *intra*molecular coupling of ketyl radicals generated from reduction of **6**, which itself is formed by reduction of dibenzoylmethane at the first

$$(C_6H_5CO)_2CH_2 \xrightarrow[H^+]{e^-} C_6H_5COCH_2\dot{C}(OH)C_6H_5$$

5

$$2\ \mathbf{5} \xrightarrow{\text{dimerization}} C_6H_5COCH_2C(OH)(C_6H_5)-C(OH)(C_6H_5)CH_2COC_6H_5$$

6

$$\mathbf{6} \xrightarrow[2H^+]{2e^-} C_6H_5\dot{C}(OH)CH_2C(OH)(C_6H_5)-C(OH)(C_6H_5)CH_2\dot{C}(OH)C_6H_5 \longrightarrow \text{1,2,4,5-tetrahydroxy-1,2,4,5-tetraphenylcyclohexane}$$

polarographic wave at pH 4.2.[10] (Dibenzoylmethane is presumably easier to reduce than **6** because of the inductive effect of the nearby carbonyl group.) Reduction of dibenzoylmethane at even more negative potentials affords a monomeric product, 1,3-diphenyl-1,3-propanediol, since at these potentials the rate of reduction of radical **5** is much higher than its rate of dimerization.

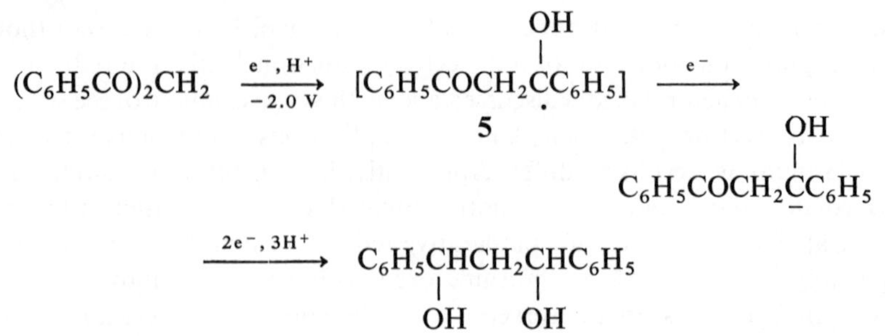

$$(C_6H_5CO)_2CH_2 \xrightarrow[-2.0\ V]{e^-,\ H^+} [C_6H_5COCH_2\dot{C}(OH)C_6H_5]\ \mathbf{5} \xrightarrow{e^-} C_6H_5COCH_2\underline{C}(OH)C_6H_5$$

$$\xrightarrow{2e^-,\ 3H^+} C_6H_5CH(OH)CH_2CH(OH)C_6H_5$$

Reduction of an unsymmetrical ketone such as ArCOR may lead to two diastereomeric pinacols, *meso* or *dl*. Although there had appeared a number of reports of stereoselective formation of one of the two diastereomers, as well as several reports that the stereochemistry can be pH dependent, this area did not receive close scrutiny until relatively recently, with the studies of Stocker and Jenevein,[1,12] as well as Evans and coworkers.[13] Stocker and Jenevein found that the *dl/meso* ratio from the reduction of acetophenone ranges from ca. 1.0 to 1.4 in acid under a variety of experimental conditions, and from 2.5 to 3.2 in alkaline media. In acid, where the pinacol arises from dimerization of neutral radicals, steric factors in the respective transition states should favor the formation of the *meso*-pinacol; the fact that the *dl*-isomer actually predominates was rationalized as arising from favorable *inter*radical hydrogen bonding in the *dl* transition state. Support for this postulate was found in studies of the electrochemical pinacolization of 2-acetylpyridine. Strong *intra*molecular hydrogen bonding in the corresponding radical **7** does result in the predominance of the *meso*-isomer expected on strictly steric grounds.

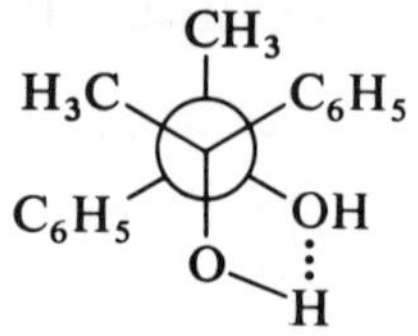

CH_3, H_3C, C_6H_5, C_6H_5, OH, O, H

dl transition state

N, $\cdot CH_3$, H–O

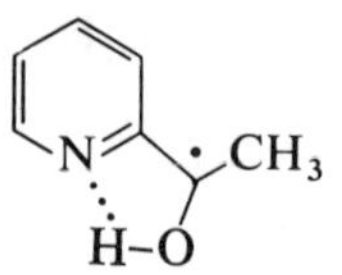

7

Stocker and Jenevein have suggested[12a] that in alkali, pinacols arise from combination of the neutral radical with the radical anion of the ketone. This postulate appears consistent with the increased preference for formation of *dl*-pinacol in alkali, since the hydrogen bond here should be stronger than that between the neutral radicals in acid.

Evans and coworkers[13] examined the *dl*/*meso* ratio of hydrobenzoins from reduction of benzaldehyde in aqueous acid. Since adsorption effects are more important for organic compounds in aqueous media, where solubilities of starting materials, intermediates, and products are low,[14] they expected surface effects to be more important than in the 80% ethanol used by Stocker and Jenevein. A variety of changes in composition of the solution that would be expected to affect the structure of the electrical double layer (i.e., changes in ionic strength, supporting electrolyte, and pH) did not change the *dl*/*meso* ratio ($\simeq$ 1.0–1.2), but addition of specifically adsorbed species, that is, iodide or tetraalkylammonium ions, lowered the *dl*/*meso* ratio considerably. (For example, addition of only 5×10^{-4}-*M* tetrabutylammonium perchlorate changed the ratio in 0.2-*M* perchloric acid from 1.16 to 0.48.) At least in these cases, dimerization must be occurring at the electrode surface, although the specific function of the adsorbed ion in changing stereochemistry is unclear. Yields were only 15–35% in aqueous media (polymeric materials are formed), compared with the 65–85% yields of Stocker and Jenevein, so that the results are not of direct synthetic use. In an interesting related experiment, Horner and Degner showed that the stereochemistry of the formation of the alcohol from a ketone can also be affected by the double-layer structure.[15] Constant-current electrolysis of acetophenone at a mercury cathode in methanol containing (−)-ephedrine hydrochloride as the supporting electrolyte results in an equimolar mixture of pinacol and α-methylbenzyl alcohol, and the latter is *optically active*. Reduction of acetophenone in

$$C_6H_5COCH_3 \xrightarrow[(-)\text{-ephedrine}\cdot HCl]{2e^-,\ CH_3OH} (R)\text{—}C_6H_5\underset{\underset{OH}{|}}{C}HCH_3 \quad (4.2\%\ \text{optical purity})$$

$$C_6H_5COCH_3 \xrightarrow[(+)\text{-ephedrine}\cdot HCl]{2e^-,\ CH_3OH} (S)\text{—}C_6H_5\underset{\underset{OH}{|}}{C}HCH_3 \quad (4.6\%\ \text{optical purity})$$

the presence of (+)-ephedrine hydrochloride produced the alcohol of opposite configuration. By analogy to the experiments of Evans, a higher degree of asymmetric induction would probably be observed in aqueous methanol. These experiments of Horner and Degner were foreshadowed by the report of Murray and Kodama that the enantiomeric copper(II) tartrates and cadmium(II) alanates exhibit differing electrochemical properties in the presence of adsorbed optically active brucine.[16]

Reduction of dialkyl ketones occurs at quite negative potentials and is often not observable, since a number of metal ions often used as electrolytes are easier to reduce than are the ketones. Using tetrabutylammonium salts in 80–90% ethanol or methanol, however, one observes a single 2-electron wave at ca. −2.5 V (*vs* S.C.E.).[17,18] The 2-electron polarographic wave, together with the isolation of alcohols (90–100% yields), indicates that the reduction follows the ECE pathway. Reduction of the neutral ketyl radical must be much faster than its dimerization, presumably because of the very negative potentials at which reduction proceeds. It has been found that bicyclic aliphatic ketones are electro-inactive in DMF containing tetraalkylammonium salts as supporting electrolytes,[19] despite the fact that more cathodic potentials can be reached in the latter system than in 80–90% ethanol used by other investigators. This may mean that reduction is facilitated in protic solvents by hydrogen bonding of the solvent to the ketone.

For a series of ketones in which the carbonyl group is located at different positions on the steroid nucleus, it has been shown that ease of reduction (half-wave potential) is related to the steric accessibility of the carbonyl group.[20] Although selective reductions of diketosteroids were not attempted in that study, others have shown that constant-current reduction of 3,12-diketocholanic acid results in selective reduction of the 3-keto group.[21a] The yield of 3-hydroxy-12-ketocholanic acid (40%) could presumably be improved by operation at constant potential. The same principle could be applied to other difunctional molecules in which one of a pair of identical groups is more sterically accessible.

Electrochemical reduction of several sugars has been examined.[21b] The carbonyl group of several aldotetroses, such as D-erythrose and D-threose, was reduced to the carbinol, but ketotetroses, such as L-*glycero*-tetrulose, underwent initial cleavage of the carbon–oxygen bond of a hydroxyl group adjacent to the carbonyl (*cf.* p. 194).

The stereochemistry of electrochemical reduction of ketones to alcohols is generally the same as that observed upon sodium–alcohol reduction. Reduction of α-methylbenzyl phenyl ketone afforded the *erythro* alcohol, at least 92% stereochemically pure, while sodium–ethanol reduction produced a 1.8:1 *erythro–threo* mixture.[22] Likewise, Kabasakalian and coworkers have found that the equatorial alcohol is apparently the exclusive product from electrochemical and sodium–alcohol reductions of a number of steroidal ketones, ketols, and ketol acetates (Section 5.1.4).[17] (It is not clear, however, that small amounts of the axial isomer would not have escaped detection.) Although it has generally been accepted that sodium–alcohol reactions afford the more stable alcohol product, recent evidence from reduction of bicyclic ketones and imines suggests that both sodium–alcohol and electrochemical reductions can, on occasion, give the less stable isomer.[23, 24] This is not a failure of electrochemical reasoning, but a demonstration that our present understanding of the stereochemical features of even rather simple and well-studied systems is still incomplete.

Reduction to the Methylene Group

A number of studies[2b, 25–28] have shown that aldehydes and ketones may be reduced completely ($>C{=}O \rightarrow CH_2$) at high overvoltage cathodes in highly acidic media:

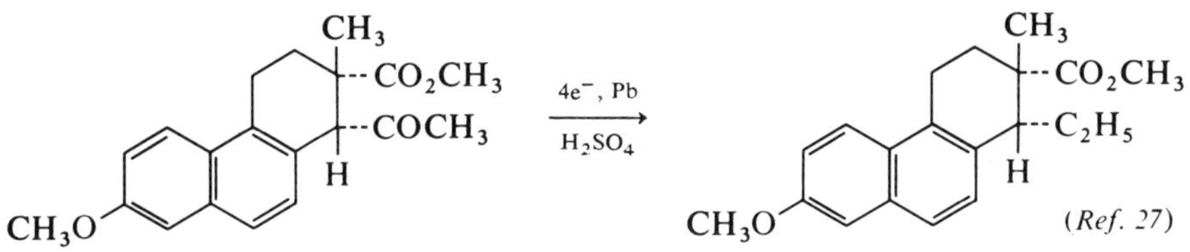

$$\text{steroid methyl ketone (}CH_3O\text{, }CH_3\text{, }CO_2CH_3\text{, }COCH_3\text{, H)} \xrightarrow[H_2SO_4]{4e^-,\ Pb} \text{(}CH_3O\text{, }CH_3\text{, }CO_2CH_3\text{, }C_2H_5\text{, H)}$$

(*Ref.* 27)

$$\text{steroidal ketone (R, =O)} \xrightarrow[D_2SO_4]{4e^-,\ Pb} \text{(R, D, D)}$$

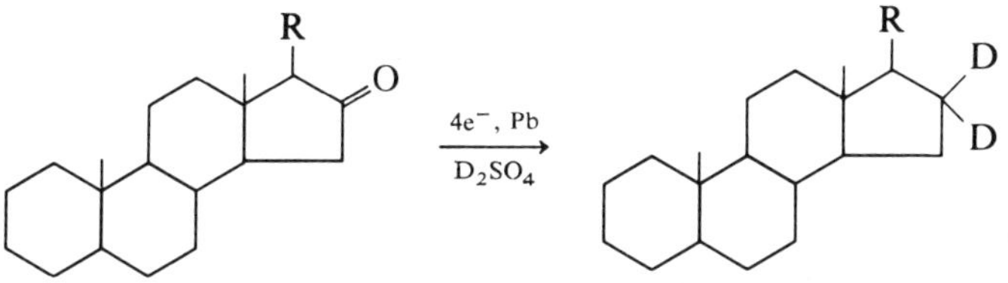

(*Ref.* 28)

Literally nothing is known about the mechanism of this reaction, since experimentation has generally been directed only towards determining empirically the best experimental conditions* for constant-current electrolysis. Mechanistic analysis by the customary electroanalytical techniques has apparently never been attempted. It is known, however, that the alcohol derived by 2-electron reduction of the carbonyl compound is not an intermediate in the reduction, since the alcohol can be recovered unchanged when submitted to the reaction conditions.[28] The cathode itself may be intimately involved in the reduction, since yields are highly dependent on the nature of the cathode and may differ substantially even at cathodes of similar hydrogen overvoltage.[2b] In fact, at mercury, with approximately the same hydrogen overvoltage as lead, little or no hydrocarbon is formed, the principal product being the dialkylmercury compound.[26] If an organolead compound were an intermediate in the reaction at a lead cathode, it would undoubtedly suffer protolysis under the strongly acid conditions of the electrolysis.[29]

6.1.2 Carboxylic Acids and Derivatives

Aromatic carboxylic acids and their derivatives are reducible at very negative potentials.[2a] Reduction apparently involves reduction to an aldehyde, which may or may not be reduced further, depending on conditions. Thus, a number of heterocyclic acids and amides are reduced only as far as the aldehyde,[30, 31] since the carbonyl of the latter is rapidly hydrated and hence electroinactive under the electrolysis conditions:

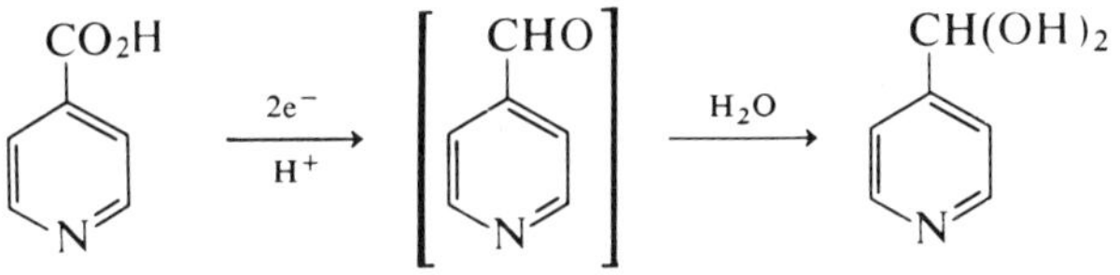

$$\text{4-pyridyl-}CO_2H \xrightarrow[H^+]{2e^-} [\text{4-pyridyl-}CHO] \xrightarrow{H_2O} \text{4-pyridyl-}CH(OH)_2$$

Aromatic acids are reduced to alcohols, on the other hand, since the corresponding aldehyde is not protected by hydration[32]:

* Throop and Tokes have reported that yields in the reduction of steroidal ketones are generally greater than 90% when the reduction is carried out at a lead cathode in a 1:1 (V/V) mixture of dioxane and 10% sulfuric acid.[28]

$$\mathrm{ArCO_2H} \xrightarrow[2\mathrm{H}^+]{2e^-} \mathrm{ArCHO} \xrightarrow[2\mathrm{H}^+]{2e^-} \mathrm{ArCH_2OH}$$

The carbinol may react intramolecularly with an appropriately placed functional group,[33a] as may be seen from the electrochemical reduction of *o*-phthalate esters to phthalide:

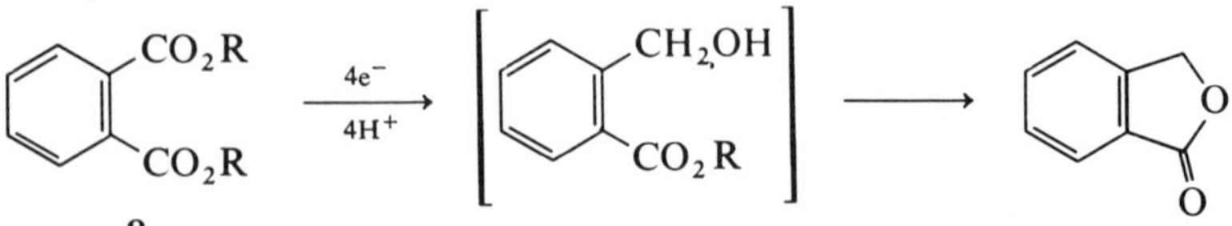

CO_2R CO_2R **8** $\xrightarrow[4H^+]{4e^-}$ [CH_2OH CO_2R] $\longrightarrow$ O O

Benkeser and coworkers were able to reduce aliphatic amides to aldehydes or alcohols (depending on conditions), using solvated electrons generated by electrolysis of a solution of lithium chloride in methylamine.[33b] The tetrachlorophthalimide **9** is reduced in two steps, the first consuming 2 electrons, and the second, at more cathodic potential, 6 electrons.[2a, 34] The reduction path is apparently

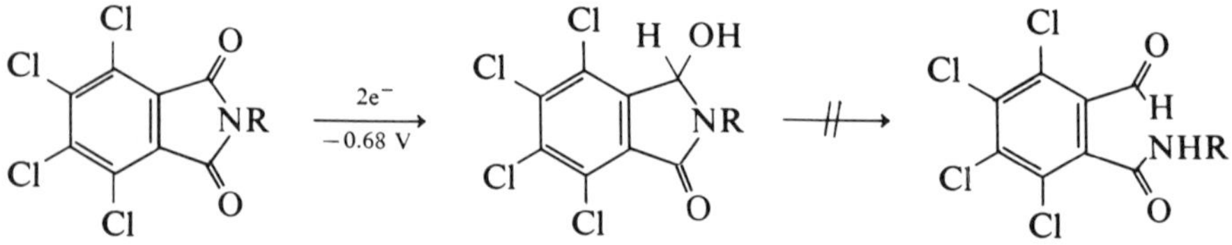

Cl Cl Cl Cl O NR O $\xrightarrow[-0.68\ V]{2e^-}$ Cl Cl Cl Cl H OH NR O $\not\longrightarrow$ Cl Cl Cl Cl O H NHR O

9, $R = CH_2CH_2N(CH_3)_2$ **11** **10**

11 $\xrightarrow[-1.1\ V]{6e^-}$ Cl Cl Cl Cl NR

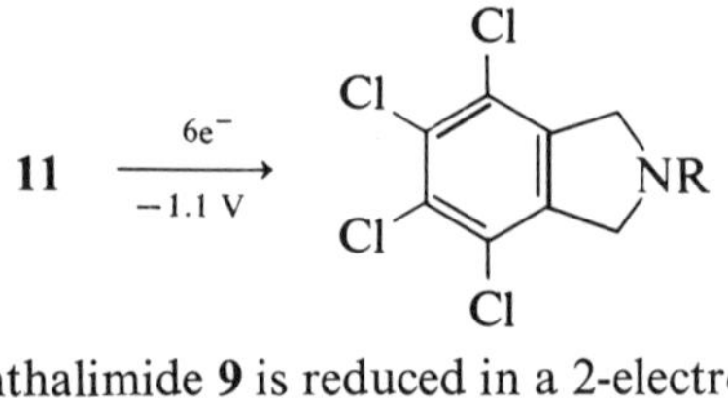

Notice that phthalimide **9** is reduced in a 2-electron step, while phthalate ester **8** is reduced in a 4-electron step. Apparently, in the former case the first product, the aldehyde **10**, is stabilized against further reduction as the carbinolamine **11**, just as it was noted that some heterocyclic aldehydes are stabilized by hydration.

6.1.3 Synthetic Scope and Limitations

The electrochemical route is one of the best methods for preparation of pinacols derived from diaryl and arylalkyl ketones, and aryl aldehydes. All three may be converted to pinacols in media of moderate acidity, and all but the diaryl ketones may also be converted to pinacols in alkaline or aprotic media. Yields are generally high (>80%). Electrolyses should always be carried out at controlled potential under conditions (pH, potential) where polarography indicates stepwise 1-electron reduction. In strong acid the initially formed pinacols may undergo dehydration or rearrangement. The quantitative yield of benzpinacolone by electrolysis of benzophenone in strong acid[4] suggests that this lability of pinacols in acid could be taken advantage of. The stereochemistry of pinacol formation is pH dependent: in acid *dl*/*meso* ratios are generally 1–1.5, while in alkali the ratio may be as high as 3 or more. This may be of interest where selective formation of one isomer is desired. The *meso*-isomer predominates when *intra*molecular hydrogen bonding can occur in the intermediate radicals.

The reduction of carbonyl compounds to alcohols in neutral or weakly alkaline ethanol or in aprotic solvents containing proton donors such as phenols is of limited synthetic interest, in view of the wide variety of chemical reagents that can also effect this conversion. There may be occasions with polyfunctional molecules where the advantages of high yield, stereoselectivity, and the degree of selectivity provided by controlled potential make the electrochemical route attractive.

The reduction of carbonyl compounds to saturated hydrocarbons in strong acid at cathodes of lead or cadmium compares favorably with procedures such as the Clemmensen and Wolff–Kishner reactions in both yield and simplicity. The latter derives from the fact that constant-current electrolysis provides excellent yields of products in this case, so that the experimental arrangement can be quite simple (Chapter 9). It is likely that further study of this reaction would provide insights into the role of the electrode, including possible formation of organometallic intermediates in some electrode reactions.

The coupling of acrylonitrile with adsorbed 2-hydroxy-2-propyl radicals generated by reduction of acetone in acid (p. 145)[35] represents a method of carbon–carbon bond formation that is very likely of considerable generality. Baizer and coworkers (Chapter 7) have effected a variety of coupling reactions using not only acrylonitrile but many other activated olefins, most of which could presumably replace acrylonitrile in this synthesis.

6.2 THE CARBON–NITROGEN DOUBLE BOND

6.2.1 Imines

The electrochemical behavior of imines (Schiff bases), $R_2C{=}NR$, changes in orderly progression as one successively makes zero, one, two, and three of the R groups aryl. Completely arylated imines, such as benzophenone anil (**12**),

$$(C_6H_5)_2C{=}NC_6H_5$$
$$\mathbf{12}$$

are easily reduced and afford relatively stable intermediates upon reduction. As aryl groups are successively exchanged for hydrogen or alkyl, ease of reduction and stability of intermediates both decrease. Thus, as with carbonyl compounds, it is not appropriate to describe the electrochemical behavior of imines without reference to the degree of aryl substitution.

Triaryl imines such as **12** closely resemble diaryl ketones in electrochemical behavior, with, however, a few modifications. Imines cannot be studied in aqueous acid because of rapid hydrolysis. Anil **12** is reasonably stable at pH 5, however, and over the pH range 5.1–7.4 it exhibits two 1-electron waves, the first of which is pH dependent, moving cathodically with increasing pH.[36] At about pH 9, the two waves finally merge into a single 2-electron wave. Recall from the preceding section that this transition occurs at about pH 5 for benzophenone: the difference is undoubtedly related to the substantially higher basicity of imines relative to carbonyl compounds. Since **12** is not basic enough to be protonated in bulk solution to more than an infinitesimal degree in the pH range 5.1–7.4, the first wave, as with benzophenone in acid, must be due to field-enhanced basicity arising from polarization of the double bond by the electrode. Controlled-potential reduction at pH 10 consumes 2 electrons/molecule and affords the expected secondary amine benzhydrylaniline as product.[36] At pH 13, the 2-electron wave has again become two 1-electron waves because of reduction of the neutral imine to its radical anion and thence to the dianion; controlled-potential electrolysis was not carried out in this pH region. In aprotic solvents **12** also exhibits two 1-electron waves, cyclic voltammetry indicates the first reduction to be reversible, and controlled-potential electrolysis at the first wave affords a solution containing the stable radical anion of **12**, which may be identified by esr spectroscopy.[37]

Diaryl imines, $Ar_2C{=}NR$ or $Ar(R)C{=}NAr$, undergo the transition from two 1-electron polarographic waves to a single 2-electron wave at about pH 13, rather than pH 9 as with **12**.[38] This behavior is undoubtedly associated with the stronger basicity of these imines relative to fully arylated species such as **12**. Because of rapid hydrolysis in acid, their electrochemical behavior can be studied only at pH ≥ 7. Controlled-potential electrolysis does not appear to have been carried out on diaryl imines in the pH range 7–11, where two polarographic waves are observed. One would expect dimeric products, *vic*-diamines, upon electrolysis at the first wave in this pH range; for example,

$$\mathrm{ArCH{=}NAr' + H^+ + e^- \longrightarrow Ar\dot{C}HNHAr' \longrightarrow \begin{array}{c} \mathrm{ArCHNHAr'} \\ | \\ \mathrm{ArCHNHAr'} \end{array}}$$

Although this prediction remains to be tested properly, that is, under controlled-potential conditions, suggestive evidence for its correctness comes from certain constant-current electrolyses carried out by Law.[39] At a high hydrogen overvoltage electrode (lead), benzylidene *p*-toluidine afforded benzyl-*p*-toluidine in high yield, and at a low overvoltage (copper) cathode the *vic*-diamine **13** (a *dl*/*meso* mixture) was obtained, along with some benzyl-*p*-toluidine. One would also expect *vic*-diamines from reduction of immonium salts if the latter could be

$$C_6H_5CH{=}N{-}Ar \xrightarrow[EtOH-H_2O]{e^-,\ Cu} \begin{array}{c} C_6H_5CHNHAr \\ | \\ C_6H_5CHNHAr \end{array} + C_6H_5CH_2NHAr$$

13

(Ar = *p*-tolyl)

reduced under nonhydrolytic conditions. Indeed, Andrieux and Saveant[40] reduced immonium salts in anhydrous aprotic solvents and obtained the expected *vic*-diamines; in some cases they even obtained esr spectra of the intermediate radicals:

$$(CH_3)_2C{=}\overset{+}{N}(C_6H_5)_2 \xrightarrow{e^-} (CH_3)_2\dot{C}{-}N(C_6H_5)_2 \xrightarrow{\text{fast}} \begin{array}{c} (CH_3)_2C{-}N(C_6H_5)_2 \\ | \\ (CH_3)_2C{-}N(C_6H_5)_2 \end{array}$$

$$(C_6H_5)_2C{=}\overset{+}{N}(CH_3)_2 \xrightarrow{e^-} \underset{\text{stable}}{(C_6H_5)_2\dot{C}N(CH_3)_2}$$

The polarographic behavior of diaryl imines in aprotic media (containing no proton donors) depends on the nature of the two aryl groups. Where the latter contain extended aromatic systems, such as in **14**, two 1-electron waves are

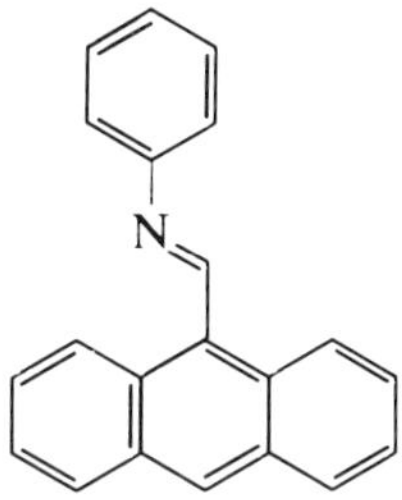

14

observed.[41a,c] Benzalanilines, on the other hand, exhibit behavior almost like the monoaryl imines to be discussed below, that is, a single 2-electron wave.[41b] Cyclic voltammetry and esr spectroscopy indicate that the stability of the radical anion formed in the first step is highly dependent on the nature of the two aryl groups.[41a] This is to be expected, since this can substantially influence the basicity of the radical anion, that is, the magnitude of the proton abstraction rate constant k_1 (Section 4.2.1).

$$ArCH{=}NAr' \xrightarrow[\substack{DMF \\ TEAB}]{e^-} Ar\dot{C}H\bar{N}Ar' \xrightarrow[\substack{TEAB \\ Et_4N^+Br^-}]{k_1} Ar\dot{C}HNHAr'$$

With highly resonance-stabilized radical anions, such as that from **14**, proton abstraction is slow on the polarographic time scale, and two 1-electron waves are observed. In most benzalanilines, proton abstraction is very fast, initiating an ECE process and thus a single 2-electron wave. Naturally, in some cases behavior intermediate between these two extremes is observed: the first wave is much larger than the second, as a consequence of partial protonation and reduction of the radical anion on the polarographic time scale.[41a, c]

Monoaryl imines (**15**) and trialkyl imines have been investigated relatively little. A single 2-electron polarographic wave is observed in protic media (at

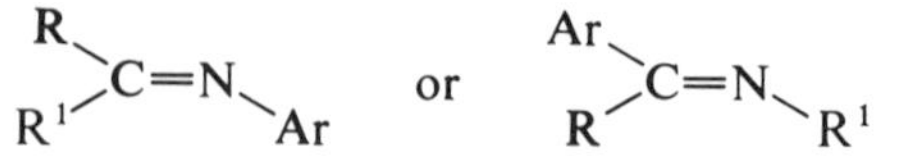

$$R(R^1)C{=}N{-}Ar \quad \text{or} \quad Ar(R)C{=}N{-}R^1$$

15a **15b**

quite negative potentials), and controlled-potential reduction affords the secondary amine.[36] This behavior indicates that these species, like dialkyl ketones, are reduced by the ECE process. Trialkyl imines, again like dialkyl ketones, are not reducible at potentials accessible in the aprotic solvent DMF, suggesting that reduction in protic solvents may be enhanced by hydrogen bonding between the solvent and the imine.[19] Monoaryl imines (**16** and **17**) have been studied in DMF,

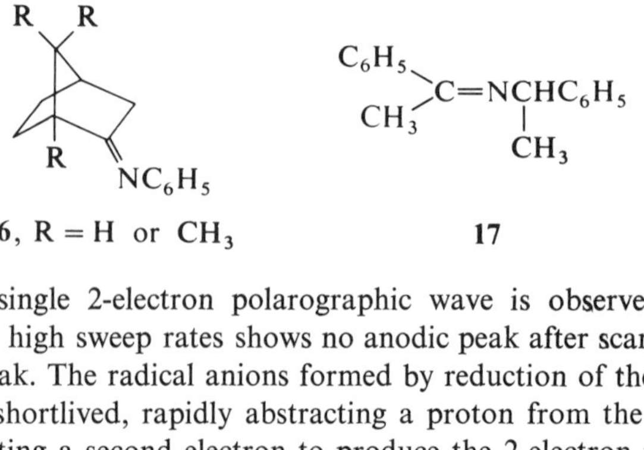

NC_6H_5

16, R = H or CH_3

$C_6H_5(CH_3)C{=}NCH(CH_3)C_6H_5$

17

however.[23] A single 2-electron polarographic wave is observed, and cyclic voltammetry at high sweep rates shows no anodic peak after scan reversal past the cathodic peak. The radical anions formed by reduction of these imines are therefore very shortlived, rapidly abstracting a proton from the surroundings and then accepting a second electron to produce the 2-electron polarographic wave. Controlled-potential electrolysis confirms the consumption of 2 electrons/molecule, and the corresponding secondary amines are isolated in high yield.[23] Reduction of all three imines affords as the principal product the same isomeric amine as is formed in sodium–alcohol reductions (Table 6.1). The results contradict the common assumption that sodium–alcohol reductions produce the most stable of a pair of epimeric products (see also Section 6.1.1).

Horner and Skaletz have carried out the constant-current electrolysis of monoaryl and diaryl imines in an ethanol:methyl acetate:water (5:3:1) mixture.[42] Products were either secondary amines or *vic*-diamines, depending on the electrolysis temperature (−10°C and 65°C, respectively). No voltammetric data were reported, making it difficult to compare their results with those of Lund[36] and others.

Table 6.1 Stereochemistry of reduction of monoaryl imines

Substrate	Mode of Reduction	Relative Percent of Products[a]	
	Bicyclic Imines[b]		
		exo-Amine	*endo*-Amine
Norcamphor anil	Sodium–ethanol	33	67
(**16**, R = H)	Catalytic hydrogenation	0	100
	Electrochemical	20	80
Camphor anil	Sodium–ethanol	0	100
(**16**, R = CH_3)	Catalytic hydrogenation	100	0
	Electrochemical	0	100
	N-α-Methylbenzylidene-α-methylbenzylamine (**17**)[c]		
		dl-(RR,SS)-Amine	*meso*-(RS)-Amine
Racemic imine	Sodium–ethanol	46	54
(±)-(R,S)	Catalytic hydrogenation	93	7
	Electrochemical	39	61
		Optically Active (–)-(RR)-Amine	*meso*-(RS)-Amine
Optically active	Sodium–ethanol	42	58
imine (–)-(R)	Catalytic hydrogenation	90	10
	Electrochemical	43	57

Source: A. J. Fry and R. G. Reed, *J. Amer. Chem. Soc.*, 91, 6448 (1969); R. G. Reed, M.A. thesis, Wesleyan University, 1969.

[a] Accuracy: ±3%.

[b] Analysis by nuclear magnetic resonance.

[c] Analysis by vapor phase chromatography on a 2 ft × ¼ in. alkaline Carbowax 20M column, operating at 160°C and flow rate of 120 ml/min.

Judging from the usual effects of proton donors on the course of electrochemical reduction generally,[43] and the results of Andrieux and Saveant[40] specifically, the prediction can be ventured that reduction of monoaryl and trialkylimines under acidic, nonhydrolytic conditions (e.g., phenol or hydrogen chloride in anhydrous ethanol or DMF) will proceed via two 1-electron steps. If this turns out to be true, it would provide a convenient procedure for conversion of these compounds to *vic*-diamines, as well as enhancing present knowledge of the electrochemical behavior of imines. There is some evidence that this postulate is true, because addition of phenol to a solution of norcamphor anil in DMF does cause a splitting of the single wave into two.[19]

6.2.2 Other Carbonyl Derivatives

Diaryl, arylalkyl, and dialkyl oximes exhibit a pH-dependent 4-electron polarographic wave in acid solution.[36,44] Coulometry at controlled potential also results in an uptake of 4 electrons/molecule of oxime, and the amine can be isolated in good yield:

$$R_2C{=}NOH + 5H^+ + 4e^- \longrightarrow R_2CHNH_3^+ + H_2O$$

The reduction has been shown to involve reduction of the conjugate acid of the oxime and also to proceed via initial cleavage of the N—O bond to form an imine, which is immediately protonated and reduced:

$$R_2C{=}NOH + 2H^+ + 2e^- \longrightarrow [R_2C{=}NH] \xrightarrow[3H^+]{2e^-} R_2CHNH_3^+$$

(*Ref. 36*)

The fact that hydroxylamines, $R_2CHNHOH$, are not reducible at potentials where oximes undergo reduction to the amine clearly demonstrates that the initial step in the latter conversion does not involve reduction of the carbon–nitrogen double bond.[36] Furthermore, a few oximes exhibit two 2-electron waves, indicating resolution of the two reduction steps, and in these cases reduction at the first wave affords the imine in good yield.[45] Above pH 6 or so, the height of the polarographic wave for oxime reduction decreases with increasing pH, and a new 4-electron wave is usually seen for reduction of the neutral oxime. (This latter wave is sometimes obscured by solvent discharge.) The question remains unanswered at this time as to why neutral oximes, $R_2C{=}NOH$, undergo this reduction in weak alkali, while no polarographic wave is observed for oxime ethers, $R_2C{=}NOR$, under the same conditions (they do undergo 4-electron reduction to the imine in acid).[38] In strongly alkaline solutions, oximes are not reducible, of course, because of conversion to the conjugate base, $R_2C{=}NO^-$. There is evidence that reduction of oximes provides the same isomeric amine as is obtained by reaction of the oxime with lithium aluminum hydride,[46] rather than that from sodium–alcohol reduction, but the reason for the different stereochemistry observed for oximes[46] and imines[23] is unclear.

Phenylhydrazones and semicarbazones are reduced in a single 4-electron step in acid, as demonstrated by polarography, coulometry, and isolation of the corresponding amine from controlled-potential electrolysis:

$$R_2C{=}NNHX + 5H^+ + 4e^- \longrightarrow R_2CHNH_3^+ + XNH_2$$

$$(X = C_6H_5 \text{ or } CONH_2)$$

Here too the reduction has been shown to proceed via initial cleavage of the N—N bond of the conjugate acid.[36,47] No reduction waves are observed in alkaline media (pH 9–10).

6.2.3 Synthetic Scope and Limitations

Conditions can generally be found for electrochemical reduction of imines to the corresponding amines. Since conditions vary depending on the degree of aryl-

Table 6.2 Electrochemical reduction of imines. Product dependence on structure and experimental conditions

Imine Type	Optimal Conditions to Produce Desired Product[a] Amine	*vic*-Diamine
Triaryl[b]	Aqueous alkali, pH 10–12[c] or Second polarographic wave, pH 7–9[d]	Aqueous alkali, pH ⩾ 13[d] or Anhydrous ethanol or DMF[e] containing a proton donor, e.g., phenol[c]
Diaryl	Aqueous alkali, pH ⩾ 12[d] or Second polarographic wave, pH 7–11[d] or DMF[e] containing a weak acid, e.g., phenol[d]	Aqueous alkali, pH 7–11[d] or Anhydrous ethanol or DMF[e] containing a strong acid, e.g., HCl[d] or Constant-current electrolysis in ethanol: methyl acetate: water (5:3:1), 65°C[c]
Monoaryl	Anhydrous ethanol or DMF[c, e] or Constant-current electrolysis in ethanol: methyl acetate:water (5:3:1), –10°C[c]	Anhydrous ethanol or DMF[e] containing a proton donor, e.g., phenol or Constant-current electrolysis in ethanol: methyl acetate: water (5:3:1), 65°C[c]
Trialkyl	Neutral ethanol[c] or Second polarographic wave in anhydrous ethanol or DMF[e] containing a proton donor, e.g., phenol[d]	Anhydrous ethanol or DMF[e] containing a proton donor, e.g., phenol[d]

[a] Controlled-potential electrolysis at the first polarographic wave, except where noted.
[b] These may be converted to radical anions by electrolysis at the first wave in aprotic media.
[c] Experimentally observed (see text).
[d] Predicted behavior (see text)
[e] DMF or other aprotic solvent.

ation, this information is presented here in tabular form, along with suggested conditions for conversion of imines to *vic*-diamines (Table 6.2). The latter are mainly conjectural, except for the work of Andrieux and Saveant,[40] but appear reasonable in view of known effects of proton donors.

Since oximes can be converted electrochemically into amines in good yield under neutral conditions, the sequence

$$R_2C{=}O \xrightarrow{H_2NOH} R_2C{=}NOH \xrightarrow{4e^-} R_2CHNH_2$$

represents an efficient procedure for the reductive amination of aldehydes and ketones. Yields are high and side reactions appear to be absent. Similarly, secondary amines can be produced by condensation of the carbonyl compound with a primary amine, followed by reduction of the imine so formed. The overall conversion of a ketone to an amine can, in fact, be effected without isolation of the intermediate imine, since imines are reducible at potentials where the corresponding carbonyl compounds are not. One dissolves the carbonyl compound in a solvent containing the amine, and reduces the imine *in situ.* For example, reduction of cyclohexanone either in ethanol containing methylamine, or in methylamine as solvent, affords methylcyclohexylamine in good yield.[36,48] In a variant on this procedure, Elving and Van Atta reduced 2-chlorocyclohexanone in the presence of ammonia. Cyclohexylamine was isolated in quantitative yield.[49] The suggested reduction sequence involved dehalogenation to cyclohexanone, condensation of the latter with ammonia, and then reduction of the imine to the observed product. Evidence favoring, but not rigorously establishing, this reaction sequence was presented.

O $\xrightarrow[CH_3NH_2]{2e^-}$ NHCH$_3$

Cl O $\xrightarrow{2e^-}$ [O] $\xrightarrow[NH_3]{2e^-}$ NH$_2$

The fact that oxime ethers, semicarbazones, and phenylhydrazones are not reducible in mild alkali (pH $\geq$ 10) suggests that one might use these derivatives as carbonyl-blocking groups when reduction of another function in the molecule is contemplated. The blocking function could also be served by ketalization of the carbonyl, but ketals must be prepared in acid, while these derivatives can be prepared under neutral or alkaline conditions.

6.3 THE NITRO GROUP

As might be expected for a functional group in such a high oxidation state, the cathodic behavior of the nitro group exhibits a wealth of detail. Because adequate reviews already exist,[3, 50] the subject will be discussed here only sufficiently to delineate principal reaction pathways, including those often observed as a result of interaction with other functional groups in the molecule. Throughout the discussion that follows, analogies to the electrochemical behavior of carbonyl compounds will often be evident. Examination of the effects of changes in experimental conditions (pH, potential, etc.) on the course of electrochemical reduction of nitro compounds will also be seen to shed a great deal of light on the variety of products produced from nitro compounds using chemical reducing agents. This is of no little significance, since it reemphasizes the point that redox processes are best studied electrochemically, and that data from electrochemical studies can often be helpful in understanding mechanisms of chemical redox reactions.

6.3.1 Aromatic Nitro Compounds

General

Nitrobenzene exhibits a single pH-dependent, irreversible 4-electron polarographic wave in aqueous media above pH 5.[50] Below pH 5, the 4-electron wave is accompanied by a second 2-electron wave at more negative potentials. Controlled-potential coulometry confirms 4 electrons/molecule. Phenylhydroxylamine is isolated when reduction is carried out in moderately acid or neutral solution, but under more vigorous conditions (stronger acid, higher temperature) the product is *p*-aminophenol, formed by rearrangement of phenylhydroxylamine:

$$C_6H_5NO_2 \xrightarrow[4H^+]{4e^-} C_6H_5NHOH \xrightarrow{H^+} HO-C_6H_4-NH_2$$

Reduction of nitrobenzene at the second wave in acidic solution affords aniline by reduction of the intermediate hydroxylamine before it can undergo rearrangement:

$$C_6H_5NO_2 \xrightarrow[4H^+]{4e^-} [C_6H_5NHOH] \xrightarrow[2H^+]{2e^-} C_6H_5NH_2$$

These reductions are undoubtedly initiated by reversible electron transfer to nitrobenzene to form a moderately stable radical anion. This may be demonstrated by controlled-potential electrolysis in an aprotic solvent[51] or even aqueous alkali[52] in the cavity of an esr spectrometer, whereupon the spectrum of the nitrobenzene radical anion can be observed. Interestingly, the second polarographic wave of nitrobenzene in acetonitrile is a 3-electron wave.[51] Apparently, proton abstraction by the nitrobenzene dianion and subsequent ejection of hydroxide to generate nitrosobenzene are fast on the polarographic time scale. Nitrosobenzene is easier to reduce than nitrobenzene, so it undergoes immediate 2-electron reduction at the potential at which it is formed, making the overall process at the second wave a 3-electron reduction.[3]

$$C_6H_5NO_2 + e^- \longrightarrow C_6H_5NO_2^{\dot{}-} \quad (\textit{first wave})$$

$$C_6H_5NO_2^{\dot{}-} + e^- \longrightarrow C_6H_5NO_2^{-2} \xrightarrow{H^+} C_6H_5N(O^-)OH$$

$$\xrightarrow{-OH^-} C_6H_5NO \xrightarrow[2H^+]{2e^-} C_6H_5NHOH \quad (\textit{second wave})$$

The 4-electron wave in protic solvents arises via rapid protonation of the nitrobenzene radical anion, reduction of the resultant radical to nitrosobenzene, and further reduction of the latter to phenylhydroxylamine.

$$C_6H_5NO_2 \xrightarrow{e^-} C_6H_5NO_2^{\dot{}-} \xrightarrow{H^+} C_6H_5\dot{N}^+(O^-)OH \xrightarrow[-OH^-]{e^-} C_6H_5NO$$

$$\xrightarrow{e^-} C_6H_5NO^{\dot{}-} \xrightarrow{H^+} C_6H_5\dot{N}OH \xrightarrow[H^+]{e^-} C_6H_5NHOH$$

Holleck and Exner have shown that in the presence of surface-active agents such as camphor, agar, tylose, or gelatin the single 4-electron wave normally observed for nitro compounds in aqueous media splits into two waves, the first a 1-electron and the second a 3-electron reduction.[53] Addition of these substances thus produces exactly the same polarographic behavior as in aprotic media. Holleck has suggested that the film of adsorbed material inhibits protonation of the radical anion at the electrode surface, at least on the polarographic time scale.

Reduction of nitroaromatics in alkali results in the formation of dimeric products. These are apparently formed by dimerization of the nitrosobenzene radical anion (**18**).

$$C_6H_5NO_2 + 2e^- + 2H^+ \longrightarrow C_6H_5NO \xrightarrow{e^-} \underset{\mathbf{18}}{C_6H_5NO^{\dot{}-}}$$

$$2\ \mathbf{18} \longrightarrow C_6H_5N(O^-)N(O^-)C_6H_5 \xrightarrow[-OH]{+H^+} \underset{\mathbf{19}}{C_6H_5N{=}N^+(O^-)C_6H_5}$$

It is well known that nitrosobenzene and phenylhydroxylamine react rapidly in alkaline ethanol to afford azoxybenzene (**19**) in yields close to quantitative.[54] This reaction presumably also involves **18**, formed by disproportionation:

$$C_6H_5NHOH + 2OH^- \longrightarrow C_6H_5NO^{-2}$$

$$C_6H_5NO^{-2} + C_6H_5NO \longrightarrow \underset{\mathbf{18}}{2C_6H_5NO^{\dot{-}}} \longrightarrow \mathbf{19}$$

Russell and his coworkers have demonstrated many other examples of radical anion formation by electron transfer from an aromatic dianion to the corresponding neutral species.[55] Observe the analogy here between reduction of aromatic nitro compounds to either hydroxylamines or azoxy compounds, depending on experimental conditions, and the corresponding reduction of aromatic carbonyl compounds to either pinacols or alcohols, also depending on experimental conditions. Except for reduction of carbonyl compounds in acid, both reductions proceed via radical anions; under conditions of high proton availability, the respective radical anions undergo the ECE pathway, that is, protonation followed by rapid reduction to afford monomeric products (alcohols or hydroxylamines). Under conditions of low proton availability, dimerization to afford pinacols or azoxy compounds is the predominant path.

It was stated above that "dimeric products" are obtained from electrochemical reduction of nitro compounds in alkaline media, but the nature of these products was not stated. This omission was intentional. There is no question that the initial reduction product is the azoxy compound. However, depending on the experimental conditions employed, the azoxy compound may or may not be reduced further. Most reductions in the literature have been run under constant-current conditions, and a great deal of effort has been invested in finding, empirically, conditions to effect a given conversion without too much attention to the reasons why a given set of conditions acts as it does. Briefly, the facts are these. Reduction of a nitro compound as a concentrated solution or suspension in hot alkali at a nickel cathode affords the azoxy compound in high (85–95%) yield, apparently because the azoxy compound precipitates from solution and hence is reduced no further.[2b, 56] If reduction is carried out under conditions (sodium acetate in hot 95% ethanol at a lead or nickel cathode) where the azoxy compound remains in solution, the corresponding azo compound is produced in 90–95% yield if the current is interrupted when hydrogen evolution begins.[57] If reduction is continued past this point, the hydrazo compound may then be isolated in 90–95% yields.[57] While the yields in these conversions are all excellent, the situation is obviously unsatisfactory with respect to an understanding of what factors are actually governing which product is formed. There is clearly a need for voltammetric and controlled-potential investigation of the electrochemical behavior of nitroarenes and their reduction products (azoxy, azo, and hydrazo compounds) in alkaline media. The isolation of azo compounds under certain conditions, as mentioned above, is particularly surprising, in view of the reports that azoxybenzene exhibits a single 4-electron polarographic wave over the pH range 2–12, and that azobenzene is *easier* to reduce than azoxybenzene by 0.3 V at room temperature.[58] The 4-electron wave apparently arises from a

2-electron reduction of azoxybenzene to azobenzene, which is then reduced immediately to hydrazobenzene in a second 2-electron step at the same potential. Controlled-potential reduction of several azoxy compounds indeed produced the hydrazo compounds in high yield.[58] One is faced with the apparent paradox of isolation of azo compounds despite the fact that they are thought to be easier to reduce than the starting azoxy compounds. The explanation may be that the isolation of azo compounds from reduction in hot alkali arises because the potential for the initial reduction of the azoxy compound to the azo stage shifts sufficiently anodically with increasing temperature for this step to be *easier* than reduction of the azo compounds.*

In acid, azoaromatics exhibit two 2-electron waves, the first for reduction to the hydrazo compound and the second due to reduction of the hydrazo compound to the aromatic amine by cleavage of the N—N bond.[59] This enables one to convert the azo compounds produced in hot alkali into any of three products: reduction at the first wave affords the hydrazo compound in weak acid and the rearranged benzidine in strong acid, and reduction at the second wave affords the amine.

Polyfunctional Compounds

The electrochemical behavior of aromatic nitro compounds is greatly influenced by inductive and resonance interactions with properly placed substituents elsewhere in the molecule, and also by condensations, dehydrations, and so on, arising through reaction of electrogenerated intermediates with other substituents. It will be useful to discuss here certain of these substituent effects on electrochemical behavior, since the lessons learned will be equally applicable to the behavior of other polyfunctional molecules. For more complete discussions the reader is referred to the reviews by Perrin[3] and Zuman.[60]

Like many other cathode processes, reduction of nitro compounds is facilitated by electron-withdrawing substituents and retarded by electron-donating groups.[3,60] As a consequence of these two effects, polynitro compounds are reduced in stepwise fashion. *m*-Dinitrobenzene, for example, exhibits two distinct 4-electron waves due to stepwise reduction of the two nitro groups. This sort of behavior, while it can be very useful, is not particularly surprising. Interactions between substituents can exert a considerably greater influence on the course of electrochemical reduction than this, however. For example, *m*-nitroaniline exhibits the expected 4-electron polarographic wave in weak acid for reduction to *m*-hydroxylaminoaniline. Under the same conditions, *p*-nitroaniline exhibits a 6-electron polarographic wave, and *p*-phenylenediamine is isolated from controlled-potential electrolysis.[50,61] Analogous behavior is observed

* Half-wave potentials for irreversible reductions (e.g., azoxy → azo) shift anodically with increasing temperature, but half-wave potentials for processes that are reversible at room temperature (e.g., azo → hydrazo)[59] remain close to E^0 as the temperature is raised (Section 2.3.2).

with the nitrophenols: the *meta*-isomer is reduced in a 4-electron step, while the *ortho*- and *para*-isomers are reduced completely to the amine.[50, 62] A number of clues point to the nature of the phenomenon responsible for this behavior. First of all, a free *ortho*- or *para*-hydroxyl or amino group is apparently necessary for 6-electron reduction to occur, since *p*-nitroanisole,[62] *p*-nitroacetanilide,[63] and *p*-dimethylaminonitrobenzene[50] follow the normal 4-electron reduction path. Second, polarographic reduction of *o*-nitrophenol is observed to consume 6 electrons only at high and low pH; in neutral solution (pH 6) the expected 4-electron wave is observed.[62] All of these facts are consistent with the assumption that in the anomalous cases the *o*- and *p*-amino and hydroxyl-substituted hydroxylamines generated in the first 4-electron process can undergo dehydration to a quinoid intermediate, which is reduced to the amine immediately upon formation:

$$HO-C_6H_4-NO_2 \xrightarrow[4H^+]{4e^-} HO-C_6H_4-NHOH \xrightarrow[\text{acid}]{\text{base or}} O{=}C_6H_4{=}NH \xrightarrow[2H^+]{2e^-} HO-C_6H_4-NH_2$$

$$\underset{\mathbf{20}}{H_2N-C_6H_4-N{=}NC_6H_5} \xrightarrow[2H^+]{2e^-} \underset{\mathbf{21}}{H_2N-C_6H_4-NHNHC_6H_5}$$

$$\mathbf{21} \longrightarrow C_6H_5NH_2 + HN{=}C_6H_4{=}NH_2$$

$$\xrightarrow{2e^-} H_2N-C_6H_4-NH_2$$

The pH dependence of the number of electrons involved in the reduction is due to the fact that dehydration is both acid and base catalyzed, but is slow in neutral media. The *pseudo*-first-order dehydration rate constant k_d can be evaluated by a number of electrochemical techniques. Alberts and Shain obtained a value of k_d for dehydration of *p*-hydroxylaminophenol of 1.3 sec^{-1} in 20% ethanol at pH 4.8,[64] while Testa and Reinmuth measured a k_d of 0.2 sec^{-1} for the *ortho*-isomer in 50% ethanol at (apparent) pH 6.2.[65]

A situation analogous to the above is observed in the electrochemical reduction of *p*-aminoazobenzene (**20**).[66] Polarography indicates a 2-electron reduction to the hydrazo compound, but upon controlled-potential electrolysis 4 electrons are consumed; the products are aniline and *p*-phenylenediamine. Apparently, elimination of aniline from the initial product **21** is slow on the polarographic time scale, but fast enough to occur in the time necessary to carry out the electrolysis. With some related compounds, elimination is fast even on the polarographic time scale, and a 4-electron polarographic wave is observed.[67] In all of these reactions, hydrolysis of the quinone imine intermediates to the corresponding quinones does not interfere because hydrolysis is much slower than reduction.[64]

Intramolecular bond formation can occur between intermediates generated electrochemically by reduction of nitro compounds and suitably located substituents. 2,2′-Dinitrobiphenyl (**22**) exhibits a 10-electron polarographic wave in alkali,[68] which presumably arises from this sequence

O_2N NO_2 (**22**) $\xrightarrow{6e^-}$ ON NHOH $\xrightarrow[^-OH]{\text{fast}}$ $N{=}N^+{-}O^-$ (**23**)

$\xrightarrow{2e^-}$ N=N (**24**) $\xrightarrow{2e^-}$ N(H)–N(H)

NO_2, CO_2H, CO_2H $\longrightarrow$ [NHOH, CO_2H, CO_2H] $\longrightarrow$ H–N–O, C=O, HOOC

Lund has shown that by proper control of pH and potential the reduction can be made to stop at **23**, **24**, or the fully reduced material.[68b] Reduction of the indicated *o*-nitrobenzoic acid results in formation of a heterocycle (quantitative yield), by dehydration of the intermediate hydroxylamine.[69] A similar reaction may be involved in the 4-electron reductions of 1,3-dinitropropane and *o*-dinitrobenzene in alkali to produce products of unknown structure.[70]

6.3.2 Aliphatic Nitro Compounds

Like their aromatic counterparts, reduction of aliphatic nitro compounds proceeds in a single 4-electron step to the hydroxylamine.[70a, 71] Aliphatic nitro compounds, however, do not exhibit a second wave in acid solution for further 2-electron reduction to the amine. The latter is apparently possible in the case of aromatic nitro compounds only because of the activating effect of the aromatic pi system.

As with nitroaromatics, the initial step in reduction of aliphatic nitro compounds involves reversible addition of an electron to afford a radical anion; this has been demonstrated by esr spectroscopy[72] and cyclic voltammetry.[73] In aprotic solvents, the radical anion appears to decompose rapidly to a free radical and nitrite ion:

$$RNO_2 + e^- \rightleftharpoons RNO_2^{\dot{-}} \xrightarrow{k_d} R\cdot + NO_2^-$$

The rate constant k_d is insensitive to changes in the radical stability of R ($k_d = 1.4$ and $3.5\ \text{sec}^{-1}$ for the radical anions of *t*-nitrobutane and 1-nitropropane, respectively), indicating that the transition state for bond breakage occurs at an early stage along the reaction coordinate.[73]

The reduction of ω-nitrostyrene (**25**) has been investigated by Masui, Sayo, and Nomura.[74a] Two polarographic waves are observed in acid solution. Controlled-potential electrolysis at the first wave consumes 4-electrons/molecule, and phenylacetaldoxime (**26**) is isolated as the product. It was reported that electrolysis at the second wave affords β-phenethylhydroxylamine (**28**).[74a] This would appear unlikely, however, since reduction of oximes generally occurs via a single 4-electron step, and the initial stage involves cleavage of the N—O bond, rather than reduction of the C=N bond.[36] If **26** were an intermediate in the reduction of **25** at the first wave, one would expect to isolate β-phenethylamine (**27**), not the hydroxylamine **28**, from reductions at the second wave. Indeed, it has recently been found that **27** *is* the product of reduction at the

$$\underset{\mathbf{25}}{C_6H_5CH{=}CHNO_2} \xrightarrow{4e^-} \underset{\mathbf{26}}{C_6H_5CH_2CH{=}NOH} \xrightarrow{4e^-} \underset{\mathbf{27}}{C_6H_5CH_2CH_2NH_2}$$

$$\mathbf{26} \longrightarrow \underset{\mathbf{28}}{C_6H_5CH_2CH_2NHOH}$$

second wave.[74b] Complete reduction of ω-nitrostyrenes to β-phenethylamines can also be effected at a lead cathode in sulfuric acid (constant current).[75] The mechanism of this reduction has not been studied.

6.3.3 Synthetic Scope and Limitations

Because of the very substantial effects of pH, potential, and temperature on the course of reduction of aromatic nitro compounds, a single substrate may be converted into any of a variety of products, depending on the choice of conditions. Although, as has been indicated already, there still exist some unanswered questions, the picture is clear enough that approximate conditions for each conversion can be suggested (Table 6.3). In practice, the choice of experimental conditions in a specific case would be aided further by polarographic data.

Reduction in acid of nitro compounds subsituted at the *ortho*- or *para*-position with groups such as hydroxyl or amino generally leads directly to the amine, as a result of elimination from the hydroxylamine to generate an easily reducible quinoid species.

One may take advantage of the stepwise nature of reduction of polynitro compounds, coupled with the usual product dependence on conditions, to convert such compounds into a multitude of useful polyfunctional substances. Tallec has reported a number of such studies,[76] of which the following sequence is representative:

CH_3, O_2N, NO_2

-0.09 V (S.C.E.), 20°C; 2-*N* H_2SO_4–EtOH (1:3) → CH_3, O_2N, NHOH

-1.0 V (S.C.E.), 20°C; 2-*N* H_2SO_4–EtOH (1:1) → CH_3, H_2N, NH_2

$+0.05$ V (S.C.E.), 80°C; 10-*N* H_2SO_4–EtOH (1:2) → CH_3, O_2N, NH_2, HO

-0.25 V (S.C.E.), 80°C; 5-*N* H_2SO_4–EtOH (1:1) → CH_3, H_2N, NH_2, HO

The extreme ease with which the nitro group is reduced relative to most other substituents allows one to effect a variety of useful selective reductions without affecting other functional groups. One such example was cited already (Section 4.1). Others may be noted:

O, O, Cl, O_2N → O, O, Cl, H_2N (*Ref. 77*)

$C_6H_5CH_2O$, $C_6H_5CH_2O$, CHOAc, $CHNO_2$, CH_3 → $C_6H_5CH_2O$, $C_6H_5CH_2O$, CHOH, CHNHAc, CH_3 (*Ref. 78*)

The yield in the first example was considerably better (90%) than could be achieved using conventional reducing agents such as tin–hydrochloric acid.[77]

Table 6.3 Electrolysis products of aromatic nitro compounds as a function of experimental conditions

	Experimental Conditions[a]		
Desired Product	Electrolysis	Temperature (°C)	pH
Amine	Controlled-potential, at second polarographic wave	25	4
	or		
	Controlled-potential reduction of hydrazo or azo compounds (see below) in acid	25	
Hydroxylamine	Controlled-potential, at first polarographic wave	25	6
p-Aminophenol	Controlled-potential, at first polarographic wave	80	0–2
	or		
	Constant current, platinum cathode	25	ca. –1[b]
Azoxy compound	Controlled-potential, at first polarographic wave	25	8
	or		
	Constant-current, water as solvent	80	8
Azo compound	Constant-current, alcohol as solvent, limited time	80	8
Hydrazo compound	Constant-current, alcohol as solvent, prolonged electrolysis	80	8
Benzidine	Reduction to hydrazo compounds (see above), followed by rearrangement of the latter	80	0–2

[a] Approximate conditions; to be modified by voltammetric data in specific cases.
[b] Concentrated sulfuric acid.

In the second example, which is notable because the benzyloxy groups survive reduction,[78] an intramolecular acyl migration of the sort often observed in the ephedrine series[79] has apparently taken place.

Finally, Harman and Cason have described a procedure for synthesis of quinones by constant-current reduction of nitro compounds under conditions where the intermediate hydroxylamine undergoes acid-catalyzed rearrangement, followed by oxidation of the *p*-aminophenol without isolation[80]:

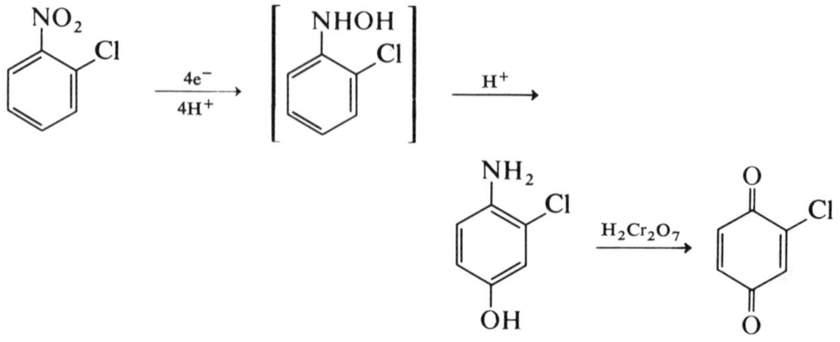

Yields could probably be improved, and the operation simplified, by electrolysis at constant potential.

6.4 OTHER MULTIPLE BONDS

While a variety of other multiple bonds have been investigated, none has been studied as intensively as those already discussed. Furthermore, most of the synthetic principles that have been deduced from inspection of the behavior of the preceding groups may be applied in fairly straightforward fashion in other systems. For this reason the following discussion is not meant to be comprehensive.

6.4.1 Azo and Nitroso Compounds

These compounds are mentioned together because the main features of the electrochemistry of each has already been outlined in the preceding section on nitro compounds. Azobenzene is reduced to hydrazobenzene in the pH range 2–6, and at pH 2 a second 2-electron polarographic wave is observed, corresponding to reduction of hydrazobenzene to aniline.[59] If electrolysis is carried out in rather strong acid (pH = 2) at the first wave, the initially formed hydrazobenzene rearranges to benzidine. Since the rearrangement has a second-order kinetic dependence on proton concentration, it is easy to arrange conditions such that the product is either the hydrazobenzene or the benzidine.[81] Nitroso compounds are converted to hydroxylamines or *p*-aminophenols at low to medium pH and to azoxy compounds in alkaline media; and at low pH they also exhibit a second 2-electron wave for reduction to the primary amine. With *ortho*- and *para*-substituents such as hydroxyl or amino, reduction is directly to the amine, via elimination from the hydroxylamine and immediate reduction of the resulting quinoid product.

$$\mathbf{29} \xrightarrow{2e^-} \mathbf{30}$$

$$\mathbf{30} \xrightarrow{H^+} \text{(N}^+H_2\text{, NH)} \xrightarrow[H^+]{2e^-} \text{(NH}_2\text{, NH}_2\text{)} \xrightarrow{H_2O} \text{cyclohexanone (=O)}$$

Lund has investigated the electrochemical behavior of 3,3-pentamethylenediazirine (**29**).[82] In acid (pH = 6), **29** exhibits a 4-electron polarographic wave; upon controlled-potential electrolysis 4 electrons/molecule are consumed, and cyclohexanone is the product. In alkali (pH = 9), a 2-electron wave is observed and the diaziridine **30** can be isolated in 80% yield from a preparative electrolysis. The results were interpreted as being due to the fact that diaziridine **30** is harder to reduce than **29** in alkali, while protonation of **30** in acid makes electrochemical cleavage of the N—N bond *easier* than the initial reduction. Hence, **30** is protonated and reduced to the *gem*-diamine immediately upon formation in acid. (This datum accounts for the failure to isolate diaziridines from some chemical reductions in acidic media.) It was suggested that diazirines generally can be reduced under alkaline conditions and, in view of the good yields obtainable, that electrochemical reduction may be the procedure of choice.

Lund and Iversen have reported use of the dianion of azobenzene as a synthetic intermediate in the Wittig reaction.[83] A mixture of azobenzene, benzyltriphenylphosphonium bromide, and benzaldehyde was electrolyzed in DMF containing lithium chloride. At the potential selected for electrolysis, the only species reduced is azobenzene, which in the presence of lithium ion is converted to its strongly basic dianion **31** (Section 4.2.2). The only proton donor in the mixture is the phosphonium salt, which reacts with **31** to generate ylide **32**. The latter then reacts with benzaldehyde to form stilbene in essentially quantitative yield.

$$C_6H_5N{=}NC_6H_5 \xrightarrow[LiCl]{2e^-} C_6H_5N(Li)N(Li)C_6H_5 \quad \mathbf{31}$$

$$\mathbf{31} + C_6H_5CH_2P^+(C_6H_5)_3Br^- \longrightarrow C_6H_5CH{=}P(C_6H_5)_3 \quad \mathbf{32}$$

$$\mathbf{32} + C_6H_5CHO \longrightarrow C_6H_5CH{=}CHC_6H_5 \quad (98\%)$$

Lund and Iversen pointed out that this procedure for "electrochemical generation of strong base" might be applied in many other base-catalyzed reactions. Its advantages include the following: (1) the use of a nucleophilic reagent such as phenyl lithium is avoided, (2) the amount of base used can be monitored by following the reaction with a coulometer in series with the cell, and (3) the steady-state concentration of base can easily be controlled by adjustment of the electrolysis current. Since azobenzene reduction takes place at a potential where many substituents are electroinactive, this clever application of simple electrochemical principles would appear capable of considerable further elaboration. Indeed, Iversen subsequently effected Stevens rearrangement of a quarternary ammonium salt using electrochemically generated base.[83b]

6.4.2 Diazo and Diazonium Compounds

Diazoacetophenone (**33**) exhibits three polarographic waves at pH 6, whose heights correspond to 6, 2, and 2 electrons, respectively[84]:

$$\underset{\mathbf{33}}{C_6H_5\underset{\substack{\| \\ O}}{C}CHN_2} \xrightarrow[\text{pH 6}]{6e^-} \underset{\mathbf{34}}{C_5H_5\underset{\substack{\| \\ O}}{C}CH_2NH_2} \xrightarrow{2e^-} C_6H_5\underset{\substack{\| \\ O}}{C}CH_3 \xrightarrow{2e^-} C_6H_5\underset{\substack{| \\ OH}}{C}HCH_3$$

Controlled-potential coulometry confirms these *n* values for the respective waves. ω-Aminoacetophenone (**34**) exhibits two polarographic waves at the same potentials as the second and third waves of **33**, and acetophenone shows a single 2-electron wave at the same potential as the third wave of **33** and second wave of **34**. Finally, addition of acetophenone results in an enhancement of the third polarographic wave of **33**. These results nicely confirm the reduction scheme outlined above. It has also been reported that reduction of **33** at pH 5 consumes only 2 electrons/molecule, and that acetophenone is the product.[85] This reaction apparently proceeds via the conjugate acid of the diazoketone:

$$C_6H_5COCHN_2 \xrightarrow{H^+} C_6H_5COCH_2N_2^+ \xrightarrow{2e^-}$$
$$N_2 + C_6H_5COCH_2^- \xrightarrow{H^+} C_6H_5COCH_3$$

The aliphatic diazoketone diazocamphor (**35**) also exhibits a single wave from pH 6 to 8.5 ($n = 6$ by polarography and coulometry), but in alkali (pH 11), this splits into an initial 2-electron wave, followed by one corresponding to 4 electrons.[86] Controlled-potential electrolysis at the first and second waves confirmed the *n*-value assignments and afforded the hydrazone (**36**) and 3-aminocamphor (**37**), respectively (no stereochemistry was suggested for the latter). Reduction of **35** proceeds directly to the amine in the pH region 6–8.5; it could not be investigated at pH lower than 6 because of the instability of **35** in acid.

$$\mathbf{35}\ (=N_2) \xrightarrow{2e^-} \mathbf{36}\ (=NNH_2) \xrightarrow{4e^-} \mathbf{37}\ (NH_2)$$

Diazonium salts in aqueous solution exhibit two polarographic waves, the first adsorption controlled. Microcoulometry[87] at a dropping mercury electrode indicates that the first wave corresponds to a 1-electron reduction; 4 electrons/molecule are consumed in reduction at the second wave.[88] Reduction products in aqueous media are either mercury compounds or phenylhydrazines, depending on potential and stirring rate.[89] In aprotic solvents, the first polarographic wave is diffusion controlled and corresponds to the process[90]

$$ArN_2^+ + e^- \longrightarrow ArN_2\cdot \longrightarrow Ar\cdot + N_2$$

There does not appear to be preparative value to reduction of diazonium salts in aqueous media. On the other hand, reduction of benzenediazonium tetrafluoborate in acetonitrile at a mercury cathode has been shown to be a useful method for free-radical phenylation of aromatic hydrocarbons.[91] Isomer ratios, total rate ratios, and partial rate factors agree with those found for phenylations with benzoyl peroxide and *N*-nitrosoacetanilide, demonstrating that there is no electrophilic component to the arylation. Yields of phenylated products are somewhat lower than with benzoyl peroxide or *N*-nitrosoacetanilide, but one is sure that the only radical generated is the phenyl radical, while complications by other species, such as benzoyloxy radical, appear with other radical precursors. The major side product in the electrode reaction is benzene, formed by hydrogen abstraction from the solvent. (Incidentally, the claim by these workers that the lower temperature limit for electrolysis at a mercury cathode is −35°C because mercury freezes at −39°C should be disregarded, since solid mercury can obviously also serve as an electrode.) Since diazonium salts are among the easiest compounds to reduce electrochemically, a large variety of aryl radicals could be generated by this technique.

6.4.3 The Sulfonyl Group

Electrochemical cleavage of sulfonamides and sulfones by tetramethylammonium amalgam has already been met in a previous chapter, as has reduction of sulfonate esters to alcohols (Section 5.2.2). Direct reduction of sulfones and sulfonamides is also feasible. Sulfones can be cleaved ($RSO_2R' \rightarrow RSO_2^- + R'H$) when at least one of the groups attached to sulfur is aryl, benzyl, or *alpha* to a carbonyl.[93, 94] This cleavage under mild neutral conditions is particularly interesting in view of the very vigorous conditions normally necessary to cleave sulfones. While the direction of cleavage of unsymmetrical sulfones by tetramethylammonium amalgam has been delineated by the work of Horner, less is known, and the situation is confused, in the case of direct electrochemical reduction. Horner[92] reports that arylalkyl sulfones ($ArSO_2R$) are cleaved by the amalgam into arylsulfinate ion and the hydrocarbon derived from the alkyl group. Bowers and Russell identified benzenesulfinate ion (quantitative yield) as the product of electrolysis of phenyl methyl sulfone in 50% ethanol containing tetramethylammonium bromide as electrolyte.[93] Manousek, Exner, and Zuman, on the other hand, found cleavage of the aryl–sulfur bond in a series of aryl methyl sulfones ($ArSO_2CH_3 \rightarrow ArH + CH_3SO_2^-$).[94a] This work

$$C_6H_5SO_2CH_3 \longrightarrow C_6H_5SO_2^- + CH_4 \qquad \textit{(Ref. 93)}$$

$$p\text{-}EtO_2C\text{-}C_6H_4\text{-}SO_2CH_3 \longrightarrow C_6H_5CO_2Et + CH_3SO_2^- \qquad \textit{(Ref. 94a)}$$

differed from that of Bowers and Russell in two ways: sodium and lithium salts were used as supporting electrolytes, and electron-withdrawing substituents were present on the aromatic ring in all cases. The differences between the two studies might therefore be due either to the fact that Bowers and Russell's work actually involved cleavage by tetramethylammonium amalgam or to a substituent effect on the direction of cleavage. Until more data are available, this discrepancy must remain unsolved. The cleavage of aryl–sulfur bonds in arylalkylsulfones cannot be effected by any chemical method, hence the reductions reported by Zuman and coworkers (who made no attempt to establish a quantitative material balance) are of obvious interest. The cleavage of the carbon–sulfur bond adjacent to the carbonyl group of β-ketosulfones, studied by Samuelsson and Lamm,[94b] may be superior to the chemical reductant, aluminum amalgam,[94c] often employed for this purpose.

6.4.4 Carbon–Carbon Multiple Bonds

It is impossible to reduce simple acetylenes and olefins by direct electron transfer at the cathode, probably because the lowest unfilled molecular orbital in these compounds is so high in energy that the necessary potentials cannot be reached before reduction of the solvent or electrolyte begins. It is possible, however, to reduce alkyl and dialkyl acetylenes cleanly to *cis*-olefins at a spongy nickel cathode in protic solvents.[95] The nature of the cathode material necessary for the conversion, and the fact that acetylenes show no voltammetric behavior at mercury even at the very negative potentials accessible in aprotic solvents, demonstrate that the reduction is more akin to catalytic hydrogenation than to reduction via electron transfer. The exact mechanism of reduction is not known, but it is notable that olefins cannot be reduced at all under the same conditions (even at a Raney nickel cathode) and that hydrogen evolution does not begin at the cathode until reduction of the acetylene is complete.[95] These facts suggest that reduction is initiated by adsorption of the acetylene upon the cathode surface.

Acetylenes can also be reduced to *cis*-olefins by hydrogenation over Lindlar catalyst,[96] but there may be reason to prefer the electrochemical procedure, in view of its experimental simplicity and the fact that reduction can be stopped cleanly at the olefin stage, without further reduction to the alkane. Clean selective reduction is not always possible using the Lindlar method.[96]

Acetylenes may also be reduced selectively to *trans*-olefins by electrochemically generated solvated electrons, that is, by electrolysis of a solution of lithium chloride in methylamine.[97] This procedure would appear to be superior to the use of alkali metals in an amine solvent for this purpose[98] when large-scale operations are intended, since only catalytic amounts of lithium chloride (and hence lithium metal) are involved.

References

[1]J. H. Stocker and R. M. Jenevein, *J. Org. Chem.*, **33**, 294 (1968).
[2a]M. J. Allen, *Organic Electrode Processes*, London, Chapman and Hall, 1958; [b]S. Swann, Jr., in A. Weissberger, ed., *Technique of Organic Chemistry*, 2nd ed., New York, Wiley, 1956, vol. 2, p. 385ff; [c]F. Fichter, *Organische Elektrochemie*, Dresden, Steinkopf, 1942 (out of print—available in Xerographic form from University Microfilms, Ann Arbor, Mich.).
[3]C. L. Perrin, *Progr. Phys. Org. Chem.*, **3**, 165 (1965).
[4]R. Pasternak, *Helv. Chim. Acta*, **31**, 753 (1948).
[5]P. J. Elving and J. T. Leone, *J. Amer. Chem. Soc.*, **80**, 1021 (1958).
[6]A. A. Pozdeeva and S. I. Zhdanov, in G. J. Hills, ed., *Polarography—1964*, New York, Wiley, 1966, vol. 2, p. 781.
[7]R. F. Michielli and P. J. Elving, *J. Amer. Chem. Soc.*, **90**, 1989 (1968).
[8]S. Wawzonek and A. Gundersen, *J. Electrochem. Soc.*, **107**, 537 (1960).
[9a]M. J. Allen, *J. Org. Chem.*, **15**, 435 (1950); [b]M. M. Baizer and J. P. Petrovich, *Progr. Phys. Org. Chem.*, **7**, 189 (1970); [c]P. Zuman, *Coll. Czech. Chem. Commun.*, **33**, 2548 (1968).
[10]D. H. Evans and E. C. Woodbury, *J. Org. Chem.*, **32**, 2158 (1967).
[11]R. M. Elofson, *J. Org. Chem.*, **25**, 305 (1960).
[12a]J. H. Stocker, R. M. Jenevein, and D. H. Kern, *J. Org. Chem.*, **34**, 2810 (1969); [b]J. H. Stocker and R. M. Jenevein, *J. Org. Chem.*, **33**, 2145 (1968); [c]J. H. Stocker and R. M. Jenevein, *J. Org. Chem.*, **34**, 2807 (1969); [d]*cf.* also E. Laviron, *Bull. Soc. Chim. Fr.*, **28**, 2325 (1961).
[13]V. J. Puglisi, G. L. Clapper, and D. H. Evans, *Anal. Chem.*, **41**, 279 (1969).
[14]E. Blomgren, J. O'M. Bockris, and C. Jesch, *J. Phys. Chem.*, **65**, 2000 (1961).

[15]L. Horner and D. Degner, *Tetrahedron Lett.*, 5889 (1968).
[16]R. W. Murray and M. Kodama, *Anal. Chem.*, **37**, 1759 (1965).
[17]P. Kabasakalian, J. McGlotten, A. Basch, and M. D. Yudis, *J. Org. Chem.*, **26**, 1738 (1961).
[18a]R. M. Powers and R. A. Day, Jr., *J. Org. Chem.*, **24**, 722 (1959); [b]W. Wiesener and K. Schwabe, *J. Electroanal. Chem.*, **15**, 73 (1967).
[19]R. G. Reed and A. J. Fry, unpublished observations.
[20]P. Kabasakalian and J. McGlotten, *Anal. Chem.*, **31**, 1091 (1959).
[21a]M. Schenck and H. Kirchhof, *Z. Physiol. Chem.*, **163**, 125 (1927); [b]M. Fedoronko, E. Fulleova, and K. Linek, *Coll. Czech. Chem. Commun.*, **36**, 114 (1971).
[22]L. Mandell, R. M. Powers, and R. A. Day, Jr., *J. Amer. Chem. Soc.*, **80**, 5284 (1958).
[23]A. J. Fry and R. G. Reed, *J. Amer. Chem. Soc.*, **91**, 6448 (1969).
[24]J. W. Huffman and J. T. Charles, *J. Amer. Chem. Soc.*, **90**, 6486 (1968).
[25]S. Swann, Jr., *Trans. Electrochem. Soc.*, **62**, 177 (1932).
[26]C. Schall and W. Kirst, *Z. Elektrochem.*, **29**, 537 (1923).
[27]W. R. J. Simpson, D. Bobbe, J. A. Edwards, and J. H. Fried, *Tetrahedron Lett.*, 3209 (1967).
[28]L. Throop and L. Tokes, *J. Amer. Chem. Soc.*, **89**, 4789 (1967).
[29]W. P. Neumann and K. Kuhlein, *Adv. Organometallic Chem.*, **7**, 241 (1968).
[30]H. Lund, *Acta Chem. Scand.*, **17**, 972 (1963).
[31a]P. E. Iversen and H. Lund, *Acta Chem. Scand.*, **21**, 279, 389 (1967); [b]P. E. Iversen, *Acta Chem. Scand.*, **24**, 2459 (1970).
[32]G. H. Coleman and H. L. Johnson, *Org. Syn.*, **21**, 10 (1941).
[33a]G. C. Whitnack, J. Reinhart, and E. St. C. Gantz, *Anal. Chem.*, **27**, 359 (1955); [b]R. A. Benkeser, H. Watanabe, S. Mels, and M. A. Sabol, *J. Org. Chem.*, **35**, 1210 (1970).
[34]M. J. Allen and J. Ocampo, *J. Electrochem. Soc.*, **103**, 452 (1956).
[35a]O. R. Brown and K. Lister, *Disc. Faraday Soc.*, **45**, 106 (1968); [b]M. M. Baizer and J. L. Chruma, *J. Electrochem. Soc.*, **118**, 450 (1971); [c]K. Sugino and T. Nonaka, *J. Electrochem. Soc.*, **112**, 1241 (1965).
[36]H. Lund, *Acta Chem. Scand.*, **13**, 249 (1959).
[37]A. J. Bard, private communication.
[38]P. Zuman and O. Exner, *Coll. Czech. Chem. Commun.*, **30**, 1832 (1965).
[39]H. D. Law, *J. Chem. Soc.*, **101**, 154 (1912).
[40a]C. P. Andrieux and J.-M. Saveant, *Bull. Soc. Chim. Fr.*, **35**, 4671 (1968); [b]C. P. Andrieux and J.-M. Saveant, *J. Electroanal. Chem.*, **26**, 223 (1970); **28**, 446 (1970).
[41a]J. M. W. Scott and W. H. Jura, *Can. J. Chem.*, **45**, 2375 (1967); [b]V. D. Bezuglyi, L. V. Kononenko, A. F. Korunova, V. N. Dmitrieva, and B. L. Timan, *J. Gen. Chem. U.S.S.R.*, **39**, 1647 (1969); [c]P. Martinet, J. Simonet, and J. Tendil, *Compt. Rend.*, **268C**, 303 (1969).
[42]L. Horner and D. H. Skaletz, *Tetrahedron Lett.*, 1103, 3679 (1970).
[43]M. J. Peover, in A. J. Bard, ed., *Electroanalytical Chemistry*, New York, Marcel Dekker, 1967, vol. 2, p. 1ff.
[44a]P. Souchay and S. Ser, *J. Chim. Phys.*, **49C**, 172 (1952); [b]H. J. Gardner and W. P. Georgans, *J. Chem. Soc.*, 4180 (1956).
[45]H. Lund, *Acta Chem. Scand.*, **18**, 563 (1964).
[46]A. J. Fry and J. Newberg, *J. Amer. Chem. Soc.*, **89**, 6374 (1967).
[47]P. Zuman and O. Exner, *Coll. Czech. Chem. Commun.*, **30**, 1832 (1965).
[48]R. A. Benkeser and S. J. Mels, *J. Org. Chem.*, **35**, 271 (1970).
[49]P. J. Elving and R. E. Van Atta, *J. Electrochem. Soc.*, **103**, 676 (1956).
[50a]J. Pearson, *Trans. Faraday Soc.*, **44**, 683 (1948); [b]I. Bergman and F. C. James, *Trans. Faraday Soc.*, **50**, 60 (1954).

[51]D. H. Geske and A. H. Maki, *J. Amer. Chem. Soc.*, **82**, 2671 (1960).
[52]L. H. Piette, P. Ludwig, and R. N. Adams, *Anal. Chem.*, **34**, 916 (1962).
[53]L. Holleck and H. J. Exner, *Z. Elektrochem.*, **56**, 46, 677 (1952).
[54a]E. Bamberger and E. Renauld, *Chem. Ber.*, **30**, 2278 (1897); [b]G. A. Russell, E. J. Geels, F. J. Smentowski, K.-Y. Chang, J. Reynolds, and G. Kaupp, *J. Amer. Chem. Soc.*, **89**, 3821 (1967).
[55]G. A. Russell, R. Konaka, E. T. Strom, W. C. Danen, K.-Y. Chang, and G. Kaupp, *J. Amer. Chem. Soc.*, **90**, 4646 (1968), and references therein.
[56a]C. Kerns, *Trans. Electrochem. Soc.*, **62**, 183 (1932); [b]C. Weygand and R. Gabler, *J. Prakt. Chem.*, **155**, 332 (1940).
[57]K. Elbs and O. Kopp, *Z. Elektrochem.*, **5**, 108 (1898).
[58a]G. Costa, *Gazz. Chim. Ital.*, **83**, 875 (1953); [b]L. Holleck and A. M. Shams El Din, *Electrochim. Acta*, **13**, 199 (1968).
[59]S. Wawzonek and J. D. Frederickson, *J. Amer. Chem. Soc.*, **77**, 3985 (1955).
[60]P. Zuman, *Substituent Effects in Organic Polarography*, New York, Plenum, 1967.
[61]O. D. Shreve and E. C. Markham, *J. Amer. Chem. Soc.*, **71**, 2993 (1949).
[62]D. Stocesova, *Coll. Czech. Chem. Commun.*, **14**, 615 (1949).
[63]M. E. Runner and E. C. Wagner, *J. Amer. Chem. Soc.*, **74**, 2529 (1952).
[64]G. S. Alberts and I. Shain, *Anal. Chem.*, **35**, 1859 (1963).
[65]A. C. Testa and W. H. Reinmuth, *J. Amer. Chem. Soc.*, **83**, 784 (1961).
[66a]G. A. Gilbert and E. K. Rideal, *Trans. Faraday Soc.*, **47**, 396 (1951); [b]H. A. Laitinen and T. J. Kneip, *J. Amer. Chem. Soc.*, **78**, 736 (1956).
[67a]G. G. McKeown and J. L. Thomson, *Can. J. Chem.*, **32**, 1025 (1954); [b]T. M. Florence and G. H. Aylward, *Austr. J. Chem.*, **15**, 65, 416 (1962).
[68a]S. D. Ross, G. J. Kahan, and W. A. Leach, *J. Amer. Chem. Soc.*, **74**, 4122 (1952); [b]H. Lund, *Preprints of Papers, Symposium on Synthetic and Mechanistic Aspects of Electro-Organic Chemistry*, Army Research Center, Durham, N.C., October 1968, p. 197; [c]H. Lund and L. G. Feoktistov, *Acta Chem. Scand.*, **23**, 3482 (1962); [d]S. Kwee and H. Lund, *Acta Chem. Scand.*, **23**, 2711 (1969); [e]H. Lund, *Disc. Faraday Soc.*, **45**, 193 (1968).
[69]K. Gleu and K. Pfannstiel, *J. Prakt. Chem.*, **146**, 129 (1936).
[70a]N. Radin and T. de Vries, *Anal. Chem.*, **24**, 971 (1952); [b]L. Holleck and H. Schmidt, *Z. Elektrochem.*, **59**, 56 (1955).
[71a]W. J. Seagers and P. J. Elving, *J. Amer. Chem. Soc.*, **72**, 5183 (1950); [b]S. Deswarte and J. Armand, *Compt. Rend.*, **258C**, 3865 (1964).
[72]A. K. Hoffmann, W. G. Hodgson, D. L. Maricle, and W. H. Jura, *J. Amer. Chem. Soc.*, **86**, 631 (1964).
[73]R. H. Gibson and J. C. Crosthwaite, *J. Amer. Chem. Soc.*, **90**, 7373 (1968).
[74a]M. Masui, H. Sayo, and Y. Nomura, *Pharm. Bull.* (*Tokyo*), **4**, 332 (1956); see *Chem. Abstr.*, **51**, 9370 (1957); [b]P. E. Iversen, private communication.
[75a]K. H. Slotta and R. Kethur, *Chem. Ber.*, **71**, 62 (1938); [b]E. Mosettig and E. L. May, *J. Amer. Chem. Soc.*, **60**, 2964 (1938).
[76a]A. Tallec, *Ann. Chim.* (*Fr.*), [14], **4**, 67 (1969); [b]A. Tallec and M.-J. Gueguen, *Compt. Rend.*, **262C**, 484 (1966); [c]A. Tallec, *Compt. Rend.*, **262C**, 1886 (1966); [d]A. Tallec, *Compt. Rend.*, **263C**, 722 (1966); [e]A. Tallec, *Compt. Rend.*, **261C**, 2915 (1965).
[77]H. Slotta and R. Kethur, *Chem. Ber.*, **71**, 335 (1938).
[78]V. Bruckner and G. von Fodor, *Chem. Ber.*, **76**, 466 (1943).
[79]L. H. Welsh, *J. Org. Chem.*, **32**, 119 (1967).
[80]R. E. Harman, *ibid.*, **17**, 1047 (1952).
[81a]D. M. Oglesby, J. D. Johnson, and C. N. Reilley, *Anal. Chem.*, **38**, 385 (1966); [b]W. M. Schwarz and I. Shain, *J. Phys. Chem.*, **69**, 30 (1965).

[82]H. Lund, *Coll. Czech. Chem. Commun.*, **31**, 4175 (1966).
[83a]P. E. Iversen and H. Lund, *Tetrahedron Lett.*, 3523 (1969); [b]P. E. Iversen, *Tetrahedron Lett.*, 55 (1971).
[84a]D. M. Coombs and L. L. Leveson, *Anal. Chim. Acta*, **30**, 209, (1964); [b]M. Bailes and L. L. Leveson, *J. Chem. Soc.*, (B), 34 (1970).
[85]A. Foffani, L. Salvagnini, and C. Pecile, *Ann. Chim.* (*Rome*), **49**, 1677 (1959).
[86]M. E. Cardinali, I. Carelli, and A. Trazza, *J. Electroanal. Chem.*, **23**, 399 (1969).
[87]L. Meites, *Polarographic Techniques*, 2nd ed., New York, Wiley, 1965, pp. 532–535.
[88]R. M. Elofson, *Can. J. Chem.*, **36**, 1207 (1958).
[89]P. Ruetschi and G. Trumpler, *Helv. Chim. Acta*, **36**, 1649 (1953).
[90]R. M. Elofson and F. F. Gadallah, *J. Org. Chem.*, **34**, 854 (1969).
[91a]F. F. Gadallah and R. M. Elofson, *J. Org. Chem.*, **34**, 3335 (1969); [b]R. M. Elofson, F. F. Gadallah, and K. F. Schulz, *J. Org. Chem.*, **36**, 1526 (1971).
[92]L. Horner and H. Neumann, *Chem. Ber.*, **98**, 1715 (1965).
[93]R. C. Bowers and H. D. Russell, *Anal. Chem.*, **32**, 405 (1960).
[94a]O. Manousek, O. Exner, and P. Zuman, *Coll. Czech. Chem. Commun.*, **33**, 3988 (1968); [b]B. Lamm and B. Samuelsson, *Acta Chem. Scand.*, **24**, 561, 3070 (1970); [c]E. J. Corey and M. Chaykovsky, *J. Amer. Chem. Soc.*, **87**, 1345 (1965).
[95]K. N. Campbell and E. E. Young, *J. Amer. Chem. Soc.*, **65**, 965 (1943).
[96]H. Lindlar and R. Dubuis, *Org. Syn.*, **46**, 89 (1966).
[97]R. A. Benkeser and C. A. Tincher, *J. Org. Chem.*, **33**, 2727 (1968).
[98]H. O. House, *Modern Synthetic Reactions*, Menlo Park, Calif., Benjamin, 1965, pp. 71–72.

7 ELECTROCHEMICAL REDUCTION OF CONJUGATED SYSTEMS

The electrochemical behavior of compounds containing extended pi systems is, in its general outlines, much like that of the functional groups discussed in the previous chapter. That is, reduction is generally initiated by injection of an electron into the lowest unfilled molecular orbital of the substrate. The fate of the resulting species (a neutral radical in acid or a radical anion in neutral, alkaline, or aprotic media) will depend on a number of circumstances such as pH, steric factors, and reduction potential. Nevertheless, there are enough differences between the electrochemical behavior of pi systems and simple unsaturated functional groups to justify separate discussion of each. Because of extensive resonance delocalization, radicals from pi systems may and often do react at more than one center. Furthermore, it appears that concerted electron transfer to two substrate molecules with generation of a dimeric intermediate in the primary electrochemical step may be energetically feasible in some instances. Dimerization is, in fact, generally more significant with extended pi systems than with simpler functional groups. This is actually quite fortunate, since the resulting dimers are often difficult to obtain by other routes.

7.1 ACTIVATED OLEFINS

The term *activated olefin* may be defined for present purposes as a compound containing a double bond to which is attached any of a variety of electronegative substituents capable of conjugation, and generally includes all compounds capable of acting as acceptors in the Michael reaction. Reduction of such com-

$$\text{CH}_2{=}\text{CH}{-}X \longrightarrow X{-}\text{CH}_2\text{CH}_2\text{CH}_2\text{CH}_2{-}X$$

1

($X = CN, COR, CO_2R, C_6H_5$, etc.)

pounds usually, but not always, results in products of formula **1**. This important carbon–carbon bond-forming reaction is known as *hydrodimerization*. There are at least two, and probably three, distinct pathways by which hydrodimerization may proceed, depending on experimental conditions and the structure of the activated olefin. In the discussions that follow, the nature of these pathways is examined and structural effects and synthetic utility of the hydrodimerization reaction are examined. It will be helpful for mechanistic analysis to organize the extensive literature in this area according to whether reduction of the activated olefin is carried out in acid, neutral and alkaline protic solvents, or aprotic media.

7.1.1 Behavior in Acid

A wide variety of acyclic and cyclic α,β-unsaturated ketones, such as cyclohexenone,[1] mesityl oxide,[2] and cortisone[3] (a cross-conjugated cyclohexadienone), have been found to exhibit a single 1-electron polarographic wave in acidic media. As the pH is increased, the height of this wave diminishes and a second wave appears at more negative potentials and becomes larger with increasing pH. In alkali, only the second wave is observed. The total wave height remains constant during this process, demonstrating that the second wave is also a 1-electron wave. The half-wave potential of the first wave is pH dependent; that of the second wave is not. Discussions of the reduction process in neutral solution will be deferred until the next section. The first wave is due to reduction of the conjugate acid of the ketone. For example, for mesityl oxide:

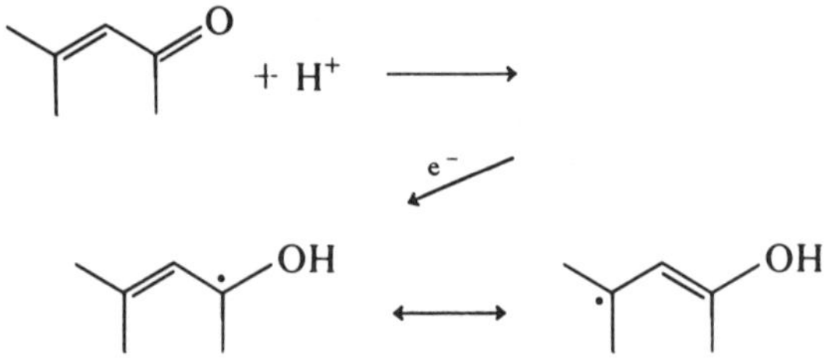

O + H^+ ⟶

e^-

OH ⟷ OH

2

It appears that the principal mode of reaction of such radicals is tail to tail, such that a diketone is initially formed. Pasternak, for example, reported the isolation of 1,6-diones from reduction of benzalacetophenone, benzalacetone, and dibenzalacetone at pH 1.3.[2] The structures of the products were not proved rigorously, however, and another of his reports has since been shown to be in error. He reported formation of 4,4,5,5-tetramethyl-2,7-octanedione (**3**) from reduction of mesityl oxide at pH 1.3, but Wiemann and Bouguerra later showed that under these conditions the initial product **3** undergoes *in situ* acid-catalyzed intramolecular aldol condensation to produce a mixture from which ketol **4** and cyclopentenone **5** can be isolated in 20 and 62% yields, respectively.[4] Ketol **4** is the

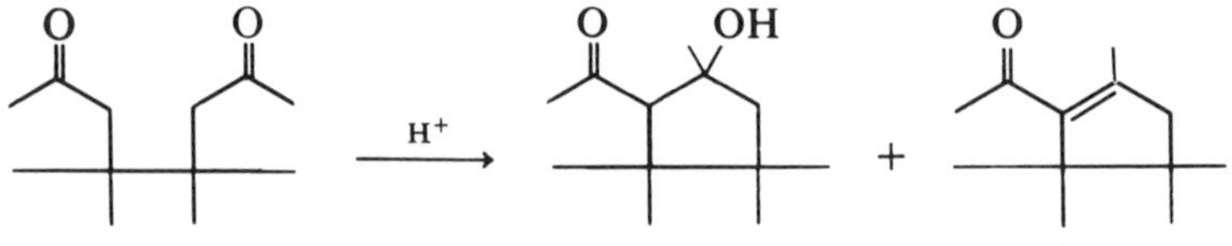

O O $\xrightarrow{H^+}$ O OH + O

3 **4** **5**

major product (62%) when reduction is carried out in weak alkali. In stronger alkali (pH 12), it is dehydrated and **5** is the major product (60%). Dione **3** can be isolated in fair yield (56%) by electrolysis in neutral solution, but a better method for its synthesis will be seen later (Section 7.1.3).

A small amount of 2,4,4,7-tetramethylocta-2,6-dien-5-one (**7**) is also isolated from reduction of mesityl oxide at pH 1.3.[4] This apparently arises through head-to-head coupling of **2** to produce a pinacol (**6**) that undergoes *in situ* acid-catalyzed rearrangement to **7**.*

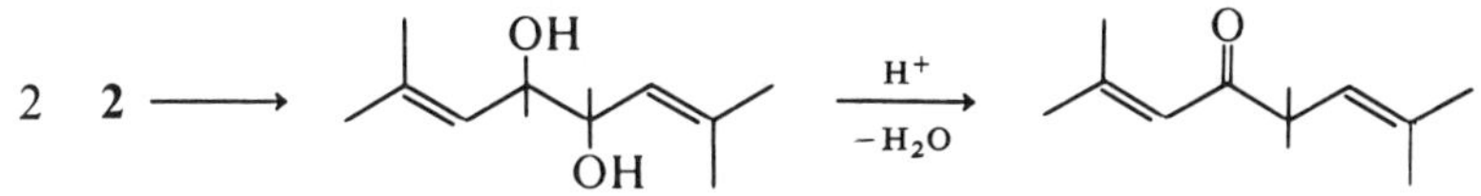

2 **2** ⟶ **6** $\xrightarrow[-H_2O]{H^+}$ **7**

It appears that pinacol formation becomes the major reaction path when tail-to-tail coupling would create severe steric crowding. A number of steroidal enones and dienones, for example, are reduced to pinacols (which can undergo acid-catalyzed dehydration),[3,8,9] and **8** is converted to its pinacol (**9**) in essentially quantitative (97%) yield.[10]

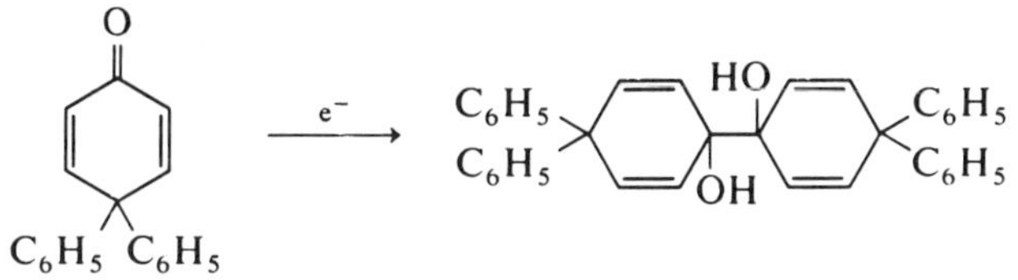

8 $\xrightarrow{e^-}$ **9**

As we have now seen, diones or their dehydration products, such as **5**, constitute the principal reduction products of α,β-unsaturated ketones in acid. Only if this mode of coupling is sterically unfavorable is pinacol formation observed. Other products have occasionally been observed, however. Holleck and Marquarding found that reduction of methyl vinyl ketone at a mercury cathode in acid (pH < 5) affords the mercury compound **10** in good yield,[11] a

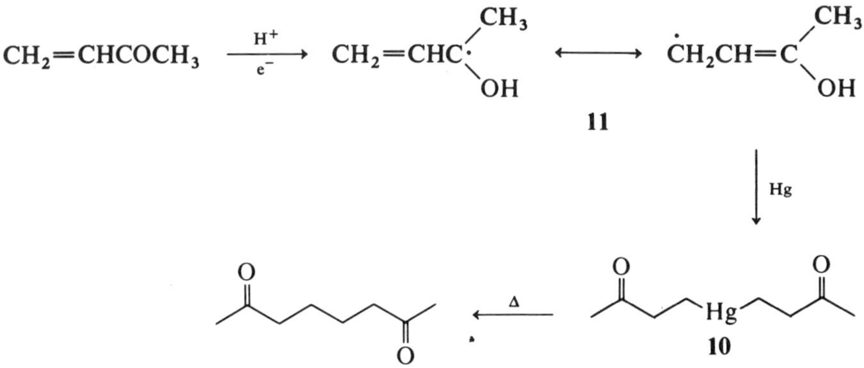

$$CH_2{=}CHCOCH_3 \xrightarrow[e^-]{H^+} CH_2{=}CH\dot{C}(CH_3)OH \longleftrightarrow \dot{C}H_2CH{=}C(CH_3)OH$$

11

$\downarrow$ Hg

10 $\xrightarrow{\Delta}$

* Preferential migration of a methyl group over a vinyl group in this pinacol rearrangement[5] is surprising. Another exception to the usual rule[6] that sp^2 carbon migrates more easily than sp^3 carbon has been reported by McMurry.[7]

result that has subsequently been confirmed by other investigators.[12,13] Under the same conditions, mesityl oxide does not yield a mercury compound; instead, as we have seen, it is converted to **4** and **5** via the diketone **3**.[4] There are several possible reasons for this difference. It may be that radical **11** is so reactive that it combines with mercury at the electrode surface immediately upon formation, while the *gem*-dimethyl groups in **2** provide the added stability to permit its escape from the electrode. The alternative possibility exists that mesityl oxide does actually form a mercury compound analogous to **10**, which decomposes to dione **3** under the reaction conditions. This possibility is suggested by the discovery that **10** decomposes thermally to 2,7-octanedione in high yield at temperatures not much over 100°C.[12,13] The corresponding mercury compound from mesityl oxide might very well be more thermally sensitive than **10**.

Another unexplained finding in this area lies in the report by Wiemann and Bouguerra that reduction of mesityl oxide in acetic acid containing acetic anhydride produces the dihydrofuran **12** in high (88 %) yield.[4] Compound **12** formally arises through head-to-tail coupling of **2**, but it is not clear why head-to-tail coupling should occur under these conditions while in aqueous acid, as mentioned above, head-to-head and tail-to-tail coupling to produce **6** and **3**, respectively, are observed. The function of acetic anhydride in the synthesis of **12** was thought to be as a water scavenger, but it may also serve to acetylate intermediates in the electrolysis. It is not obvious, at any rate, why the presence of water should affect the mode of coupling of **2**.

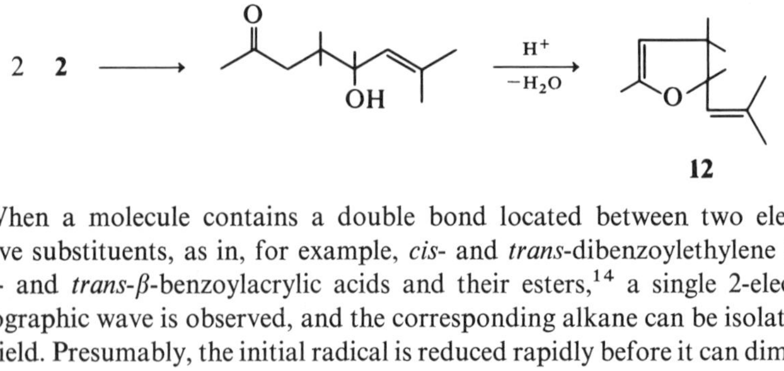

2 **2** ⟶ O ... OH $\xrightarrow[-H_2O]{H^+}$ O

12

When a molecule contains a double bond located between two electronegative substituents, as in, for example, *cis*- and *trans*-dibenzoylethylene (**13**)[2] or *cis*- and *trans*-β-benzoylacrylic acids and their esters,[14] a single 2-electron polarographic wave is observed, and the corresponding alkane can be isolated in high yield. Presumably, the initial radical is reduced rapidly before it can dimerize because of its location adjacent to an electronegative substituent, that is, for **13**,*

$$C_6H_5COCH{=}CHCOC_6H_5 \xrightarrow[e^-]{H^+}$$

13

$$C_6H_5CO\dot{C}HCH_2COC_6H_5 \xrightarrow[(2)\ H^+]{(1)\ e^-\ (fast)} (C_6H_5COCH_2)_2$$

* The reduction might also proceed via a *bis*-enol:

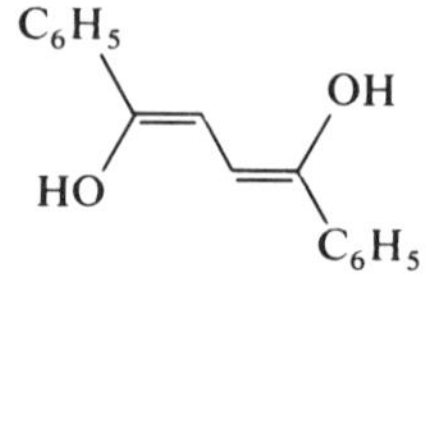

C_6H_5 OH HO C_6H_5

It should be noted parenthetically here that in every such case studied the *trans*-isomer is easier to reduce than the *cis*-isomer. This generalization appears to be firm enough to be used as a criterion for assigning configurations to an isomeric pair of unknown geometry. The stereochemistry of the reduction is of some interest. Elving and coworkers have reported that dimethylmaleic and dimethylfumaric acids are reduced stereospecifically to *dl*- and *meso*-dimethylsuccinic acids, respectively.[15] If true (product analysis was only by determination of the melting point of the electrolysis products), these experiments represent overall stereospecific *trans*-addition of hydrogen to the double bond, but the factors controlling stereochemistry are not obvious. In the only other study in this area, it was found that cyclohexene-1,2-dicarboxylic acid is converted to the *cis*-diacid upon electrochemical reduction.[16]

Reduction of activated olefins takes a completely different course in acid when the pi system contains a cyano group. For example, reduction of *p*-cyanocinnamic acid or its ethyl ester produces the primary amine in acid.[17] In alkali, reduction of the double bond takes place.[17] Cinnamic acid itself and its esters, on the other hand, undergo hydrodimerization in acid, as expected.[18,19,44] The neighboring pi system assists reduction of the cyano group, since saturated aliphatic nitriles are not electroactive under the same conditions. Smirnov and Tomilov took advantage of this fact to effect selective reduction of the conjugated cyano group in 1,4-dicyano-1-butene at a lead cathode in strong acid.[20]

The electrochemical reduction of derivatives of conjugated ketones and aldehydes is of some synthetic interest. The oximes of crotonaldehyde, mesityl oxide, cinnamaldehyde, and benzalacetone exhibit a single polarographic wave, which is diffusion controlled at low pH and kinetically controlled in solutions ranging from weakly acid to weakly alkaline.[21] (In strong alkali, no wave is observed because of conversion of the oximes to their electroinactive conjugate bases.) Reduction of the oximes of cinnamaldehyde and benzalacetone in strong acid consumed 4 electrons/molecule and afforded the corresponding amines as the principal product, along with a small amount of unidentified saturated carbonyl compounds, possibly hydrodimers of the corresponding unsaturated parent carbonyl compounds:

$$C_6H_5CH{=}CH\underset{\displaystyle R}{\underset{|}{C}}{=}\overset{+}{\underset{\displaystyle H}{\underset{|}{N}}}OH \xrightarrow{2e^-}$$

$$C_6H_5CH{=}CH\underset{\displaystyle R}{\underset{|}{C}}{=}NH \xrightarrow{2e^-} C_6H_5CH{=}CH\underset{\displaystyle R}{\underset{|}{C}}HNH_2$$

$$\xrightarrow{H_2O} C_6H_5CH{=}CH\underset{\displaystyle R}{\underset{|}{C}}{=}O \xrightarrow[H^+]{e^-} \text{hydrodimer}$$

Evidence for this mechanism comes from the oximes of carvone and testosterone propionate, for which in strong acid the usual 4-electron wave is resolved into two 2-electron waves.[21] Reduction of testosterone propionate oxime at the first wave afforded testosterone propionate in good yield, indicating that the first step in reduction is cleavage of the N—O bond. The hydrolysis of imine intermediates to carbonyl compounds could presumably be averted by electrolysis in anhydrous acidic solvents, such as methanol or DMF containing hydrogen chloride. Reduction of the oximes in alkali is less interesting—amines are formed only in low yield, and the principal products are unidentified saturated carbonyl compounds, presumably hydrodimers and the saturated monomers. On the other hand, the semicarbazones of cinnamaldehyde and benzalacetone exhibit considerably different behavior in alkali.[21] Two electrons per molecule are consumed, and the *saturated* semicarbazone can be isolated in good yield. This reaction may be of interest for reducing the double bond of α,β-unsaturated

$$C_6H_5CH{=}CH\underset{\displaystyle R}{\underset{|}{C}}{=}NNHCONH_2 \xrightarrow{2e^+} C_6H_5CH_2CH_2\underset{\displaystyle R}{\underset{|}{C}}{=}NNHCONH_2$$

carbonyl compounds when operations are being conducted on a small scale, since the product semicarbazones are nicely crystalline and hence easily isolable. Notice the contrast between the behavior of conjugated oximes in acid, where amines are formed, and conjugated semicarbazones in alkali, where reduction of the double bond takes place. Presumably, reduction of the semicarbazones in acid (anhydrous) would also afford the corresponding amine.

7.1.2 Neutral and Alkaline Media

Mechanisms

The mechanisms of electrochemical reduction of activated olefins in neutral and weakly alkaline media have been the object of considerable attention in recent years, excited by the realization that not only is the hydrodimerization reaction a powerful and versatile carbon–carbon bond-forming reaction in the organic laboratory, but also that it can be of commercial importance. Large amounts of the synthetic intermediate adiponitrile are presently prepared through electrochemical hydrodimerization of acrylonitrile at the Monsanto Chemical Co., where a research group headed by M. M. Baizer has contributed much to current knowledge of the mechanism of hydrodimerization.

Certain features of the hydrodimerization reaction in weakly alkaline and neutral media demonstrate its mechanistic complexity. Acrylonitrile exhibits a single 2-electron polarographic wave under such conditions, suggesting an ECE mechanism, that is, stepwise addition of an electron, proton, electron, and finally a second proton to yield propionitrile (**15**).[22] Beck has demonstrated, however, that a molecule of water is also involved in the rate-determining reduction step,[23] indicating that the initial electron and proton transfers are concerted:

$$CH_2=CHX + e^- + H_2O \longrightarrow \left[\begin{matrix} H_2C \text{---} CHX \\ \quad\quad\; \vdots \\ \quad\quad H\text{---}OH \end{matrix}\right]^{\dot{-}} \longrightarrow \bar{O}H + \cdot CH_2CH_2X$$

14

$$\cdot CH_2CH_2X + e^- \longrightarrow \bar{C}H_2CH_2X \xrightarrow{H_2O} CH_3CH_2X$$

15, X = CN

In the face of this apparently well-established pathway of polarographic reduction of acrylonitrile, it is surprising to find that the major product isolated from preparative-scale electrolysis of acrylonitrile in aqueous media is not propionitrile, but adiponitrile (**16**). The reason for the difference lies, of course, in the more concentrated solutions employed in the preparative electrolysis. Under these conditions a pathway which might appear attractive is

$$CH_2{=}CHCN \xrightarrow{e^-} \bar{C}H_2\dot{C}HCN \xrightarrow{CH_2=CHCN} \begin{matrix} CH_2\bar{C}HCN \\ | \\ CH_2\dot{C}HCN \end{matrix}$$

17 **18**

$$\mathbf{18} \xrightarrow[2H_2O]{e^-} NC(CH_2)_4CN$$

16

Petrovich, Anderson, and Baizer pointed out a major flaw in this mechanism, however: it requires anion **17** to react preferentially with acrylonitrile, while anion **18** must react preferentially with water.[24] A way out of this paradox was suggested by the experiments that these investigators carried out with *bis*-activated olefins (**19**). It was found that, depending on the nature of Z, certain

$$Z\begin{matrix} \diagup CH{=}CX_2 \\ \diagdown CH{=}CX_2 \end{matrix} \longrightarrow Z\begin{matrix} \diagup CHCHX_2 \\ \diagdown CHCHX_2 \end{matrix} \text{(ring)} + Z\begin{matrix} \diagup CH_2CHX_2 \\ \diagdown CH_2CHX_2 \end{matrix} + \text{oligomers}$$

19 **20** **21**

bis-activated olefins produced only cyclic products (**20**), others produced a mixture of **20** and tetrahydro compound **21**, and from others no cyclic product could be isolated. The important discovery was made that polarographic half-wave potentials for compounds producing cyclic product **20** in high yield (*intramolecular hydrocyclization*) are shifted *anodically relative to those of model compounds.* Furthermore, compounds affording a mixture of **20** and **21** upon preparative-scale reduction exhibit two polarographic waves, and the relative heights of the first and second waves are in the same ratio as the ratio of **20** to **21** produced upon electrolysis. The anodic shift is clearly associated with cyclization. It does not arise from an inductive effect of one activated olefin upon the other as in, for example, *m*-dinitrobenzene (Chapter 6), since it is not a function of chain length. Rather, the anodic shift (and high yields of **20**) arise only when it is possible to align the two double bonds of **19** parallel and with the β-carbon atoms ca. one carbon–carbon bond length apart.[24] (It is therefore possible to predict the likelihood of efficient *intra*molecular hydrodimerization from examination of molecular models.) These experiments led Petrovich, Anderson, and Baizer to suggest a reduction mechanism involving electron transfer concerted with bond formation:

$$\mathbf{19} \xrightarrow{e^-} \left[Z\begin{matrix} CH_2 \overset{...}{=} CX_2 \\ \vdots \\ CH \overset{...}{=} CX_2 \end{matrix} \right]^{\dot{-}} \longrightarrow Z\begin{matrix} CH\bar{C}X_2 \\ | \\ CH\dot{C}X_2 \end{matrix} \xrightarrow{e^-} Z\begin{matrix} CH\bar{C}X_2 \\ | \\ CH\bar{C}X_2 \end{matrix}$$

22 **23**

$$\mathbf{23} \xrightarrow{2H^+} \mathbf{20}$$

The energy gained by delocalization in transition state **22** is responsible for the anodic shift and is analogous to the anodic shift in half-wave potential observed for *vic*-dihalides that undergo concerted reduction (Section 5.1). In *bis*-olefins **19** which cannot assume the proper geometry for this concerted transition state, the double bonds are reduced independently at more negative potentials via transition state **14**, and **21** is the product. It was suggested that these experiments are relevant to the mechanism of bimolecular hydrodimerization. The apparent preference for dimerization relative to saturation of the double bond can be understood if the transition state for reduction of an activated olefin is lowered more through participation by a second olefin molecule (transition state **24**) than through participation by water (transition state **14**). In dilute aqueous solution,

$$2CH_2{=}CHX \xrightarrow{e^-} \left[\begin{matrix} CH_2 \overset{...}{=} CHX \\ \vdots \\ CH_2 \overset{...}{=} CHX \end{matrix} \right]^{\dot{-}} \longrightarrow \begin{matrix} CH_2\bar{C}HX \\ | \\ CH_2\dot{C}HX \end{matrix} \xrightarrow{e^-} \begin{matrix} CH_2\bar{C}HX \\ | \\ CH_2\bar{C}HX \end{matrix}$$

24 **25**

$$\mathbf{25} \xrightarrow{2H^+} \mathbf{1}$$

24 should be disfavored relative to **14**, and reduction to the saturated compound should be observed. As has been noted already, this is exactly what happens when acrylonitrile is examined in the very dilute (ca 1 m*M*) solutions used for polarography.[22] At very high olefin concentrations, the water concentration near the electrode will be low, and anions such as **23** or **25** can initiate anionic polymerization of the starting material. This side reaction appears to be the source of the oligomers that often accompany hydrodimerization products. One is therefore faced with the experimental problem of electrolyzing at high enough olefin concentrations for dimerization to compete successfully with saturation of the double bond, while still maintaining a high enough water content in the solvent to protonate carbanions as formed. Baizer and his coworkers have solved this problem by the use of the so-called *hydrotropic solvents* or *hydrotropes*, that is, certain concentrated aqueous salt solutions that are excellent solvents for organic compounds (Chapter 9).[25]

Baizer and his coworkers have considerably enhanced the synthetic usefulness of the hydrodimerization reaction through introduction of the concept of *mixed* reductive coupling of two different activated olefins.[26] In principle, such reductions can afford three different products:

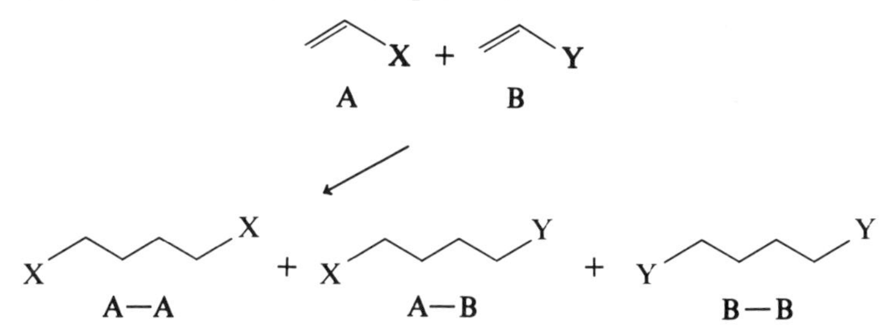

However, if A is easier to reduce than B by at least 0.2 V and electrolysis is carried out at the reduction potential of A, then only A—A and A—B are formed. The yield of A—B relative to A—A can be enhanced by employing B in large excess. Ease of coupling is related to the tendency of B to act as an acceptor in the Michael reaction, not to reduction potential, hence no A—B is observed if B is a poorer Michael acceptor than A. For example, reduction of acrylonitrile (−1.9 V) and styrene (−2.3 V) at a potential sufficient to reduce only acrylonitrile results in formation of adiponitrile and acrylonitrile oligomers, but only a trace of 5-phenylvaleronitrile, the mixed product.[27] Conversely, the best yields of mixed dimer are to be expected when the more easily reduced olefin (the donor) is the poorer Michael acceptor of the pair. This is exemplified by reduction of a mixture of diethyl fumarate (−1.2 V) and methyl vinyl ketone (−1.4 V), which affords **26** as the principal product in good yield.[26b]

$$CH_3COCH_2CH_2\underset{\displaystyle CO_2Et}{\underset{|}{C}}HCH_2CO_2Et$$

26

Structural Effects

It may be helpful to examine at this point some trends in the electrochemical reduction of activated olefins as a function of structure, as was done for simple ketones in Chapter 6. This analysis will clarify some limitations on the hydrodimerization reaction.

Cyclohexenones and cross-conjugated cyclohexadienones, such as **8** and a number of $\Delta^{1,4}$- and Δ^4-3-keto steroids (cortisone, hydrocortisone, prednisolone, prednisone, cholestenone, etc.), exhibit a 1-electron wave in neutral and alkaline media.[1, 3, 9] As in acid, pinacols, rather than hydrodimers, are the products from preparative electrolysis. Apparently, steric crowding at the β-carbon does not permit the usual hydrodimerization. The mechanism of pinacol formation has not been investigated. It could involve a concerted bimolecular reduction; thus, for reduction of **8** in alkali,

$$2\ \mathbf{8} + e^- \longrightarrow \left[(C_6H_5)_2C\langle\text{ring}\rangle C(O)\cdots C(O)\langle\text{ring}\rangle C(C_6H_5)_2 \right]^{\dot{-}}$$

27

As a result of its extensive electronic delocalization transition state, **27** should possess an energetic advantage similar to that gained in delocalized transition state **22** proposed by Baizer for hydrodimerizations. Alternatively, one might choose as a mechanistic model Stocker's studies of the formation of pinacols from alkylaryl ketones in alkali.[28] In this case, pinacol formation from unsaturated ketones could take place by reaction between a monomeric radical anion and neutral radical:

$$\underset{\mathbf{28}}{RCH{=}CHCOR^1} \xrightarrow{e^-} \underset{\mathbf{29}}{RCH{=}\dot{C}H\langle^{O^-}_{R^1}}$$

$$\mathbf{29} \xrightarrow{H^+} \underset{\mathbf{30}}{RCH{=}CH\dot{C}\langle^{OH}_{R^1}}$$

$$\mathbf{29} + \mathbf{30} \longrightarrow RCH{=}CH\overset{R^1}{\underset{HO}{C}}{-}\overset{O^-}{\underset{R^1}{C}}CH{=}CHR \longrightarrow \text{pinacol}$$

$$\mathbf{30} \xrightarrow{e^-} \mathrm{RCH{=}CH\dot{C}(R^1)(OH)^-} \longrightarrow \mathrm{RCH{=}CHCH(R^1)OH} \quad \text{or} \quad \mathrm{RCH_2CH_2COR^1}$$

The first of these two possible mechanisms, that is, concerted reduction, has the distinct advantage, however, of accounting for the absence in these reductions of products in which the olefinic bond is saturated, since one would expect, at least in neutral solution, rapid reduction of neutral radicals such as **30**. The second mechanism would be preferred where transition states **24** and **27** are both sterically unlikely and/or the monomeric radical anion (akin to **29**) is especially stable. As will be seen later, this appears to be the case with highly arylated unsaturated ketones.

Pinacols formed in alkali from steroidal ring *A* ketones generally differ stereochemically from those formed in acid.[9] Three isomers are possible in principle: referring to the respective configurations of the two hydroxyl groups, the isomeric pinacols may be designated as diequatorial, diaxial, or mixed (equatorial–axial). Lund has assigned stereochemistry to the pinacols arising from reduction of androsta-1,4-diene-17-β-ol-3-one in acid and base on semi-intuitive grounds.[9] Bladon, Cornforth, and Jaeger[8] have criticized Lund's stereochemical assignments, however, and it appears that the only steroidal pinacol whose stereochemistry may be regarded as established is that from cholestenone. This pinacol was prepared by these investigators through constant-current electrolysis of cholestenone in ethanol containing sodium acetate. Chemical degradation showed it to be the diequatorial pinacol.[8] The same pinacol could be prepared by reduction of cholestenone with sodium amalgam, but the electrochemical method was found to be preferable for large-scale preparations, since the use of large amounts of sodium amalgam was thereby avoided. This advantage of electrochemical methods over dissolving-metal reductions has already been pointed out (Section 5.2).

A variety of aryl-substituted α,β-unsaturated carbonyl compounds, such as phenyl vinyl ketone,[29] benzalacetone,[2] dibenzalacetone,[2] benzalacetophenone,[2,30] substituted γ-pyrones,[31] and cinnamic acids,[32] exhibit polarographic behavior very similar to that observed with aromatic ketones, that is, two 1-electron waves in acid that merge into a single 2-electron wave as the pH is increased, sometimes splitting again into two 1-electron waves in strong alkali.[30] Apparently, aryl substitution can retard dimerization sufficiently, both sterically and electronically, to permit reduction of these substrates to a radical anion that can then suffer protonation and further reduction to the saturated carbonyl compound in neutral or weakly alkaline media. It is possible, nevertheless, to hydrodimerize at least some of these compounds. Pasternak obtained a mixture of hydrodimer (4,5-diphenyl-2,7-octanedione) and benzylacetone from controlled-potential electrolysis of benzalacetone at pH 8.6,[2] but Baizer and Anderson were able to obtain the hydrodimer in 74% yield with no detectable formation of benzylacetone.[26b] Mesityl oxide was hydrodimerized in 83% yield,[26b] but the dimer **3** underwent subsequent partial aldol condensation to the same complex mixture of **3**, **4**, and **5** isolated under similar conditions by Wiemann and Bouguerra.[4] In coumarin (**31**), the steric bulk of the aryl group is minimized by its incorporation into a ring and by the *cis*-olefinic linkage. As a result, dimerization is not hindered; this is evidenced by isolation of a mixture of hydrodimers (*dl* and *meso*) in good yield even in quite dilute solution.[33] The alkyl group R in alkyl

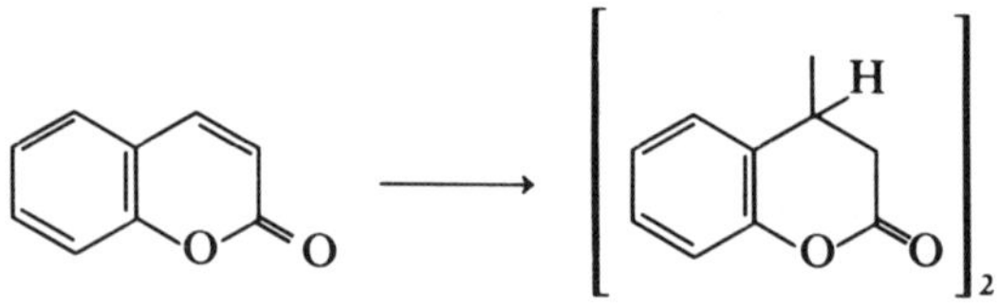

31

vinyl ketones of structure $RCOCH{=}CH_2$ can apparently sterically inhibit both hydrodimerization and pinacolization. When R is methyl, hydrodimerization is the exclusive reduction pathway in neutral solution. When R is *t*-butyl, however, the product of reduction is *t*-butyl ethyl ketone; mixtures of hydrodimer and saturated ketone are observed for R groups of intermediate size.[34]

Very little study has been made of the factors governing the mode of dimerization of activated dienes (**32**). Coupling may occur at the β or δ positions, and so, neglecting stereochemistry, three possible products (β–β, β–δ, and δ–δ) can arise. Hydrodimerization of 1-cyanobutadiene yields mainly the product of δ–δ

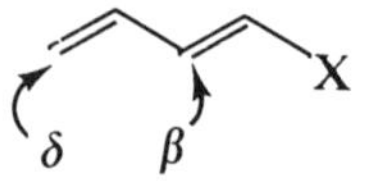

32

coupling,[35] but ethyl β-styryl acrylate affords a mixture of stereoisomers arising from β–β coupling.[27] Analysis of such mixtures could conceivably be complicated by concomitant thermal electrocyclic rearrangement[36] of initial products.

7.1.3 Aprotic Media

Reduction of activated olefins in rigorously aprotic solvents often results in polymer formation. The reason for this will be evident from the mechanistic discussion of the preceding section, that is, that carbanions such as **24** which are formed by hydrodimerization can initiate polymerization of the starting material if not protonated.* In a mechanistically revealing study, Simonet examined a series of ketones of general formula **33** in DMF.[37] When R is methyl, three waves are observed; the first and third are almost equal in height, and the second is quite small. When R is isopropyl, however, it is the third wave that is small, while the first and second are almost equal in height. When R is *t*-butyl, no third

$$C_6H_5CH{=}CHCOR$$
33

wave is observed at all. With mesityl oxide, only one wave is observed. The three waves observed for the first two compounds were assigned by Simonet to reduction of the ketone to a radical anion, further reduction of the radical anion to a dianion, and reduction of the hydrodimer, respectively.[37] When R is methyl, dimerization is fast and the third wave is larger than the second. When R is isopropyl, dimerization is slowed sterically, and the height of the second wave is enhanced relative to the third. When R is *t*-butyl, dimerization is so slow on the polarographic time scale that the third wave is not observed. Dimerization can also be slowed electronically: *p*-aminobenzalacetone fails to exhibit a third polarographic wave, apparently because its radical anion is highly stabilized by resonance. With mesityl oxide, dimerization is so fast that the second wave is not observed; since the product is electroinactive, the third wave is also absent. The lability of radical anions derived from aliphatic conjugated carbonyl compounds is substantiated by the report of Tolles and Moore that acrolein, methyl vinyl ketone, and crotonaldehyde exhibit no esr signal upon electrochemical reduction, even at −80°C.[38a] Likewise, Wawzonek and Gundersen were able to trap the radical anions of benzalacetophenone and benzalacetone with carbon dioxide, but could not intercept those from crotonaldehyde and cinnamaldehyde.[39] Simonet substantiated his interpretation of the polarographic behavior of compounds **33** through stationary-electrode voltammetry. At high sweep rates, the second wave for all compounds of structure **33** is larger than the third; at slow sweep rates, where dimerization has had time to proceed, the third wave becomes larger than the second. The yield of hydrodimer is also related to the size of R: for R = hydrogen, methyl, *n*-propyl, isopropyl, and *t*-butyl, yields of hydrodimer are 0, 0, 25, 65, and 95% respectively.[37] Apparently, subsequent Michael

* Evans has recently found that yields of hydrodimer from the reduction of α,β-unsaturated ketones are much higher in dimethyl sulfoxide when the electrolyte is lithium perchlorate than when the electrolyte is tetrabutylammonium perchlorate. Apparently, coordination with lithium lowers electron density on the radical anion sufficiently to permit rapid dimerization, thus subverting polymerization by attack of the radical anion on neutral starting materials. (D. H. Evans, private communication.)

reactions of the intermediate dianion **34** are slowed relative to proton abstraction as the size of R increases.

$$\mathrm{RCO\overset{-}{C}H\underset{\substack{|\\ H_5C_6}}{C}H\underset{\substack{|\\ C_6H_5}}{C}H\overset{-}{C}HCOR}$$

34

Where hydrodimers could be obtained by Simonet from **33**, it was generally possible to separate them into the corresponding *dl*- and *meso*-isomers by fractional crystallization.[12] The *dl* : *meso* ratio, which might shed some light on the preferred geometry for hydrodimerization, was not reported. Relative to this point, however, it has been reported that reduction of *trans*-**35** proceeds stereospecifically to the *dl*-hydrodimer under all conditions, while the *cis*-isomer affords *meso*-hydrodimer at −78°C and the *dl*-hydrodimer at −35°C. Apparently, at the higher temperature the *cis*-radical anion can isomerize before coupling.[40]

t-BuCH=CHCO*t*-Bu

35

Other experiments by Simonet have suggested a method of hydrodimerization of α,β-unsaturated ketones that would normally undergo rapid subsequent *intra*molecular aldol condensation. It will be recalled from previous discussions that mesityl oxide is a notorious example of such behavior: its hydrodimer **3** can only be isolated under neutral conditions, and even then only in fair yields.[4] Simonet reports, however, that reduction of mesityl oxide can be stopped successfully at hydrodimer **3** by reduction in DMF containing an equivalent of a proton donor (phenol or acetic acid) to protonate carbanions before they can initiate anionic polymerization or aldol condensation.[12]

7.1.4 Synthetic Scope and Limitations

A wide variety of activated olefins can be hydrodimerized successfully. The principal requirement is that the activating group be sufficiently electronegative to stabilize incipient adjacent carbanions. Thus, ease of hydrodimerization tends to be related to the tendency of the olefin to act as an acceptor in the Michael reaction. This requirement is not particularly stringent, however, as can be seen by the fact that even styrene and 2- or 4-vinylpyridines can be electrochemically hydrodimerized in good yields.[41] The preferred conditions for hydrodimerization consist of electrolysis of a concentrated solution of the olefin in a mildly alkaline hydrotrope, such as a concentrated aqueous solution of a tetraalkylammonium tosylate, sometimes with an organic cosolvent.[25, 42] When the olefin or its hydrodimer are sensitive to base, however, it is preferable to effect dimerization in an aprotic solvent such as DMF or acetonitrile, in the presence of a proton donor to protonate carbanions.[12] Of course, it may sometimes be desirable to take synthetic advantage of the lability of certain hydrodimers toward acid or base, as witnessed by the electrolytic conversion of mesityl oxide to **5** at pH 1.3 or 12 and to **4** under mild alkaline conditions.[4]

The synthetic utility of the hydrodimerization reaction has been broadened considerably by its adaptation to the synthesis of mixed hydrodimers.[26] There are limitations, however, on the kinds of mixed dimers that can be prepared. The best yields of mixed dimer can be expected when the two olefins differ by at least 0.2 V in half-wave potential and the more difficulty reduced olefin is the better Michael acceptor. Yields of mixed dimer can also be improved by employing a large excess of the more difficultly reduced olefin. The polyfunctional molecules that can be prepared by mixed electrolytic hydrodimerization are of obvious interest in complex synthetic schemes. In principle, a wide variety of such mixed hydrodimers could be prepared. A few examples include

$$\textit{trans}\text{-}EtO_2CCH{=}CHCO_2Et + CH_2{=}CHCOCH_3 \longrightarrow \underset{\displaystyle CH_2CH_2COCH_3}{EtO_2CCH_2\overset{|}{C}HCO_2Et} \qquad (\textit{Ref. 26b})$$

$$(CH_3)_2C{=}CHCOCH_3 + CH_2{=}CHCN \longrightarrow \underset{\displaystyle CH_2CH_2CN}{(CH_3)_2\overset{|}{C}CH_2COCH_3} \qquad (\textit{Ref. 26b})$$

$$CH_2{=}CHCH{=}CHCN + CH_2{=}CHCO_2Et \xrightarrow[(2)H_2,\ Pd]{(1)\ e^-} NC(CH_2)_6CO_2Et \qquad (\textit{Ref. 26a})$$

$$\underset{\displaystyle CO_2CH_3}{CH_2{=}\overset{|}{C}NHCOCH_3} + CH_2{=}CHCN \longrightarrow \underset{\displaystyle \mathbf{36}}{\underset{\displaystyle CO_2CH_3}{NCCH_2CH_2CH_2\overset{|}{C}HNHCOCH_3}} \qquad (\textit{Ref. 26a})$$

The last example is the first step in a convenient three-step synthesis of the common amino acid lysine; to complete the synthesis it is merely necessary to subject **36** to hydrogenation and hydrolysis.

As Anderson, Baizer, and Petrovich have pointed out, *intra*molecular hydrodimerization of *bis*-activated olefins of general formula **19** can be a useful route to cyclic molecules.[24] Yields are best when three-, five-, or six-membered rings are formed and when the β-positions of the two double bonds are sterically unencumbered. (Incidentally, unlike *inter*molecular hydrodimerization, the *intra*-molecular reaction is independent of the concentration of electroactive substance.) An important corollary of mechanistic studies carried out by Anderson, Baizer, and Petrovich is the fact that one can usually determine the likelihood of successful *intra*molecular hydrodimerization (hydrocyclization) simply by examination of a molecular model of the *bis*-olefin.

When hydrodimerization of α,β-unsaturated ketones is sterically difficult, the dominant mode of reaction is pinacol formation. Where this is desired, reductions should be carried out at pH > 5, since reduction in stronger acid can result in acid-catalyzed rearrangement or dehydration of the pinacol.[3,8]

Reduction of the semicarbazones of α,β-unsaturated ketones in neutral or alkaline solution affords the semicarbazone of the corresponding saturated ketone in high yield.[21] On the other hand, reduction of the oximes (and probably also semicarbazones), in anhydrous acidic solvents, such as methanol or DMF containing hydrogen chloride or benzoic acid, should result in formation of the allylic amine, as was noted in Section 7.1.1.

The electrochemical reduction of methyl *p*-chlorophenylpropiolate (**37**) has already been discussed in another context (Chapter 4). It will be recalled that this molecule, and presumably related activated acetylenes, can be reduced either to the cinnamate or to the dihydrocinnamate, depending on whether the supporting electrolyte is a tetramethylammonium or a tetra-*n*-butylammonium salt.[43] This selectivity is of obvious synthetic advantage. Also, since cinnamates can be hydrodimerized in acid,[18,19] it should also be possible to convert **37** into **39** by

$$\text{Cl}-C_6H_4-C{\equiv}CCO_2CH_3$$

37

$$\underset{\mathbf{38}}{ArCH{=}CHCO_2CH_3} \xleftarrow[Me_4N^+]{2e^-} \mathbf{37} \xrightarrow[Bu_4N^+]{4e^-} ArCH_2CH_2CO_2CH_3$$

$$\mathbf{38} \xrightarrow[Me_4N^+]{e^-,\ H^+} \underset{\mathbf{39}}{\begin{array}{c} ArCHCH_2CO_2CH_3 \\ | \\ ArCHCH_2CO_2CH_3 \end{array}}$$

$$Ar = p\text{-}Cl-C_6H_4$$

electrolysis in acidic media containing tetramethylammonium salts. This conjecture is supported by the discovery of Ono that electrolysis of phenylpropiolic acid at pH < 3 affords *trans*-cinnamic acid, 3,4-diphenyladipic acid, or dihydrocinnamic acid, depending on the cathode potential.[44]

7.2 AROMATIC COMPOUNDS AND RELATED SPECIES

7.2.1 Aromatic Compounds

Since the essential features of the electrochemical reduction of polycyclic aromatic compounds have already been discussed in Chapter 4 in connection with the effects of proton donors, they need not be repeated at length here. A short summary will be helpful, though.[45] Recall that in aprotic solvents the initial step is transfer of an electron to the hydrocarbon R, generating a stable radical anion $R^{\dot{-}}$. At more negative potentials, the radical anion can be reduced further to a dianion, which is generally shortlived, usually reacting by proton abstraction from the solvent to afford a dihydro derivative of the starting hydrocarbon. If reduction is carried out in the presence of a proton donor—this might be by reduction either in a protic solvent or in an aprotic solvent containing varying amounts of an added proton source such as phenol—the radical anion is generally quickly protonated to afford a radical RH·. If the starting material R was an even alternate hydrocarbon (as polycyclic benzenoid aromatic hydrocarbons indeed are), then RH· will always be an odd alternate hydrocarbon. Hoijtink has pointed out an important consequence of this latter property: the lowest unfilled orbital of R is always antibonding, but because odd alternate hydrocarbons have a nonbonding molecular orbital, RH· will always be easier to reduce than R, and a second electron will be added to RH· as soon as it is formed.[46a] This sequence, a so-called ECE process, is evidenced polarographically by a 2-electron wave in proton-donating media. Santhanam and Bard have contributed a model study illustrating such behavior, that of 9,10-diphenylanthracene in DMF.[47]

It is interesting to note that the second polarographic wave in aprotic solvents is generally about 0.4 V cathodic of the first wave, irrespective of the structure of the aromatic hydrocarbon.[46b] This is generally ascribed to the fact that both electrons are being added to the same orbital of R, and to a first approximation the amount of energy (0.4 V or 9.2 kcal/mole) necessary to overcome electronic charge repulsion in the addition of the second electron will be independent of the structure of the hydrocarbon. Hoijtink has pointed out, however, that for conjugated polyenes this electronic repulsion term should decrease in energy as the length of the pi system increases. At the same time, the rate of protonation of the radical anion should also decrease, so that for very long conjugated systems a direct 2-electron reduction of R to R^{-2} may occur even in protic media.[48]

Wawzonek and Laitinen have shown that it is possible to effect clean stepwise reduction of polycyclic aromatic hydrocarbons in protic media.[49] Naphthalene exhibits a single 2-electron polarographic wave in 75% dioxane. Since 1,2-dihydronaphthalene would be detected polarographically if formed, the product must be 1,4-dihydronaphthalene; this has been confirmed by preparative-scale electrolysis.[50] Likewise, the fact that anthracene exhibits a single 2-electron wave under the same conditions indicates that it is reduced to 9,10-dihydroanthracene, which having only isolated benzene rings would be expected to be electroinactive. Phenanthrene (**40**) exhibits two 2-electron waves. The first reduction must afford 9,10-dihydrophenanthrene, since the second polarographic wave is at the same potential as the single wave of biphenyl (which is presumably reduced to 3-phenyl-1,4-cyclohexadiene.) Continuing this progression, it is found that pyrene (**41**) exhibits three polarographic waves. The second two are at the same potentials as those of **40**, hence the first wave must involve saturation of the 4,5 double bond of **41** to generate a bridged phenanthrene. Deductions similar to these may be

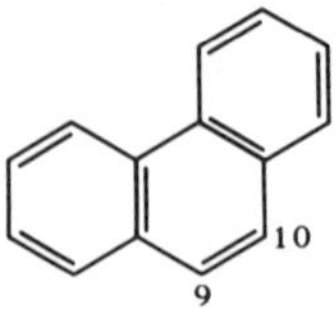

40 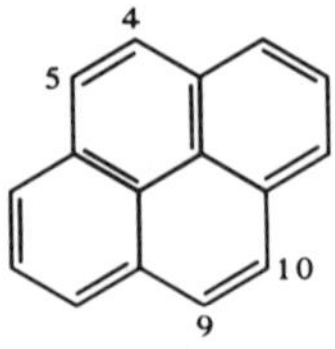 **41**

made for the other polycyclic aromatics studied by Wawzonek and Laitinen. As was discussed already in Chapter 4, one may also use electrochemically generated lithium in methylamine to effect reduction of benzenoid compounds to either 1,4-dihydro or tetrahydro derivatives, depending on the experimental conditions employed.[51] Reduction of polycyclic hydrocarbons such as **40** or **41** would be better carried out at controlled potential in aqueous dioxane or ethanol, however, since one would not expect the lithium–methylamine procedure to be very selective with such compounds.

Except for the lithium–methylamine procedure, which does not involve direct electron transfer from the electrode to the aromatic, it is generally not possible to reduce the benzene nucleus electrochemically. Electrolysis of phthalic acid in alkali at a mercury cathode produces cyclohexene-2,3-dicarboxylic acid with remarkable efficiency (78% yield), however.[52] This may mean that the presence of electronegative substituents can lower the energy of the lowest unfilled benzenoid molecular orbital sufficiently to permit reduction.

Diamond and Soffer have shown that β-naphthyl methyl ethers (**42**) can be reduced in high yield to the corresponding enol ethers (**43**), which are in turn readily hydrolyzed in high yield to 2-tetralones.[53] Electrochemical reduction was found to be definitely preferable to sodium–alcohol reduction, which provided

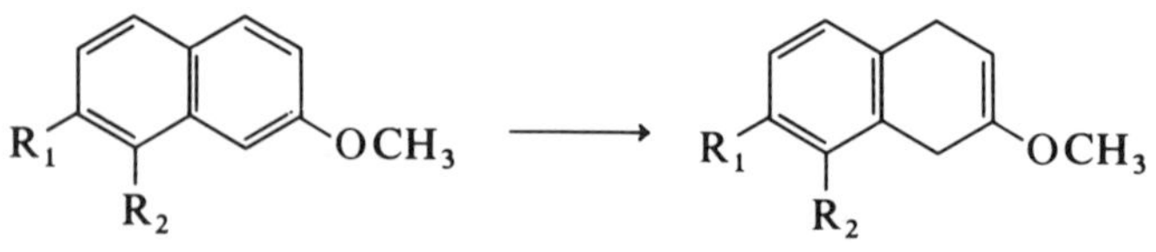

42a, $R_1 = R_2 = H$
42b, $R_1 = OCH_3$, $R_2 = H$
42c, $R_1 = R_2 = OCH_3$

43a, 96%
43b, 80%
43c, 95%

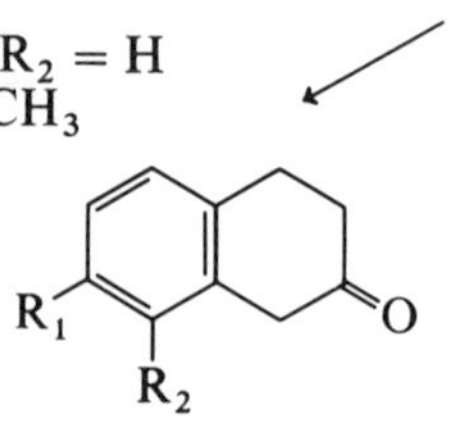

a, 89%
b, 93%
c, 89%

the ethers **43a–c** in only 30–50% yields. The strongly alkaline conditions involved in sodium–alcohol reduction presumably isomerize the double bond of enol ether **43** into conjugation with the ring to produce a 1,2-dihydronaphthalene, which can then undergo undesirable further reduction.

It is not necessary that the proton be the agent used to intercept the initial radical anion $R^{\overset{\bullet}{-}}$. Any electrophilic reagent that is electroinactive at the potentials used can be employed. The use of carbon dioxide for this purpose has already been mentioned in Chapter 4: it is, for example, possible to prepare 1,4-dihydro-1,4-dicarboxynaphthalene and 9,10-dihydro-9,10-dicarboxyphenanthrene in good yields merely by bubbling carbon dioxide through the solution during electrolysis of the hydrocarbon in an aprotic solvent.[54] This procedure is capable, in principle, of making available a variety of specifically substituted dihydroarene dicarboxylic acids (and, after dehydrogenation, arene diacids). The data of Wawzonek and Laitinen[49] on the position of protonation of aromatic radical anions would be of considerable help in such studies.

Aromatic radical anions are, in general, capable of reaction both as nucleophiles (as in reactions with protons or carbon dioxide) and as electron-transfer agents. For naphthalene radical anion and methyl chloride this dichotomy can be expressed as

H CH_3

nucleophilic displacement → products

+ CH_3Cl

electron transfer → + $CH_3\cdot$ → → CH_4

It appears that the electron-transfer path, which results in reduction of the electrophilic reagent, is the preferred pathway with the radical anions of those hydrocarbons that are reducible only at very negative potentials, such as naphthalene, phenanthrene, and triphenylene [$E_{1/2}$ (*vs* S.C.E.) = −2.5, −2.5, and −2.4 V, respectively].[55] Nucleophilic properties of the sort that are of interest in the present context are more characteristic of radical anions of more readily reduced aromatic hydrocarbons, such as anthracene ($E_{1/2}$ = −2.0 V). A variety of electrophilic reagents, such as acetic anhydride, alkyl halides, and alkyl sulfonates and sulfates, can be used to effect efficient carbon–carbon bond and even carbocyclic ring formation with radical anions of the latter type; Lipkin and coworkers have reported a number of such conversions, for example,[55d]

+ $Cl(CH_2)_3Cl$ → →

7.2.2 Nonbenzenoid and Pseudoaromatic Compounds

Electrochemical reduction of triphenylcyclopropenyl cation (**44**) consumes 1 electron/molecule of **44** and leads to the dimer **45**.[56,57] Evidence that this reaction proceeds via the triphenylcyclopropenyl radical comes from both cyclic voltammetry (reversible 1-electron reduction at high scan rates, but no anodic

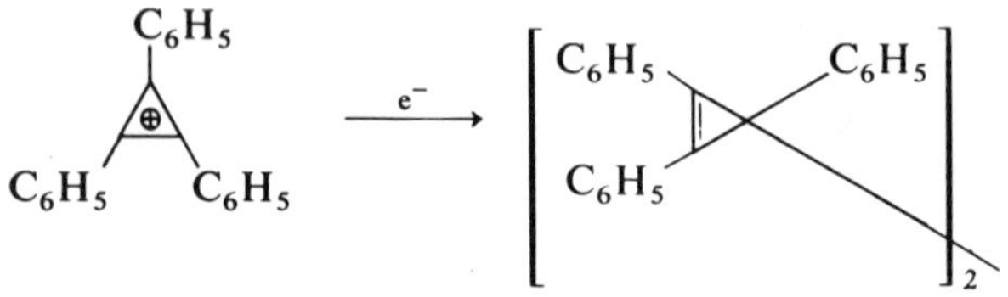

C_6H_5
C_6H_5
C_6H_5
e⁻
C_6H_5
C_6H_5
C_6H_5
2

44 **45**

peak at low scan rates)[57] and electrolysis in the presence of hydrogen atom donors, such as cyclohexene or chloroform, after which triphenylcyclopropene can be found in the products.[56] Reduction of diphenylcyclopropenyl cation **46** affords 1,2,4,5-tetraphenylbenzene (**47**, 39 %) and the amide **48** (24 %).[56] Isolation of **47** indicates that the (presumed) initial intermediate, the dimer of diphenylcyclopropenyl radical, must be less thermally stable than **45**, which rearranges to hexaphenylbenzene only at elevated temperatures.[58] Amide **48** could conceiv-

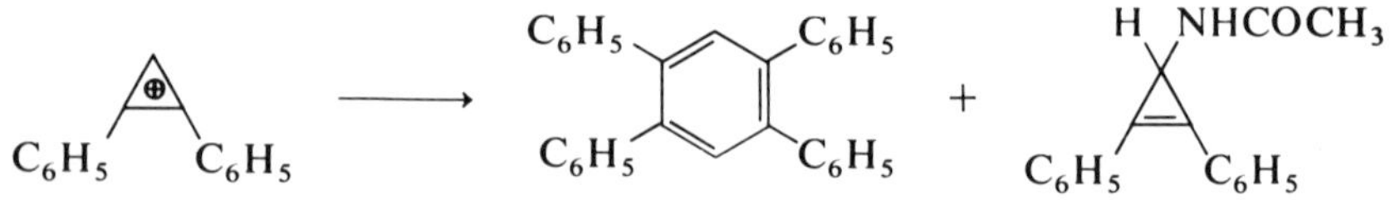

C_6H_5
C_6H_5
C_6H_5
C_6H_5
C_6H_5
C_6H_5
+
H NHCOCH$_3$
C_6H_5
C_6H_5

46 **47** **48**

ably have arisen by a Ritter reaction between **46** and the solvent (and traces of water in the solvent),[59] but it was stated that **48** is not formed unless current is allowed to flow, suggesting that it too arises via the diphenylcyclopropenyl radical. In a study akin to these, Zhdanov and Frumkin showed that tropylium ion is converted to ditropyl upon electrochemical reduction, a reaction presumably involving the unstable cycloheptatrienyl radical.[60] Tropyl alcohol and ditropyl ether are converted first to tropylium ion and thence to ditropyl upon reduction in acid.[60]

Trityl cation (**49**) exhibits a reversible 1-electron polarographic wave. The product of reduction, triphenylmethyl radical (**50**) is stable, and hence cyclic voltammetry indicates a reversible reduction even at low scan rates. It is possible to observe **49** and **50** by electrochemical measurements on triphenylcarbinol (**51**) in strongly acidic solution, that is, by *in situ* generation of **49**.[61] In methanesulfonic acid or sulfuric acid of low water content, **51** exhibits a 2-electron irreversible wave, which splits into two 1-electron waves upon addition of small

$$\underset{\mathbf{51}}{(C_6H_5)_3COH} \xrightarrow[-H_2O]{H^+} \underset{\mathbf{49}}{(C_6H_5)_3C^+} \underset{}{\overset{e^-}{\rightleftharpoons}} \underset{\mathbf{50}}{(C_6H_5)_3C\cdot} \xrightarrow{e^-} (C_6H_5)_3C^-$$

amounts of water.[62] Wawzonek and coworkers have suggested that the 2-electron wave is due to rapid disproportionation of **50** in very strong acid.

$$\mathbf{49} + e^- \longrightarrow \mathbf{50}$$

$$2\,\mathbf{50} + H^+ \longrightarrow \mathbf{49} + (C_6H_5)_3CH$$

Diphenylcyclopropenone (**51**) was found to exhibit two 2-electron polarographic waves in aqueous ethanol, but preparative electrolysis was not carried out.[63] It is not obvious what processes are occurring at each wave. It was suggested without supporting evidence that the first wave corresponds to saturation

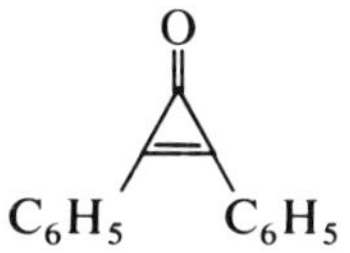

O

C_6H_5 C_6H_5

51

of the carbon–carbon double bond (which would afford a cyclopropanone), but this wave might equally well correspond to reduction of the carbonyl function or cleavage of the carbon–carbon single bond. In acid, **51** exhibits a 1-electron polarographic wave. It was demonstrated from the pH dependence of the electrochemical behavior of **51** in acid that reduction proceeds via its conjugate acid, hydroxydiphenycyclopropenyl cation. Presumably, the cation is reduced to a cyclopropenyl radical which then dimerizes, but once again the electrolysis products were not investigated.

Pseudoaromatic species may be defined here as those compounds that possess closed conjugated pi systems but that gain no resonance stabilization from this conjugation. The best-known class of pseudoaromatics consists of monocyclic conjugated systems containing 4*n* pi-electrons, and the most extensively investigated member of this class is cyclooctatetraene (**52**). Its behavior is instructive as a guide to the behavior to be expected from other pseudoaromatics. In aprotic solvents, **52** exhibits two 1-electron polarographic waves.[64] Electron transfer to aromatic hydrocarbons is generally very fast in aprotic solvents, since all that is involved is injection of an electron into the lowest unfilled molecular orbital of the hydrocarbon, with no change in geometry.[45] The first electron transfer to **52** is relatively slow, however, apparently because flattening of **52** to a partly planar radical anion is well advanced in the transition state for electron transfer.[66]

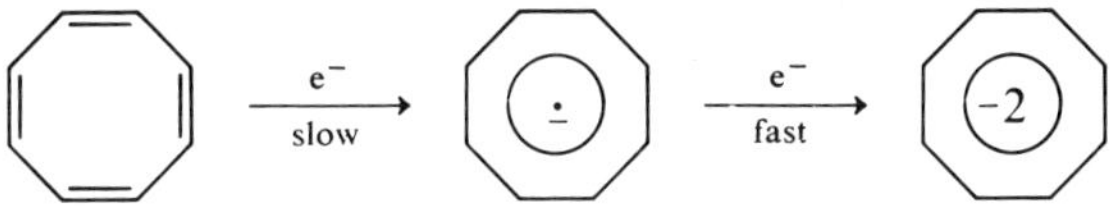

e⁻ slow → (·−) e⁻ fast → (−2)

52

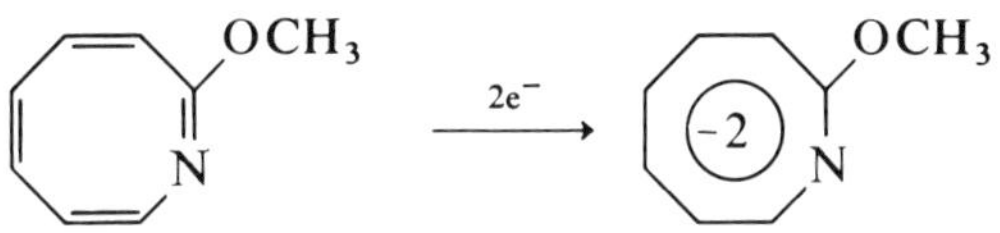

OCH_3 N $\xrightarrow{2e^-}$ (−2) OCH_3 N

53

The second electron transfer involves complete flattening to a planar dianion.[66] Controlled-potential electrolysis of **52** in aprotic solvents affords principally 1,3,5-cyclooctatriene[64]; in protic solvents the major product is 1,3,6-cyclooctatriene.[65]

The electrochemical behavior of 2-methoxyazocine (**53**) and several of its derivatives differs from cyclooctatetraene in that a single 2-electron polarographic wave is observed.[66] Cyclic voltammetry demonstrates this wave to be highly irreversible. Slow electron transfer to **53** was ascribed to flattening of the radical anion during the electron-transfer step. The radical anion of **53** takes up a second electron immediately upon formation. It should be simple to prepare electrochemically dianions of **52**, **53**, and related species.

7.2.3 Heterocyclic Compounds

As a consequence of both limitations of space and the personal interests of the author, an extensive survey of the electrochemical behavior of heterocyclic compounds will not be presented here. (Perrin has discussed such compounds in some detail.[67]) The electrochemical behavior of heterocyclic compounds such as **53** can generally be rather well understood by application of principles that have already been elucidated in this and preceding chapters. A few examples may be helpful to illustrate typical behavior.

Pyrylium and pyridinium ions, such as **54** and **55**, exhibit a reversible 1-electron wave in aqueous media. The usual reaction path for such species is

CN

N^+ — C_2H_5

54 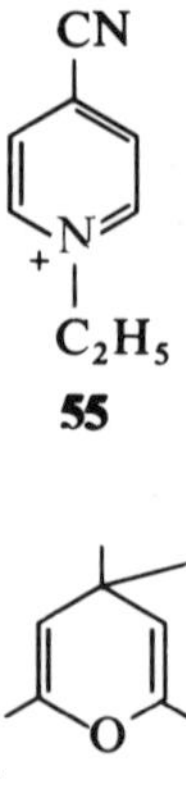 **55**

54 $\xrightarrow{1e^-}$ $[\ldots]_2$

56

55 $\xrightarrow{e^-}$ Et—N ... (CN)(CN) ... N—Et $\xrightarrow{-2CN^-}$ Et—N^+ ... N^+—Et

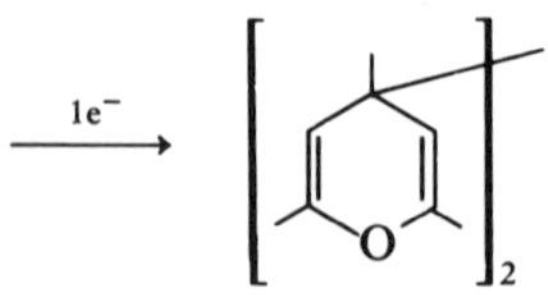

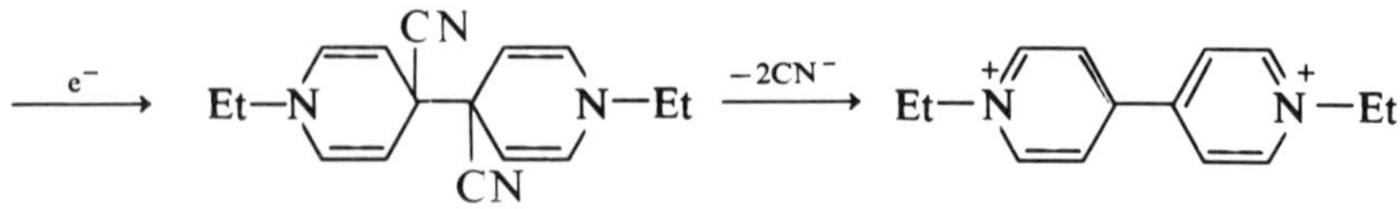

57 $\xrightarrow{e^-}$ radical cation 57

reduction to a neutral radical, which then dimerizes. Thus, reduction of **54** at pH > 5 consumes 1 electron/molecule and affords **56**.[68] Reduction of **55** is more complex, however, since the initial dimer rather rapidly extrudes cyanide ion to give dication **57**, which is further reducible to a stable radical cation.[69] The reversible 1-electron nature of the first step can be demonstrated by rapid-scan cyclic voltammetry, however. The nicotinamide derivative **58** exhibits a pair of polarographic waves at pH > 7.[70a] Controlled-potential electrolysis at the second wave produces a 1,4-dihydropyridine (**59**), which had also been prepared by Karrer and coworkers through reduction of **58** with sodium dithioniite.[71] Electrochemical reduction of **58** is the method of choice for the preparation of **59**, however, since the product is obtained in quantitative yield and unaccompanied

$CONH_2$ N^+ CH_3 **58**

$2e^-$, −1.8 V (S.C.E.) → $CONH_2$ N CH_3 **59**

$1e^-$, −1.2 V (S.C.E.) → H_2NCO $CONH_2$ N N CH_3 CH_3 **60**

by sulfur-containing contaminants. Reduction of **58** at the potential of the first wave results in a dimer, also in quantitative yield, assigned structure **60**. Steric factors may account for dimerization at the 6 position, rather than the 4 position as in reduction of **55**. Replacement of the N-methyl group of **58** with a propyl group results in a 4,4′-dimer.[70a] (On the other hand, Volke and coworkers reported finding only 4,4 dimerization upon reduction of unsubstituted N-alkylpyridinium salts, irrespective of the nature of the substituent on nitrogen.[70b] The situation remains confused.)

Reduction patterns become more complex as the number of heteroatoms in a ring system is increased. Comparisons with acyclic model compounds are often helpful in predicting behavior. 4-Methylcinnoline (**61**) exhibits two 2-electron polarographic waves at pH 3.[72] Reduction at the first wave affords the 1,4-dihydrocinnoline **62** in 85% yield. This compound may be regarded as a phenylhydrazone; it will be recalled (Section 6.2.2) that such compounds undergo cleavage of the N—N bond upon electrolytic reduction in acid.[73, 21] It is not surprising, therefore, to find that reduction of **61** at the second wave affords skatole (**63**) in good (83%) yield.[72]

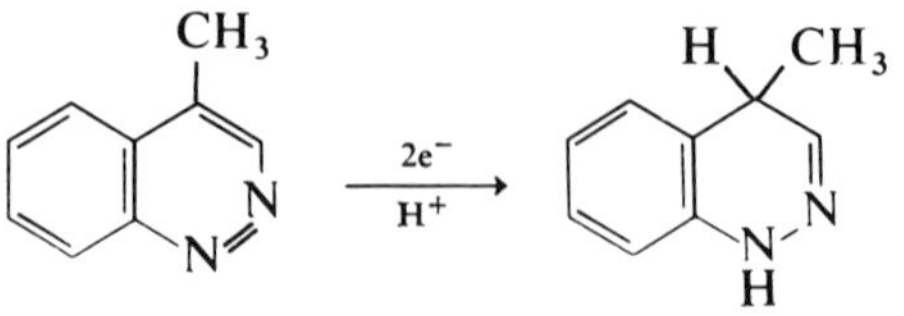

7.3 CONJUGATED OLEFINS AND ACETYLENES

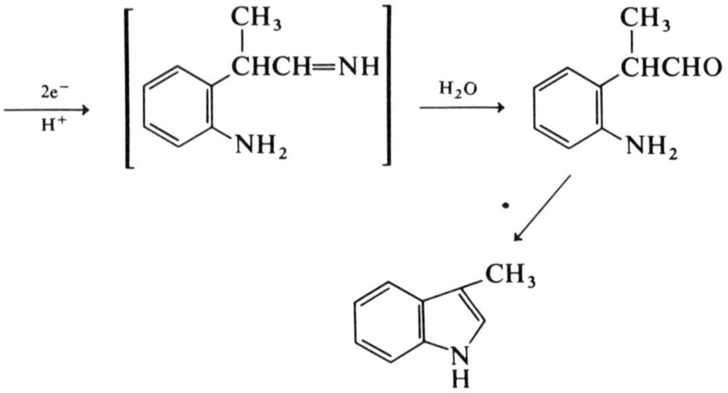

Conjugation lowers the energy of the lowest unfilled molecular orbital of olefins and acetylenes sufficiently to permit direct electrochemical reduction. In protic solvents, such compounds undergo saturation of the multiple bond via the usual ECE process. Acetylenes are generally harder to reduce than the corresponding olefins, and hence reduction cannot be stopped at the latter. Thus, phenylacetylene ($E_{1/2} = -2.37$ V *vs* S.C.E.) exhibits a 4-electron polarographic wave in 75% dioxane, since it is first reduced to styrene ($E_{1/2} = -2.34$ V) which then undergoes immediate reduction to ethylbenzene.[74] These reductions are of little synthetic importance, since other methods already exist for saturation of multiple bonds. A more useful feature lies in the fact that olefinic bonds conjugated with an aromatic ring can be hydrodimerized by proper choice of electrolysis conditions, thus affording in one step compounds whose synthesis might otherwise be complicated. For example, styrene, 9-benzalfluorene, and the 2- and 4-vinylpyridines have all been hydrodimerized in good yields[41,75]; the hydrodimer of 4-vinylpyridine, 1,4-di(4-pyridyl)-butane, had previously been available only by inefficient multistep syntheses. As was mentioned in Section 7.1, these olefins can also be utilized successfully in mixed hydrodimerizations.[41]

Conjugated dienes, unlike the aryl olefins just mentioned, are difficult to hydrodimerize. Although little evidence is available, it appears that 1,4 reduction is preferred over hydrodimerization. 2-Phenylbutadiene, under the usual hydrodimerization conditions, affords mainly 2-phenyl-2-butene (85%) but only 8.3% of the hydrodimer.[75] Likewise 1,4-diphenyl-1,3-butadiene exhibits only a single 2-electron polarographic wave in 75% dioxane, indicating that the product must be 1,4-diphenyl-2-butene.[74] Controlled-potential reduction of conjugated olefins in aprotic solvents in the presence of carbon dioxide has been reported to afford mono- and diacids. Butadiene was converted in this manner to $HO_2CCH_2CH{=}CHCH_2CO_2H$ in >50% yield.[76] The reaction presumably involves an ECE reaction initiated by the radical anion of butadiene. Similarly, reduction of *cis*- and *trans*-stilbene in DMF saturated with carbon dioxide affords mixtures of *dl*- and *meso*-diphenylsuccinic acid.[77] Isomerization of *cis*-stilbene radical anion to that of *trans*-stilbene could be shown to possess a half-life of greater than 15 sec, and the *dl*/*meso* ratio from *cis*-stilbene is different from that obtained from *trans*-stilbene (1.4 and 2.7, respectively). The pseudo-first-order rate constant for reaction of *trans*-stilbene radical anion with excess carbon dioxide was estimated to be ca. 700 sec^{-1}.

In a reaction of considerable interest, butadiene was converted into *trans,trans,trans*-hexadeca-1,6,10,14-tetraene in good (50%) yield by electrolysis in glyme or ethanol containing (*tetrakis*pyridine)nickel(II) perchlorate or nickel(II) chloride.[78] In the presence of (*bis*-triphenylphosphine)nickel(II) chloride, butadiene was converted instead into octa-1,3,7-triene. These reactions, for which no mechanistic information was reported, presumably are initiated by coordination of butadiene to nickel. It is likely that much useful chemistry could arise through the use of metal ions as templates to govern the course of electrochemical reaction, as they here, for example, control the extent of oligomerization of the diene. A considerable amount of mechanistic information is obviously required before the synthetic utility of such reactions can be assessed.

Dietz, Peover, and Wilson have reported a detailed study of the electrochemical behavior of tetraphenylallene (**64**) and 1,1,3,3-tetraphenylpropene (**65**).[79] A 2-electron polarographic wave is observed for **64** in aprotic media.

$$(C_6H_5)_2C{=}C{=}C(C_6H_5)_2$$

64

$$(C_6H_5)_2C{=}CHC(C_6H_5)_2 \text{ (with X on the central C)}$$

65, X = H
66, X = OEt

$$(C_6H_5)_2CHCH_2CH(C_6H_5)_2$$

67

If phenol is added, a second wave appears at more negative potentials. This wave is at the same potential as the 2-electron wave observed for **65**. Controlled-potential reduction of **64** or **66** affords a deep red solution; addition of water quenches the red color and **65** can then be isolated. Treatment of **65** with strong base results in the same red solution. Controlled-potential reduction of **65** also results in a red solution from which, after quenching with water, there is obtained an equimolar mixture of 1,1,3,3-tetraphenylpropane (**67**) and the starting material **65**. These data were interpreted according to the following scheme:

$$\mathbf{64} \xrightarrow{e^-} (C_6H_5)_2\dot{C}{-}\bar{C}{=}C(C_6H_5)_2 \xrightarrow[k_1]{H^+} (C_6H_5)_2\dot{C}CH{=}C(C_6H_5)_2$$

$$(C_6H_5)_2\dot{C}CH{=}C(C_6H_5)_2 \underset{}{\overset{e^-}{\rightleftharpoons}} (C_6H_5)_2\bar{C}CH{=}C(C_6H_5)_2$$

$$\mathbf{66} \xrightarrow{2e^-} \underset{\mathbf{68}}{(C_6H_5)_2\bar{C}CH{=}C(C_6H_5)_2}$$

$$\mathbf{68} \underset{KO{-}t\text{-}Bu}{\overset{H_2O}{\rightleftharpoons}} \mathbf{65}$$

$$\mathbf{65} \xrightarrow{e^-} (C_6H_5)_2\dot{C}\bar{C}HCH(C_6H_5)_2 \xrightarrow[k_2]{H^+} \underset{\mathbf{69}}{(C_6H_5)_2\dot{C}CH_2CH(C_6H_5)_2}$$

$$\mathbf{69} \xrightarrow{e^-} (C_6H_5)_2\bar{C}CH_2CH(C_6H_5)_2 \xrightarrow{H^+} \mathbf{67}$$

The red color is that of carbanion **68**; its stability in solution accounts for the observation that in polarography of **64**, the wave for reduction of **65** does not appear until phenol is added. Unlike planar aromatic hydrocarbons, **64** and **65** afford unstable radical anions: potentiostatic measurements indicated a lower limit for k_1 of 10^3 sec^{-1} and a value of ca. 200 sec^{-1} for k_2. The proton source is presumably the quarternary ammonium ion used as electrolyte, as with the strongly basic radical anions generated from some Schiff bases.[80] As an interesting sidelight, solutions of anion **68** can be oxidized to a yellow solution, presumably of the stable tetraphenylallyl radical; **68** can be regenerated by controlled-potential reduction of the yellow solution.

A similar, but less thorough, study has been reported by Kemula and Kornacki, who examined the electrochemical behavior of 1,1,4,4-tetraphenyl-1,2,3-butatriene (**70**).[81] Cumulene (**70**) exhibits three polarographic waves. The first is due to saturation of the central bond to afford 1,1,4,4-tetraphenyl-1,3-

$$\underset{\mathbf{70}}{(C_6H_5)_2C{=}C{=}C{=}C(C_6H_5)_2}$$

butadiene. The latter is reduced to a mixture of 1,1,4,4-tetraphenyl-2-butene and 1,1,4,4-tetraphenyl-1-butene. The third wave arises from reduction of the latter compound to the fully saturated tetraphenylpropane.

References

[1]E. J. Denney and B. Mooney, *J. Chem. Soc.*, (B), 1410 (1968).
[2]R. Pasternak, *Helv. Chim. Acta*, **31**, 753 (1948).
[3]P. Kabasakalian and J. McGlotten, *J. Amer. Chem. Soc.*, **78**, 5032 (1956).
[4]J. Wiemann and M. L. Bouguerra, *Ann. Chim.*, [14], **2**, 35 (1967).

[5]Y. Deux, *Compt. Rend.*, **208**, 522 (1939).
[6]H. O. House, E. J. Grubbs, and W. F. Gannon, *J. Amer. Chem. Soc.*, **82**, 4099 (1960).
[7]J. E. McMurry, *J. Amer. Chem. Soc.*, **91**, 3676 (1969).
[8]P. Bladon, J. W. Cornforth, and R. H. Jaeger, *J. Chem., Soc.*, 863 (1958).
[9]H. Lund, *Acta Chem. Scand.*, **11**, 283 (1957).
[10]A. Mazzenga, D. Lomnitz, J. Villegas, and C. J. Polowczyk, *Tetrahedron Lett.*, 1665 (1969).
[11]L. Holleck and D. Marquarding, *Naturwissenschaften*, **49**, 468 (1962).
[12]J. Simonet, *Compt. Rend.*, **267C**, 1548 (1968).
[13a]J. Wiemann and M. L. Bouguerra, *Ann. Chim.*, [14], **3**, 215 (1967); [b]A. Ryvolova-Kejharova, *J. Electroanal. Chem.*, **21**, 197 (1969).
[14a]S. Wawzonek and J. H. Fossum, *Proceedings of the First International Congress on Polarography at Prague*, **1**, 548 (1951); [b]S. Wawzonek, R. C. Reck, W. W. Vaught, Jr., and J. W. Fan, *J. Amer. Chem. Soc.*, **67**, 1300 (1945).
[15]I. Rosenthal, J. R. Hayes, A. J. Martin, and P. J. Elving, *J. Amer. Chem. Soc.*, **80**, 3050 (1958).
[16]C. T. Camilli, Ph.D. dissertation, Ohio State University, 1953.
[17]M. J. D. Brand and B. Fleet, *J. Electroanal. Chem.*, **19**, 157 (1968).
[18a]C. L. Wilson and K. B. Wilson, *Trans. Electrochem. Soc.*, **84**, 153 (1943); [b]E. P. Goodings and C. L. Wilson, *Trans. Electrochem. Soc.*, **88**, 77 (1945).
[19a]J. P. Petrovich, M. M. Baizer, and M. R. Ort, *J. Electrochem. Soc.*, **116**, 743, 749 (1969); [b]S. Wawzonek, A. R. Zigman, and G. R. Hansen, *J. Electrochem. Soc.*, **117**, 1351 (1970); [c]J. Wiemann and M. L. Bouguerra, *Compt. Rend.*, **265C**, 751 (1967).
[20]Yu. D. Smirnov and A. P. Tomilov, *Zh. Org. Khim.*, **5**, 864 (1969); see *Chem. Abstr.*, **71**, 222 (1969).
[21]H. Lund, *Acta Chem. Scand.*, **13**, 249 (1959).
[22]I. G. Sevastyanova and A. P. Tomilov, *J. Gen. Chem. U.S.S.R.*, **33**, 2741 (1963).
[23]F. Beck, *Chem. Ing. Tech.*, **37**, 607 (1965).
[24a]J. P. Petrovich, J. D. Anderson, and M. M. Baizer, *J. Org. Chem.*, **31**, 3897 (1966); [b]J. D. Anderson, M. M. Baizer, and J. P. Petrovich, *J. Org. Chem.*, **31**, 3890 (1966).
[25]R. H. McKee, *Ind. Eng. Chem.*, **38**, 382 (1946).
[26a]M. M. Baizer, *J. Org. Chem.*, **29**, 1670 (1964); [b]M. M. Baizer and J. D. Anderson, *J. Org. Chem.*, **30**, 3138 (1965); [c]M. M. Baizer, J. P. Petrovich, and D. Tyssee, *J. Electrochem. Soc.*, **117**, 173 (1970).
[27]J. H. Wagenknecht and M. M. Baizer, *J. Org. Chem.*, **31**, 3885 (1966).
[28]J. H. Stocker, R. M. Jenevein, and D. H. Kern, *J. Org. Chem.*, **34**, 2810 (1969).
[29]P. Zuman and J. Michl, *Nature*, **192**, 655 (1961).
[30a]A. Ryvolova-Kejharova and P. Zuman, *J. Electroanal. Chem.*, **21**, 197 (1969); [b]P. Carsky, P. Zuman, and V. Horak, *Coll. Czech. Chem. Commun.*, **30**, 4316 (1965).
[31a]C. Parkanyi and R. Zahradnik, *Coll. Czech. Chem. Commun.*, **27**, 1355 (1962); [b]E. Knobloch, *Advan. Polarography*, **3**, 875 (1960); [c]T. A. Geissman and S. L. Friess, *J. Amer. Chem. Soc.*, **71**, 3893 (1949).
[32]M. J. D. Brand and B. Fleet, *J. Electroanal. Chem.*, **16**, 341 (1968).
[33]A. J. Harle and L. E. Lyons, *J. Chem. Soc.*, 1575 (1950).
[34]J. Kossanyi, *Bull. Soc. Chim. Fr.*, **32**, 714 (1965).
[35]M. M. Baizer and J. D. Anderson, *J. Electrochem. Soc.*, **111**, 226 (1964).
[36]R. B. Woodward and R. Hoffmann, *Conservation of Orbital Symmetry*, Weinheim Bergstrasse, Verlag Chemie, 1970, chap. 5.
[37]J. Simonet, *Compt. Rend.*, **263C**, 1546 (1966).
[38a]W. M. Tolles and D. W. Moore, *J. Chem. Phys.*, **46**, 2102 (1967); [b]also see numerous references in G. A. Russell and G. R. Stevenson, *J. Amer. Chem. Soc.*, **93**, 2432 (1971).
[39]S. Wawzonek and A. Gundersen, *J. Electrochem. Soc.*, **111**, 324 (1964).

[40a]K. W. Bowers, R. W. Giese, J. Grimshaw, H. O. House, N. H. Kolodny, K. Kronberger, and D. K. Roe, *J. Amer. Chem. Soc.*, **92**, 2783 (1970); [b]R. W. Giese, J. Grimshaw, H. O. House, K. Kronberger, and D. K. Roe, *Extended Abstracts, 135th National Meeting*, The Electrochemical Society, New York, May 1969, p. 342.
[41]J. D. Anderson, M. M. Baizer, and E. J. Prill, *J. Org. Chem.*, **30**, 1645 (1965).
[42]M. M. Baizer, *J. Electrochem. Soc.*, **111**, 215 (1964).
[43]S. R. Missan, E. I. Becker, and L. Meites, *J. Amer. Chem. Soc.*, **83**, 58 (1961).
[44]S. Ono, *Nippon Kagaku Zasshi*, **77**, 665 (1956); see *Chem. Abstr.*, **52**, 9020 (1958).
[45a]M. E. Peover, in A. J. Bard, ed., *Electroanalytical Chemistry*, New York, Marcel Dekker, 1967, vol. 2, p. 1; [b]G. J. Hoijtink, in P. Delahay and C. W. Tobias, eds., *Advances in Electrochemistry and Electrochemical Engineering*, New York, Wiley, 1970, vol. 7, p. 221.
[46a]G. J. Hoijtink, J. van Schooten, E. de Boer, and W. Y. Aalbersberg, *Rec. Trav. Chim.*, **73**, 355 (1954); [b]G. J. Hoijtink, E. de Boer, P. H. van der Meij, and W. P. Weijland, *Rec. Trav. Chim.*, **75**, 487 (1956).
[47]K. S. V. Santhanam and A. J. Bard, *J. Amer. Chem. Soc.*, **88**, 2669 (1966).
[48]G. J. Hoijtink, *Rec. Trav. Chim.*, **73**, 895 (1954).
[49]S. Wawzonek and H. A. Laitinen, *J. Amer. Chem. Soc.*, **64**, 2365 (1942).
[50a]I. Bergman, in G. J. Hills, ed., *Polarography—1964*, New York, Wiley, 1966, vol. 2, p. 925; [b]A. Misono, T. Osa, and T. Yamagishi, *Bull. Chem. Soc. Japan*, **41**, 2921 (1968).
[51a]R. A. Benkeser and E. F. Kaiser, *J. Amer. Chem. Soc.*, **85**, 2858 (1963); [b]R. A. Benkeser, E. M. Kaiser, and R. F. Lambert, *J. Amer. Chem. Soc.*, **86**, 5272 (1964).
[52a]F. Fichter and C. Simon, *Helv. Chim. Acta.*, **32**, 1414 (1949); [b]S. Ono, *Nippon Kagaku Zasshi*, **75**, 1195 (1954); see *Chem. Abstr.*, **51**, 12704 (1957).
[53]G. B. Diamond and M. D. Soffer, *J. Amer. Chem. Soc.*, **74**, 4126 (1952).
[54]S. Wawzonek and D. Wearring, *J. Amer. Chem. Soc.*, **81**, 2067 (1959).
[55a]I. Bergman, *Trans. Faraday Soc.*, **50**, 829 (1954); [b]S. Wawzonek and J. Fan, *J. Amer. Chem. Soc.*, **68**, 2541 (1946); [c]G. J. Hoijtink and J. van Schooten, *Rec. Trav. Chim.*, **72**, 691, 903 (1953); [d]D. Lipkin, F. R. Galiano, and R. W. Jordan, *Chem. Ind.*, 1657 (1963).
[56]T. Shono, T. Toda, and R. Oda, *Tetrahedron Lett.*, 369 (1970).
[57a]R. Breslow, W. Bahary, and W. Reinmuth, *J. Amer. Chem. Soc.*, **83**, 1763 (1961); [b]R. Breslow and K. Balasubramanian, *J. Amer. Chem. Soc.*, **91**, 5183 (1969).
[58]R. Breslow, P. Gal, H. W. Chang, and L. J. Altman, *J. Amer. Chem. Soc.*, **87**, 5139 (1965).
[59]L. I. Krimer and D. J. Cota, *Organic Reactions*, **17**, 329 (1969).
[60]S. I. Zhdanov and A. N. Frumkin, *Proc. Acad. Sci. U.S.S.R.*, **122**, 659 (1958).
[61]M. Feldman and W. C. Flythe, *J. Amer. Chem. Soc.*, **91**, 4577 (1969).
[62]S. Wawzonek, R. Berkey, and D. Thomson, *J. Electrochem. Soc.*, **103**, 513 (1956).
[63]S. I. Zhdanov and M. I. Polievktov, *J. Gen. Chem. U.S.S.R.*, **31**, 3607 (1961).
[64]R. D. Allendoerfer and P. H. Rieger, *J. Amer. Chem. Soc.*, **87**, 2336 (1965).
[65]L. E. Craig, R. M. Elofson, and I. J. Ressa, *J. Amer. Chem. Soc.*, **75**, 480 (1953).
[66]L. B. Anderson, J. F. Hansen, T. Kakihana, and L. A. Paquette, *J. Amer. Chem. Soc.*, **93**, 161 (1971).
[67]C. L. Perrin, *Progr. Phys. Org. Chem.*, **3**, 165 (1965).
[68]A. T. Balaban, C. Bratu, and C. N. Rentea, *Tetrahedron*, **20**, 265 (1964).
[69]W. M. Schwarz, E. M. Kosower, and I. Shain, *J. Amer. Chem. Soc.*, **83**, 3164 (1961).
[70a]J. N. Burnett and A. L. Underwood, *J. Org. Chem.*, **30**, 1154 (1965); [b]J. Volke, M. Naarova, and V. Volkova, *Extended Abstracts, 21st Meeting*, Comite Internationale de Thermodynamique et de Cinetique Electrochimiques, Prague, Czechoslovakia, September 1970, p. 33.

[71]P. Karrer, G. Schwarzenbach, F. Benz, and U. Solmssen, *Helv. Chim. Acta*, **19**, 811 (1936).
[72]H. Lund, *Acta Chem. Scand.*, **21**, 2525 (1967).
[73]P. Zuman and O. Exner, *Coll. Czech. Chem. Commun.*, **30**, 1832 (1965).
[74]H. A. Laitinen and S. Wawzonek, *J. Amer. Chem. Soc.*, **64**, 1765 (1942).
[75]M. M. Baizer and J. D. Anderson, *J. Org. Chem.*, **30**, 1348 (1965).
[76]J. W. Loveland, U.S. Patent No. 3,032,489 (May 1, 1962); see *Chem. Abstr.*, **57**, 4470 (1962).
[77]R. Dietz and M. E. Peover, *Disc. Faraday Soc.*, **45**, 154 (1968).
[78]N. Yamazaki and S. Murai, *Chem. Commun.*, 147 (1968).
[79]R. Dietz, M. E. Peover, and R. Wilson, *J. Chem. Soc.*, (B), 75 (1968).
[80]A. J. Fry and R. G. Reed, *J. Amer. Chem. Soc.*, **91**, 6448 (1969).
[81]W. Kemula and J. Kornacki, *Roczniki Chem.*, **36**, 1835, 1857 (1962); see *Chem. Abstr.*, **59**, 218 (1963).

8 OXIDATION PROCESSES

At present, electrochemical oxidation of organic compounds is not nearly as valuable a technique as is electrochemical reduction. To a large degree this is related to the fact that until recently the mechanisms of anodic processes had been studied far less than had cathodic processes. That state of affairs is changing rapidly, however, and given the undoubted fact that much interesting and useful anodic chemistry is already known, it is clear that electrochemical oxidation will be an organic synthetic tool of steadily increasing value.

Somewhat paradoxically, in view of the fact that electrochemical oxidations of organic compounds have been studied a good deal less than have reductions, it happens that oxidation processes have been reviewed in the literature much more extensively in recent years than have the cathodic processes discussed in Chapters 5–7 of this book. For example, there are now available several good reviews of the Kolbe reaction, both from the point of view of its mechanism and its many attractive synthetic applications.[1–3] Furthermore, Adams has discussed the mechanisms of electrochemical oxidation of aromatic hydrocarbons and amines,[4] and Weinberg and Weinberg have compiled a valuable comprehensive survey of electroorganic oxidations, with emphasis on products and mechanisms.[5] (The latter review contains, incidentally, the best available compilation of oxidation potentials of organic compounds.) Because of the existence of these several reviews, which themselves attest to the increasing vitality of this area in the last decade, the present chapter need not present as extensive a discussion as might otherwise have been necessary. Rather than presenting a survey of the oxidative behavior of a broad spectrum of structural types, there will be outlined here instead the principal pathways encountered upon oxidation of a few important compound types. As in previous chapters, the emphasis will be on common mechanistic elements and the more synthetically attractive applications of each process. Some recent mechanistic advances in this area, not covered in the above-mentioned reviews, are also included because of their synthetic implications.

8.1 CARBOXYLIC ACIDS

It has been known for considerably more than a century that electrochemical oxidation of salts of carboxylic acids results in dimeric products[6]:

$$2RCO_2^- \xrightarrow{[O]} 2CO_2 + RR$$

This reaction is known as the *Kolbe reaction* after the early investigator who, though not its discoverer, studied it in sufficient detail to outline its scope. Perhaps no other organic reaction has been studied as intensively as has the Kolbe reaction; indeed, investigations into its subtler aspects have continued right up to the present day. While some mechanistic details still remain unclear, the preparative methodology and mechanism are sufficiently well established that synthetic applications are straightforward. It has also been discovered in recent years that under certain conditions the oxidation can afford carbonium ions:

$$RCO_2^- \xrightarrow{[O]} R^+$$

This reaction and its relation to the Kolbe reaction will be of interest in the following mechanistic and synthetic discussions.

8.1.1 The Kolbe and Related Reactions

Mechanism

Weedon has discussed optimal experimental procedures for the Kolbe reaction.[1] Best yields (50–90%) of dimer RR are obtained when the carboxylic acid is only partially neutralized, then oxidized at low applied potential at a platinum anode in methanol[1] or DMF.[7] Under these conditions, the carboxylate anion suffers 1-electron oxidation to an acyloxy radical (**1**), which loses carbon dioxide to give radical **2**, which then dimerizes.

$$RCO_2^- - e^- \longrightarrow \underset{\mathbf{1}}{RCO_2\cdot} \xrightarrow{-CO_2} \underset{\mathbf{2}}{R\cdot}$$

$$2\ \mathbf{2} \longrightarrow RR$$

The review by Vijh and Conway may be consulted for convincing evidence that the Kolbe oxidation is, in fact, a reaction of *adsorbed* radicals.[2] A number of its characteristic features point to this; perhaps the most striking from an organic chemical point of view is the high ratio of radical coupling to disproportionation products as compared to the reactions of the same radicals generated in homogeneous solution by other methods, such as thermolysis of acyl peroxides.[2,8] It is this very feature, of course, that makes the Kolbe reaction of synthetic utility. Another surprising aspect of the Kolbe is apparently also associated with the adsorption of intermediates on the anode surface. This is the fact that the Kolbe oxidation often proceeds at a *more positive* anode potential than does oxidation of the solvent itself. For example, one can measure the anodic decomposition potential of slightly alkaline water, that is, the potential at which oxygen evolution is noticeable. If one then adds a carboxylate salt to the same solution, it is found that a more positive anode potential can be reached before gas evolution occurs, and, furthermore, the gas is now found to be carbon dioxide rather than oxygen. Apparently, a film of adsorbed intermediates inhibits oxidation of the solvent and permits a potential to be attained where the Kolbe reaction can then occur. This feature is of interest synthetically because it results in considerable experimental simplification. Since anodic potential control (to avoid oxidation of the solvent) is rendered unnecessary by this adsorption process, one may use constant-current, rather than controlled-potential, electrolysis. As Weedon has pointed out, the experimental apparatus for the Kolbe reaction can be quite simple, merely a beaker containing both the solution and a pair of platinum electrodes connected to a source of direct current (see Chapter 9).[1]

Eberson[9] has suggested that electron transfer from the carboxylate ion may be concerted with carbon–carbon bond breaking, that is, that the reaction may be represented as a one-step process,

$$RCO_2^- - e^- \longrightarrow R\cdot + CO_2$$

without formation of the acyloxy radical **1**. It appears, however, that at least in some cases the acyloxy radical is actually a discrete intermediate. For example,

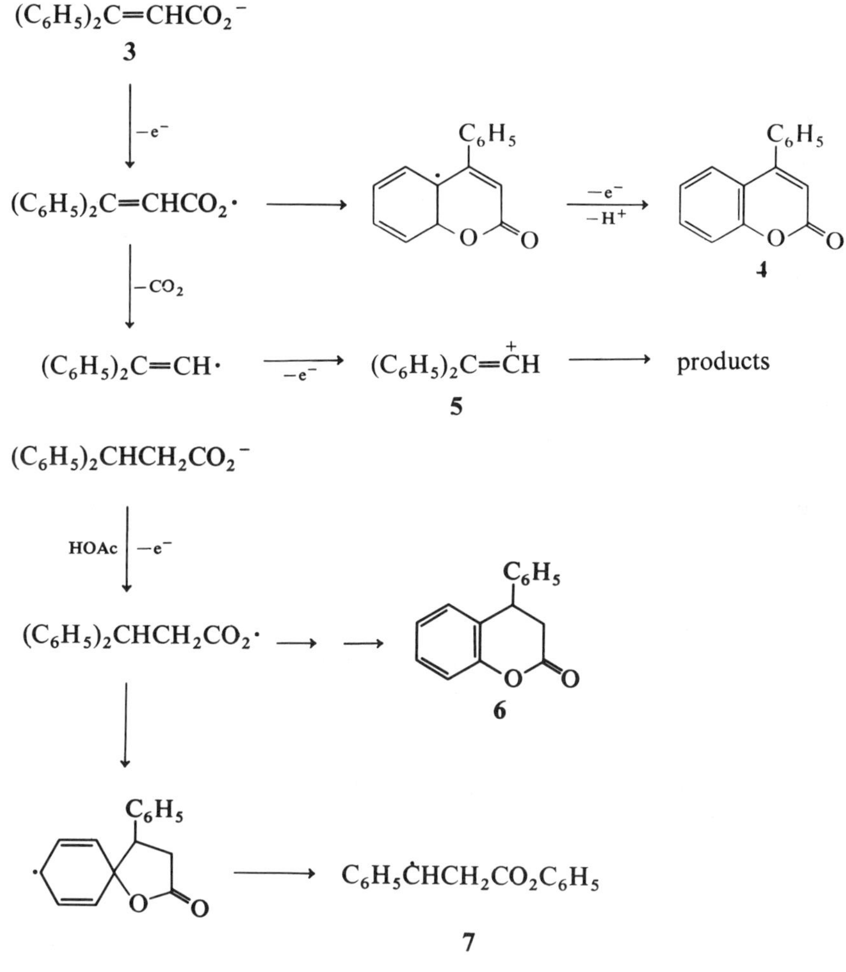

$$(C_6H_5)_2C{=}CHCO_2^- \quad \mathbf{3}$$

$$\downarrow -e^-$$

$$(C_6H_5)_2C{=}CHCO_2\cdot \longrightarrow \text{(radical, } C_6H_5\text{)} \xrightarrow[-H^+]{-e^-} \text{(}C_6H_5\text{ coumarin) } \mathbf{4}$$

$$\downarrow -CO_2$$

$$(C_6H_5)_2C{=}CH\cdot \xrightarrow[-e^-]{} (C_6H_5)_2C{=}\overset{+}{C}H \ \mathbf{5} \longrightarrow \text{products}$$

$$(C_6H_5)_2CHCH_2CO_2^-$$

$$\downarrow \text{HOAc}, -e^-$$

$$(C_6H_5)_2CHCH_2CO_2\cdot \longrightarrow \longrightarrow \text{(}C_6H_5\text{ dihydrocoumarin) } \mathbf{6}$$

$$\downarrow$$

$$\text{(}C_6H_5\text{ spiro lactone radical)} \longrightarrow C_6H_5\dot{C}HCH_2CO_2C_6H_5 \quad \mathbf{7}$$

$$\mathbf{7} \xrightarrow{-e^-} C_6H_5\overset{+}{C}HCH_2CO_2C_6H_5$$

$$\downarrow$$

$$\underset{\mathbf{8}}{C_6H_5CH{=}CHCO_2C_6H_5} + \underset{\mathbf{9}}{C_6H_5\underset{}{\overset{\overset{\displaystyle OAc}{|}}{C}}HCH_2CO_2C_6H_5}$$

electrochemical oxidation of 3,3-diphenylacrylic acid (**3**) in acetic acid affords 4-phenylcoumarin (**4**), along with several other products apparently derived from the vinyl cation **5**.[10] Similarly, oxidation of 3,3-diphenylpropanoic acid in acetic acid affords the dihydrocoumarin **6** along with products (**8** and **9**) that apparently arise by 1,4-phenyl migration.[11] Apparent phenyl migrations have also been observed[12] during electrochemical oxidation of 3,3,3-triphenylpropanoic acids:

$$(p\text{-}X{-}C_6H_4)_3CCH_2CO_2^- \xrightarrow[CH_3OH]{-e^-} (p\text{-}X{-}C_6H_4)_2\underset{\underset{\displaystyle OCH_3}{|}}{C}CH_2CO_2C_6H_4{-}X\text{-}p \qquad (X = H, t\text{-Bu})$$

Other cyclizations involving acyloxy radicals during attempted Kolbe oxidation include that of biphenyl-2-carboxylic acid to the lactone of 2-hydroxybiphenyl-2′-carboxylic acid,[13] and those of benzophenone-2-carboxylic acids to arylalkyl phthalates.[14] All of these cyclizations may appear surprising in view of the short lifetimes characteristic of acyloxy radicals in homogeneous solution, but as Vijh and Conway have pointed out,[2] nothing is known about the lifetimes of radicals adsorbed on noble metal surfaces. Indeed, as we have already seen, radicals have a much higher tendency to couple (relative to disproportionation) when adsorbed as opposed to their bulk-solution chemistry, and it would not be surprising to find that other chemical properties were also affected by adsorption.

Synthetic Scope and Limitations

Kolbe dimers are formed in 50–90% yield when the intermediate radical is either primary[1] or is substituted by electron-withdrawing groups,[15] that is,

$$R\dot{C}H_2 \quad \text{or} \quad R\dot{C}HX \quad \text{or} \quad R\dot{C}X_2 \qquad (X = CO_2R', CONR_2', \text{etc.})$$

When the α position of the acid contains a substituent capable of stabilizing a carbonium ion (X = alkyl, alkoxyl, halogen, aryl, etc.), yields of the Kolbe dimer are very low (0–10%). In such cases, the initially formed radical is oxidized further to a carbonium ion:

$$RCO_2^- - e^- \xrightarrow{-CO_2} R\cdot \xrightarrow{-e^-} R^+ \longrightarrow \text{products}$$

This reaction will be discussed in more detail in a later section (Section 8.1.2). There are a few other cases where Kolbe yields are low. Aryl[16] and α,β-unsaturated acids[17] afford little or no dimer, for example. The isolation of benzene from the oxidation of benzoic acid,[16] or **4** and other products from oxidation of **3**[10] show, however, that the carboxylate anions are oxidizable, hence the failure to obtain dimers must be because the intermediate radicals are diverted into other paths. β,γ-Unsaturated acids undergo Kolbe coupling, but afford mixtures of products derived from the intermediate allylic radicals[18]:

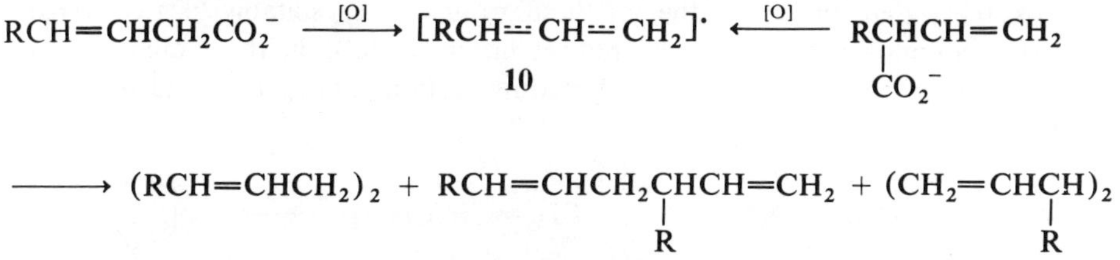

$$RCH{=}CHCH_2CO_2^- \xrightarrow{[O]} \underset{\mathbf{10}}{[RCH{\cdots}CH{\cdots}CH_2]^\cdot} \xleftarrow{[O]} \underset{\underset{CO_2^-}{|}}{RCH}CH{=}CH_2$$

$$\mathbf{10} \longrightarrow (RCH{=}CHCH_2)_2 + RCH{=}CHCH_2\underset{\underset{R}{|}}{C}HCH{=}CH_2 + (CH_2{=}CH\underset{\underset{R}{|}}{C}H)_2$$

Other than these, there are really very few restrictions on the nature of the R group in a carboxylic acid (RCO_2H) that is to be submitted to the Kolbe reaction. Hydroxy, acetoxy, alkoxy, keto, carboalkoxy, cyano, amido, halo, nitro, olefinic, acetylenic, and ketal acids have all been successfully dimerized via the Kolbe reaction.[1] Some interesting examples are shown below.

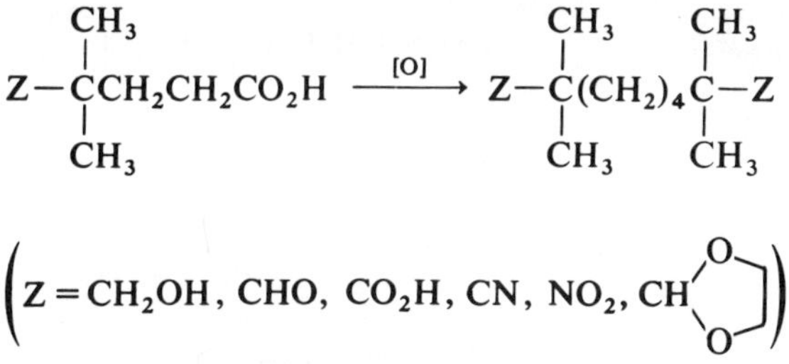

$$Z{-}\underset{CH_3}{\overset{CH_3}{C}}CH_2CH_2CO_2H \xrightarrow{[O]} Z{-}\underset{CH_3}{\overset{CH_3}{C}}(CH_2)_4\underset{CH_3}{\overset{CH_3}{C}}{-}Z \qquad (Ref.\ 19)$$

$$\left(Z = CH_2OH,\ CHO,\ CO_2H,\ CN,\ NO_2,\ CH\langle\text{1,3-dioxolan-2-yl}\rangle\right)$$

$$\underset{\mathbf{11}}{RO_2C(CH_2)_nCO_2H} \xrightarrow{[O]} \underset{\mathbf{12}}{RO_2C(CH_2)_{2n}CO_2R} \qquad (Ref.\ 1)$$

$$CH_3CH_2CO(CH_2)_4CO_2H \xrightarrow{[O]} CH_3CH_2CO(CH_2)_8COCH_2CH_3 \qquad (Ref.\ 20)$$

$$CH_3CONH(CH_2)_5CO_2H \xrightarrow{[O]} CH_3CONH(CH_2)_{10}NHCOCH_3 \qquad (Ref.\ 21)$$

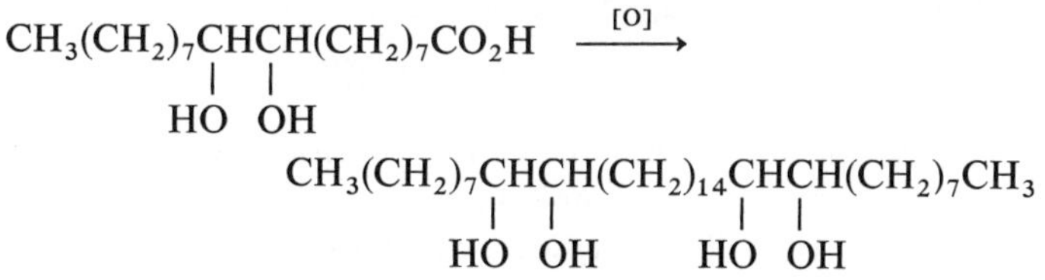

$$CH_3(CH_2)_7\underset{HO}{\underset{|}{C}}H\underset{OH}{\underset{|}{C}}H(CH_2)_7CO_2H \xrightarrow{[O]} CH_3(CH_2)_7\underset{HO}{\underset{|}{C}}H\underset{OH}{\underset{|}{C}}H(CH_2)_{14}\underset{HO}{\underset{|}{C}}H\underset{OH}{\underset{|}{C}}H(CH_2)_7CH_3 \qquad (Ref.\ 22)$$

Notice in the first example that a diacid ($Z = CO_2H$) can be selectively dimerized at the primary carboxyl group; this may be a steric effect. The oxidation of **11** to **12** is shown here to illustrate the usual method of selective dimerization of dicarboxylic acids: anodic oxidation of the half acid-ester of the diacid, the so-called *Brown–Walker reaction*, to an α,ω-diester.[1] The last example is notable because it demonstrates the fact that even the readily oxidized *vic*-diol moiety is stable to Kolbe conditions. Of course, the fact that many functional groups are unchanged under the conditions of the Kolbe reaction does not preclude the possibility that one or another of these groups, when suitably located, might react intermolecularly with the radical intermediate(s) in the Kolbe, as, for example, in the reactions of the preceding section leading to **4** and **6**, or the following example:

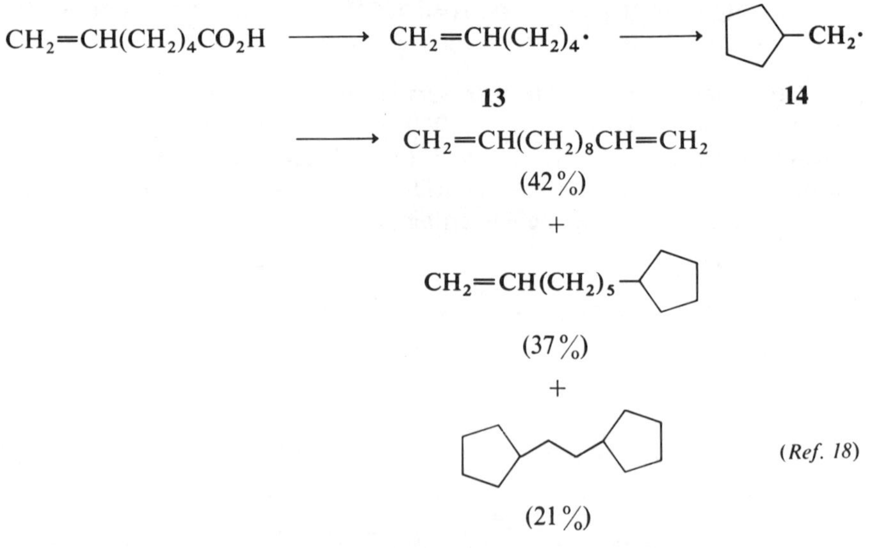

$$CH_2{=}CH(CH_2)_4CO_2H \longrightarrow \underset{\mathbf{13}}{CH_2{=}CH(CH_2)_4\cdot} \longrightarrow \underset{\mathbf{14}}{\text{(cyclopentyl)}{-}CH_2\cdot}$$

$$\longrightarrow \underset{(42\%)}{CH_2{=}CH(CH_2)_8CH{=}CH_2} + \underset{(37\%)}{CH_2{=}CH(CH_2)_5{-}\text{(cyclopentyl)}} + \underset{(21\%)}{\text{(cyclopentyl)}{-}CH_2CH_2{-}\text{(cyclopentyl)}}$$

(*Ref. 18*)

It is interesting that oxidation of cyclopentylacetic acid gives no product derived from the linear radical **13**,[18] indicating that cyclization of **13** to **14** is irreversible at the electrode surface. The fact that a mixture of products derived from both **13** and **14** is formed in the above reaction demonstrates that the rates of product formation by radical coupling and cyclization of **13** are comparable in magnitude.

A valuable extension of the Kolbe reaction was first reported by Wurtz.[23] He found that oxidation of a mixture of two different carboxylic acids results in a mixture of the three possible coupling products:

$$RCO_2H + R'CO_2H \xrightarrow[-CO_2]{-e^-} RR + RR' + R'R'$$

The fact that the three are formed in more or less statistical ratio need not be a deterrent, for by proper choice of the molecular size of R and R′ the products can be made to differ substantially in physical properties, thus making separation easy, and use of a large excess of the cheaper acid increases the relative yield of the desired mixed product RR′. For example, the second of the following routes to *n*-nonane (**15**) would be clearly superior, because the three products differ to a much greater degree in boiling point than those from the first route, and also because one could use a large excess of the inexpensive component propionic acid to improve the conversion of octanoic acid to **15**.

$$n\text{-}C_5H_{11}CO_2H + n\text{-}C_4H_9CO_2H \longrightarrow n\text{-}C_{10}H_{22} + \underset{\mathbf{15}}{n\text{-}C_9H_{20}} + n\text{-}C_8H_{18}$$

$$n\text{-}C_7H_{15}CO_2H + CH_3CH_2CO_2H \longrightarrow n\text{-}C_{14}H_{30} + \mathbf{15} + n\text{-}C_4H_{10}$$

(One common application of this *crossed Kolbe* coupling is the oxidation of a carboxylic acid in the presence of a large excess of acetate ion, thus resulting in efficient conversion of the acid RCO_2H to the alkane RCH_3 in one step.[10–12] This reaction would be useful for introduction of a labeled methyl group.)

The crossed Kolbe coupling reaction has the dual advantages of being able to accommodate a wide variety of functional groups and of enabling quick assembly of molecular subunits whose coupling by other routes might be lengthy. In these respects the reaction bears a close operational resemblance to the silver ion promoted coupling reaction of organoboranes reported by Brown and coworkers[24]:

$$RCH{=}CH_2 + R'CH{=}CH_2 \xrightarrow[(2)\ AgOH]{(1)\ BH_3} R(CH_2)_4R + R(CH_2)_4R' + R'(CH_2)_4R'$$

Which of these coupling reactions would be selected in a given situation would depend on the relative availability of the two olefins *vis-à-vis* the two carboxylic acids. In some cases, the Kolbe reaction will possess certain advantages over borane coupling, such as for the synthesis of dineopentyl structures (**16**), and the

$$\underset{\mathbf{16}}{R_3C(CH_2)_2CR_3}$$

fact that yields of disproportionation products are very low in the Kolbe reaction. On the other hand, coupling of secondary radicals can only be carried out via borane coupling.

A useful procedure for chain lengthening of carboxylic acid derivatives was first reported by von Miller and Hofer.[25] They carried out crossed coupling reactions between a carboxylic acid (**17**) and the half-ester of a diacid. Focusing attention on the mixed dimer (**18**), it will be noted that the reaction has the net effect of extending the chain length of **17** by any desired number of carbon atoms, depending only on the value of *n* in the half acid-ester. The products are generally easily separated, and of course an excess of the cheaper reagent can be used to improve the efficiency of conversion of **17** to **18**. This procedure for

$$\underset{\mathbf{17}}{RCO_2H + HO_2C(CH_2)_nCO_2R'} \xrightarrow[-CO_2]{-e^-} RR + \underset{\mathbf{18}}{R(CH_2)_nCO_2R'} + R'O_2C(CH_2)_{2n}CO_2R'$$

chain lengthening has been used extensively by investigators in the field of terpene and lipid synthesis, for example, in syntheses of phytol,[1] muscone,[26,27] erucic and nervonic[22,28] acids, (−)-tuberculostearic acid (**20**),[29] and humulane (**26**).[30] The latter two syntheses incorporate novel and potentially very useful features. The synthesis of **20** involves two successive crossed Kolbe reactions; this procedure is particularly useful for assembling larger molecules where the necessary smaller fragments are readily available. It has found frequent application in synthesis of long branched-chain structures. Sorm and coworkers used this

$$CH_3(CH_2)_6CO_2H + HO_2CCH_2\underset{\displaystyle CH_3}{\underset{|}{C}}HCH_2CO_2CH_3 \xrightarrow[\text{(2) hydrolysis}]{\text{(1) crossed Kolbe}} \underset{\mathbf{19}}{CH_3(CH_2)_7\underset{\displaystyle CH_3}{\underset{|}{C}}HCH_2CO_2H}$$

$$\mathbf{19} + HO_2C(CH_2)_7CO_2CH_3 \xrightarrow[\text{(2) hydrolysis}]{\text{(1) crossed Kolbe}} \underset{\mathbf{20}}{CH_3(CH_2)_7\underset{\displaystyle CH_3}{\underset{|}{C}}H(CH_2)_8CO_2H}$$

principle in their synthesis of humulane (**26**).[30] A more notable feature of this synthesis, however, is the initial crossed coupling of a methyl half-ester (**21**) with a benzyl half-ester (**22**). The product (**23**) could be selectively hydrogenolyzed to a new half-ester (**24**), which was subjected to a second crossed

$$\underset{\mathbf{21}}{CH_3O_2C\underset{\displaystyle CH_3}{\underset{|}{C}}HCH_2CH_2CO_2H} + \underset{\mathbf{22}}{HO_2CCH_2\overset{\displaystyle CH_3}{\overset{|}{\underset{\displaystyle CH_3}{\underset{|}{C}}}}CH_2CO_2CH_2C_6H_5} \xrightarrow{[O]}$$

$$\underset{\mathbf{23}}{CH_3O_2C\underset{\displaystyle CH_3}{\underset{|}{C}}H(CH_2)_3\overset{\displaystyle CH_3}{\overset{|}{\underset{\displaystyle CH_3}{\underset{|}{C}}}}CH_2CO_2CH_2C_6H_5}$$

$$\mathbf{23} \xrightarrow[\text{catalyst}]{H_2} CH_3O_2CCH(CH_3)(CH_2)_3C(CH_3)_2CH_2CO_2H \quad \mathbf{24}$$

$$\mathbf{24} + HO_2CCH_2CH(CH_3)CH_2CO_2CH_3$$

$$\xrightarrow{[O]} CH_3O_2CCH(CH_3)(CH_2)_3C(CH_3)_2CH_2CH_2CH(CH_3)CH_2CO_2CH_3 \quad \mathbf{25}$$

$$\mathbf{25} \longrightarrow \longrightarrow \mathbf{26}$$

Kolbe reaction with methyl hydrogen β-methylglutarate. The resulting diester (**25**) was then converted to **26** by standard methods (acyloin ring closure followed by reduction).

In recent years, several electrolytic oxidations of carboxylic acids that do not lead to Kolbe dimers have been reported. Corey and coworkers, for example, observed skeletal rearrangement upon electrochemical oxidation of β-hydroxy acids[31]:

$$\text{cyclic ketone } (CH_2)_n \xrightarrow[\text{(2) hydrolysis}]{\text{(1) Reformatsky}} \text{1-hydroxy-1-}(CH_2CO_2H)\text{ cycloalkane } (CH_2)_n \xrightarrow[-CO_2]{-e^-} \text{cyclic ketone } (CH_2)_{n+1}$$

$$(n = 0, 1, 2)$$

$$(C_6H_5)_2C(OH)CH(C_6H_5)CO_2H \xrightarrow[-CO_2]{-e^-} C_6H_5C(=O)CH(C_6H_5)_2$$

Rand and Rao investigated the oxidation of β-hydroxy acids in more detail, and found it to be conformationally unselective: both isomers of **27** give the same mixture of ketones **28** and **29**.[32] Ring expansion and expulsion of carbon dioxide are probably concerted.

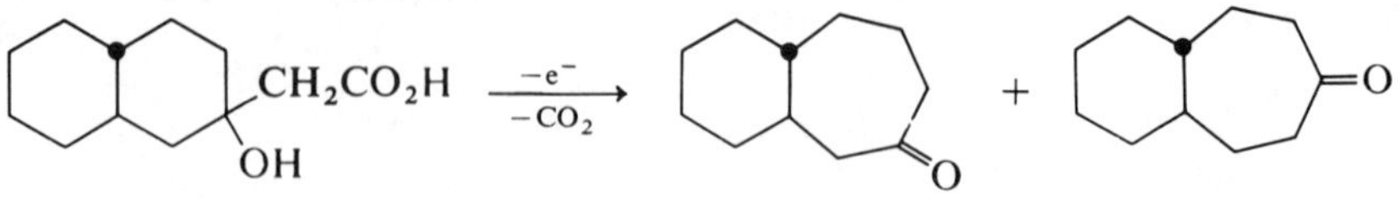

CH_2CO_2H, OH $\xrightarrow[-CO_2]{-e^-}$ O + =O

27 **28** **29**

Oxidation of γ-hydroxy acids has been found to result in carbon–carbon bond cleavage, and sometimes in rearrangement[33]:

$$C_6H_5CH_2C(CH_3)(OH)CH_2CH_2CO_2H \xrightarrow[-CO_2]{-2\,e^-} C_6H_5CH_2C(CH_3)(OH)CH_2CH_2^{+} \longrightarrow$$

$$C_6H_5CH_2\overset{O}{\overset{\|}{C}}CH(CH_3)_2 + CH_3\overset{O}{\overset{\|}{C}}CH(CH_2C_6H_5)CH_3 + C_6H_5CH_2\overset{O}{\overset{\|}{C}}CH_3$$

$$+ CH_2{=}CH_2$$

Wharton and coworkers used this reaction to prepare *trans*-5-cyclodecenone in fair yield.[34] The conditions used were clearly inappropriate, however. Such syntheses are probably best effected using carbon anodes, high applied voltage, and alkaline media, that is, conditions favoring formation of carbonium ions (*vide infra*). Where the normal Kolbe product is desired from a γ-hydroxy acid, these cleavages and rearrangements can be eliminated simply by acetylating the hydroxyl before oxidation.[33]

A number of research groups have found that electrochemical oxidation of *vic*-dicarboxylic acids represents a convenient olefin synthesis; for example,

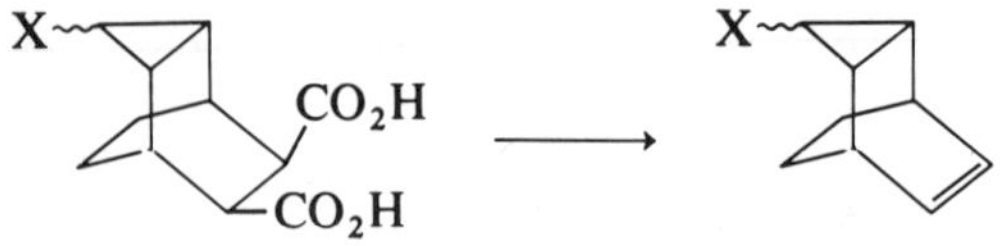

X, CO_2H, CO_2H $\longrightarrow$ X (*Ref. 35*)

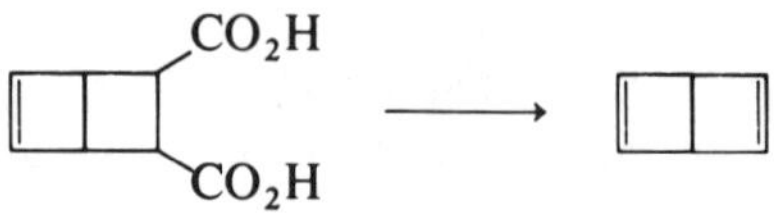

CO_2H, CO_2H $\longrightarrow$ (*Ref. 36*)

Although the same conversion can be effected by the use of lead tetra-acetate, yields are considerably better using the electrochemical method. (The two carboxyl groups are presumably not lost simultaneously, judging from the fact that *dl*- and *meso*-2,3-diphenylsuccinic acids both afford *trans*-stilbene upon electrochemical oxidation.[31a]) It has been pointed out that a sequence consisting of cycloaddition of maleic anhydride followed by hydrolysis and oxidation is tantamount to overall cycloaddition of acetylene.[35b]

8.1.2 Carbonium Ion Formation

As was noted in the preceding section, the yield of Kolbe dimer is very low from oxidation of a carboxylic acid containing an electron-supplying group at the α-position. Typical products from such acids are alcohols (the *Hofer–Moest* reaction), ethers, and olefins. Furthermore, it is not uncommon to find that these products possess rearranged carbon skeletons. In other words, the products are just those that would be expected if a carbonium ion were generated in the oxidation. It is now generally believed that such products do arise from cationic intermediates formed by further oxidation of the initial radical:

$$RCO_2^- \xrightarrow[-CO_2]{-e^-} R\cdot \xrightarrow{-e^-} R^+ \longrightarrow \text{products}$$

An approximate thermochemical analysis[9] carried out by Eberson several years ago led to the conclusion that those radicals whose ionization potential is less than 8 eV, as measured by the electron-impact method, will be oxidized further to carbonium ions when generated under Kolbe conditions, thus restricting the Kolbe dimerization to those radicals of ionization potential greater than 8 eV. Although Vijh and Conway[2] have rightly criticized Eberson's analysis on the grounds that it does not take heats of adsorption of intermediates into account, it does appear to correlate the available data fairly well. For example, the ionization potentials of primary and negatively substituted radicals all lie well above 8 eV (methyl, 9.95; ethyl, 8.78; butyl, 8.64; CH_2CN, 10.87; $(CH_3)_2CCN$, 9.15),[37] and for such structures R, it is known that RCO_2H affords Kolbe dimers, generally in good yields. Conversely, secondary, tertiary, halo, oxygenated, and so on, radicals have low ionization potentials (*i*-propyl, 7.90; *t*-butyl, 7.42)[37] and oxidation of the corresponding acids results in products derived from carbonium ions. This approximate correlation between ionization potential and ease of Kolbe dimerization applies, however, only to those conditions that are known to be optimal for carrying out the Kolbe reaction, that is, methanol as solvent, the carboxylic acid only partially neutralized, a smooth platinum anode, and low applied voltage.[1] It appears, on the other hand, that even primary radicals can be converted to carbonium ions by oxidation in highly alkaline media and at high applied voltage. Conversely, phenylacetic acid[38] and even *p*-methoxyphenylacetic acid[39] both afford some dimer under optimal Kolbe conditions, even though the ionization potential of the benzyl radical (7.76 eV) suggests that no dimer should be formed. Furthermore, the nature of the electrode surface can affect the course of reaction: carbonium ion products are formed to a considerably greater degree at a carbon anode than at platinum, all other conditions being equal.[40] Finally, use of DMF as solvent enhances the yield of Kolbe product even from acids in which the radical has a low ionization potential: diphenylacetic acid affords no dimer (instead, there is obtained an 80% yield of benzhydryl methyl ether) when oxidized in methanol, but a 24% yield of dimer in DMF, and the yield of 2,3-diphenylbutane from α-phenylpropionic acid rises from 21 to 41% upon going from methanol to DMF.[7a] Thus, it would appear that by proper manipulation of experimental conditions most acids may be converted either into their Kolbe dimers or into the corresponding carbonium ions. At the very least, it should be possible to tip the balance sufficiently to make either process account for a significant fraction of the product mixture.

The electrochemical generation of carbonium ions from carboxylic acids exhibits several features that make it of synthetic interest. Eberson and Nyberg, for example, prepared acetyl-*t*-butylamine (**30**) by oxidation of pivalic acid in wet acetonitrile[41]:

$$(CH_3)_3CCO_2H \xrightarrow{[O]} (CH_3)_3C^+ \xrightarrow[H_2O]{CH_3CN} \underset{\mathbf{30}}{(CH_3)_3CNHCOCH_3}$$

The amide **30** can also be prepared by a Ritter reaction between *t*-butanol and acetonitrile in strong mineral acid (a reaction also involving the *t*-butyl cation), but these authors point out that the electrochemical synthesis of amides would be preferable when the carboxylic acid RCO_2H is more readily accessible than the alcohol ROH, or where the group R is itself acid labile, for example,

$$\begin{matrix} (CH_3)_2CCO_2H \\ | \\ (CH_3)_2CCO_2CH_3 \end{matrix} \xrightarrow[CH_3CN,\ H_2O]{[O]} \begin{matrix} (CH_3)_2CNHCOCH_3 \\ | \\ (CH_3)_2CCO_2CH_3 \end{matrix}$$

In a similar type of reaction, Władislaw and Ayres showed that oxidation of α-alkoxy acids in alcoholic solvents affords ketals in good yield via the intermediate α-alkoxycarbonium ions.[42] The reaction can be used to make the rare and otherwise difficult-to-prepare mixed ketals and acetals, such as **31**. These investigators studied only substituted phenylacetic acids, but the reaction should also be applicable to aliphatic acids.

$$\begin{matrix} OCH_3 \\ | \\ C_6H_5CCO_2H \\ | \\ R \end{matrix} \xrightarrow[CH_3OH]{[O]} \begin{matrix} C_6H_5C(OCH_3)_2 \\ | \\ R \end{matrix} \qquad R = H, C_6H_5$$

$$\begin{matrix} OC_2H_5 \\ | \\ C_6H_5CCO_2H \\ | \\ R \end{matrix} \xrightarrow[CH_3OH]{[O]} \begin{matrix} OC_2H_5 \\ | \\ C_6H_5C\text{-}OCH_3 \\ | \\ R \end{matrix}$$

31a, R = H, 71%
31b, R = C_6H_5, 74%

A number of studies have demonstrated that electrochemically generated carbonium ions are "hot"; that is, their reactions usually resemble those of carbonium ions generated by diazotization of amines rather than those formed by solvolysis. Thus, (1) oxidation of medium-ring cycloalkanecarboxylic acids results in a much higher percentage of bicyclic products than does solvolysis,[43] (2) oxidation of butyric acid, particularly at a carbon anode, affords substantial amounts of cyclopropane,[44] and (3) oxidation of allylacetic, cyclopropylacetic, and cyclobutanecarboxylic acids affords products derived from incompletely equilibrated cations.[45] On the other hand, anodic oxidation of *exo*- or *endo*-norbornanecarboxylic acids in methanol affords only *exo*-norbornyl methyl ether, and the optically active *endo*-acid affords racemic ether.[31b] The cationic intermediate in the latter oxidations resembles that in solvolysis, *not* deaminations. The question of the nature of anodically generated carbonium ions must, therefore, be considered unsettled at present. Part of the confusion in this area is no doubt due to the variety of experimental conditions that have been used. Comparisons between individual studies of this type would doubtless be considerably easier if all were run under similar conditions of substrate concentration, solvent, pH, electrode material, current density, and so on.

Oxidation of cyclopropanecarboxylic acids usually yields products formally derived from allyl cations such as

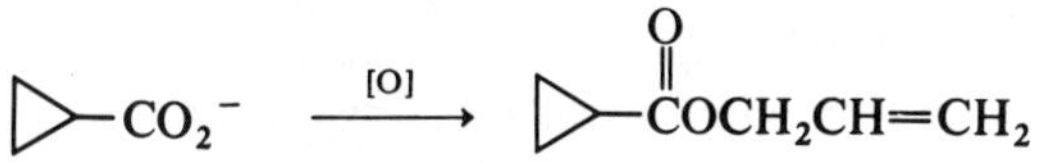

$$\triangleright\!-CO_2^- \xrightarrow{[O]} \triangleright\!-\overset{O}{\overset{\|}{C}}OCH_2CH{=}CH_2 \qquad (Ref.\ 46)$$

Shono, Nishiguchi, Yamane, and Oda have made the interesting discovery that ring opening proceeds stereospecifically[47]:

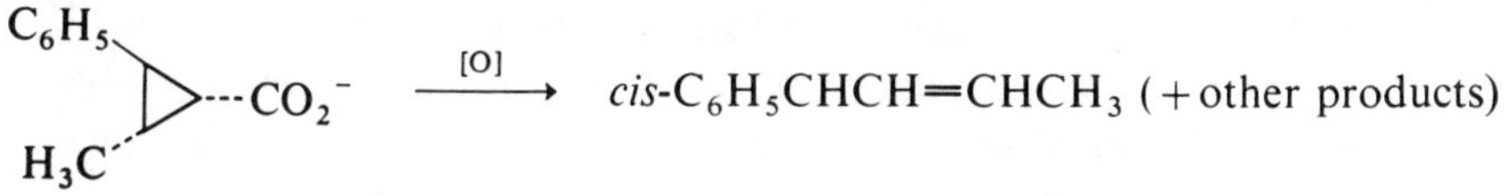

$$\mathbf{32}\ (C_6H_5,\ H_3C\text{-substituted cyclopropane}{-}CO_2^-) \xrightarrow{[O]} cis\text{-}C_6H_5CHCH{=}CHCH_3 \ (+\text{other products})$$

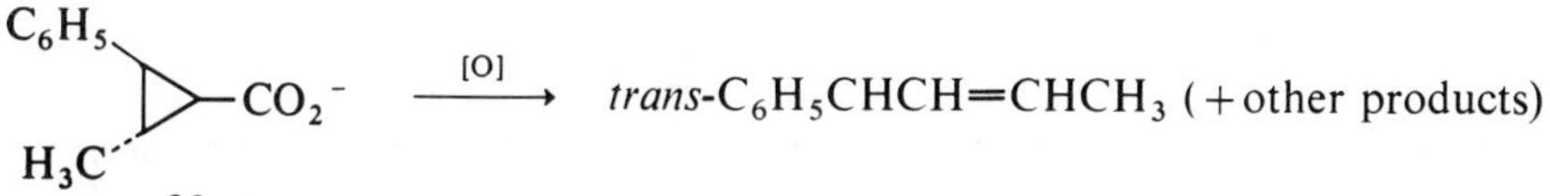

$$\mathbf{33}\ (C_6H_5,\ H_3C\text{-substituted cyclopropane}{-}CO_2^-) \xrightarrow{[O]} trans\text{-}C_6H_5CHCH{=}CHCH_3 \ (+\text{other products})$$

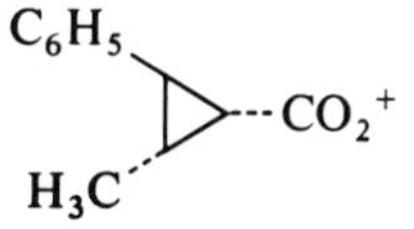

$$\mathbf{34}\ (C_6H_5,\ H_3C\text{-substituted cyclopropane}{-}CO_2^+) \qquad \mathbf{35}\ (C_6H_5,\ H_3C\text{-substituted cyclopropane}{-}CO_2^+)$$

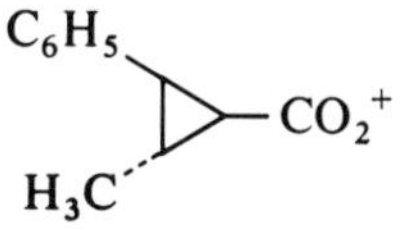

It was argued that cyclopropyl radicals or cations could not be intermediates in the oxidation, since these should open nonstereospecifically. It was therefore suggested that the reactions proceed via oxidation to the acyloxonium ions **34**

and **35**, followed by loss of carbon dioxide concerted with ring opening. Orbital symmetry considerations[48] predict that the latter process should be stereospecific. As the authors pointed out, acyloxonium ions might be more likely in these reactions than in oxidation of other carboxylates, since the acyloxy radicals formed by 1-electron oxidation of **32** and **33** could conceivably decarboxylate more slowly than acyclic acyloxy radicals.

8.2 AROMATIC COMPOUNDS

Along with the Kolbe reaction, the most intensively studied organic anode processes in recent years have been the oxidations of aromatic hydrocarbons, amines, and phenols. This area exhibits a wealth of mechanistic detail as well as a number of experimentally challenging problems. The newer experimental techniques, particularly cyclic voltammetry, controlled-potential electrolysis, the rotated-disk electrode, and use of dipolar aprotic solvents have, however, contributed greatly to advances in understanding of the mechanisms of oxidation of such compounds.

8.2.1 Hydrocarbons

Oxidation Mechanisms

A number of reactions can happen when one oxidizes a solution containing an aromatic hydrocarbon. Which actually does take place is a function of the structure of the aromatic hydrocarbon and the environment (solvent and other components of the medium) in which the oxidation is performed. Although much still remains to be learned concerning which of a variety of possible oxidation paths will be followed in a specific case, the broad outlines are clear. Scrutiny of the mechanistic complexities of hydrocarbon oxidation will be rewarding not only as a means of predicting the best experimental conditions for carrying out reactions of interest but also in a wider context, since the chemical principles learned here can be applied to all oxidative processes.

The most common reactions observed in the electrochemical oxidation of aromatic compounds can be classified generally as side-chain substitution, nuclear substitution, addition, and dimerization. Typical examples are shown below.

Side-Chain Substitution

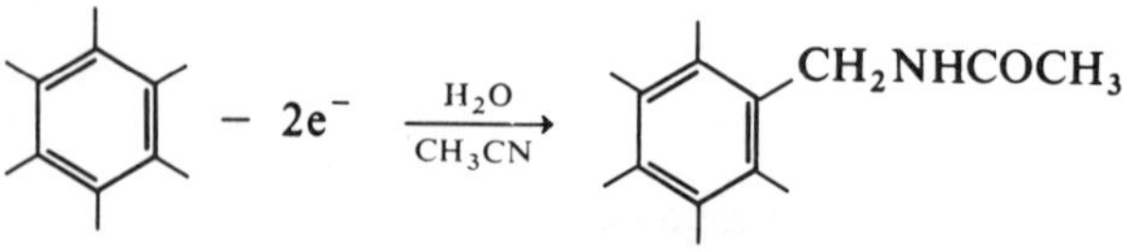

$$\mathbf{36} \; - \; 2e^- \xrightarrow[CH_3CN]{H_2O} \mathbf{37}\,(\text{Ar}CH_2NHCOCH_3)$$

(*Ref. 49*)

Nuclear Substitution

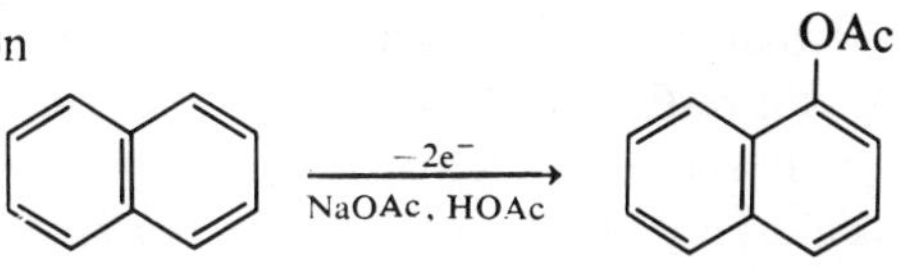

$$\text{naphthalene} \xrightarrow[\text{NaOAc, HOAc}]{-2e^-} \text{1-acetoxynaphthalene (OAc)}\ \mathbf{38} \qquad (Ref.\ 50)$$

Addition

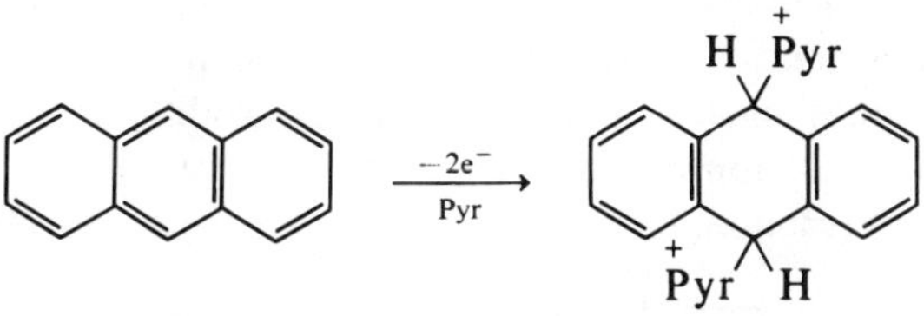

$$\mathbf{39}\ (\text{anthracene}) \xrightarrow[\text{Pyr}]{-2e^-} \mathbf{40}\ (\text{9,10-dihydro-9,10-bis}(\overset{+}{\text{Pyr}})\text{anthracene; H, }\overset{+}{\text{Pyr}}) \qquad (Ref.\ 51)$$

(Pyr = pyridine)

Dimerization

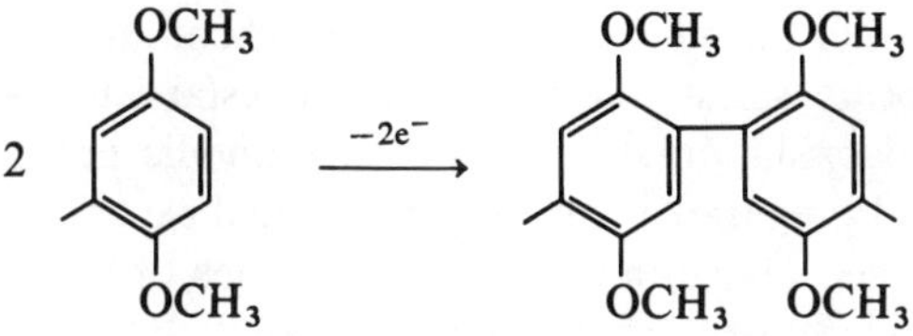

$$2\ \mathbf{41}\ (OCH_3,\ OCH_3) \xrightarrow{-2e^-} \mathbf{42}\ (OCH_3\ OCH_3,\ OCH_3\ OCH_3) \qquad (Ref.\ 52)$$

After some preliminary discussions have cleared the way for a proper analysis, it will be seen that the course of an oxidation can often be shifted to favor one or another of these processes through proper choice of experimental conditions.

In any electrochemical study, the first question that must be answered is which component of the solution is actually undergoing electron transfer. The conversion of naphthalene to **38** shown above, and in fact all anodic nuclear substitutions, were long thought to involve attack on the hydrocarbon by an electrochemically generated reagent, in this case the acetoxyl radical. It is now known from voltammetric data, however, that naphthalene is more easily oxidized than acetate ion, and that the primary electrochemical step in this reaction involves oxidation of naphthalene.[50b] It is also known that oxidation of the organic substrate is the primary electrochemical step in the other three examples shown above. Voltammetric data alone, however, do not suffice as proof of the identity of the electroactive species. Also, the conclusion ought not to be made, based on these examples, that anodic reactions are *always* initiated by oxidation of the organic substrate. The first of these two cautions may be illustrated by the studies of Parker and Burgert, and also Andreades and Zahnow, on the mechanism of nuclear cyanation of aromatic compounds, that is,

$$ArH + CN^- - 2e^- \longrightarrow ArCN \qquad (Ref.\ 53)$$

These studies showed that cyanide is considerably *easier* to oxidize than the aromatic hydrocarbons investigated. Nevertheless, electrolysis at a potential at which cyanide is the only species oxidized does not result in nuclear cyanation; the latter process occurs only at potentials where the aromatic substrate itself is oxidized. Regarding the second warning above, it should be noted that while it now appears that most anodic reactions of aromatic compounds involve initial oxidation of the organic substrate, this is by no means always the case. For example, electrochemical oxidation of mesitylene in acetic acid containing sodium acetate does indeed proceed by initial oxidation of mesitylene, but replacement of acetate ion by the very readily oxidized nitrate ion results in a different course of reaction, initiated in this case by the radical **43.**[54] In the light

$$NO_3^- - e^- \longrightarrow \cdot NO_3 \quad (\mathbf{43})$$

$$ArCH_3 + \mathbf{43} \longrightarrow HNO_3 + ArCH_2\cdot \longrightarrow \text{products}$$

of present knowledge of the mechanisms of electrochemical oxidation of organic compounds, it appears safest, on the whole, to establish experimentally the nature of the initial oxidation step for any new anodic process, for example, by voltammetric and preparative controlled-potential experiments.

As was mentioned above, most anodic reactions of aromatic compounds are initiated by oxidation of the hydrocarbon itself. It is appropriate at this point to inquire into the nature of this initial oxidation step. Adams has summarized the experimental evidence,[4] but a brief review will be helpful here. Until fairly recently, it was thought that most oxidations involved removal of 2 electrons from the aromatic hydrocarbon to afford a dication. For example, **37** and **40** were suggested to arise in the following manner:

$$\mathbf{36} \xrightarrow{-2e^-} [\text{dication}] \xrightarrow{-H^+} \mathbf{44}\ (\overset{+}{C}H_2) \xrightarrow[H_2O]{CH_3CN} \mathbf{37}$$

$$\mathbf{39} \xrightarrow{-2e^-} [\text{dication}] \xrightarrow{2Pyr} \mathbf{40}$$

It now appears, however, that in most cases the first step is a 1-electron oxidation to the radical cation of the aromatic hydrocarbon. This species usually undergoes rapid chemical reaction to generate a new species, which then loses a second electron; that is, what were formerly thought to be direct 2-electron oxidations are in most cases ECE processes, quite analogous in many ways to the 2-electron reductions observed for aromatic hydrocarbons in the presence of proton donors (Chapters 4 and 7). Thus, the conversion of **36** to **37** probably actually proceeds as follows:

$$\mathbf{36} \xrightarrow{-e^-} [\text{radical cation}] \xrightarrow{-H^+} [CH_2\cdot \text{ radical}] \xrightarrow{-e^-} \mathbf{44} \xrightarrow[H_2O]{CH_3CN} \mathbf{37}$$

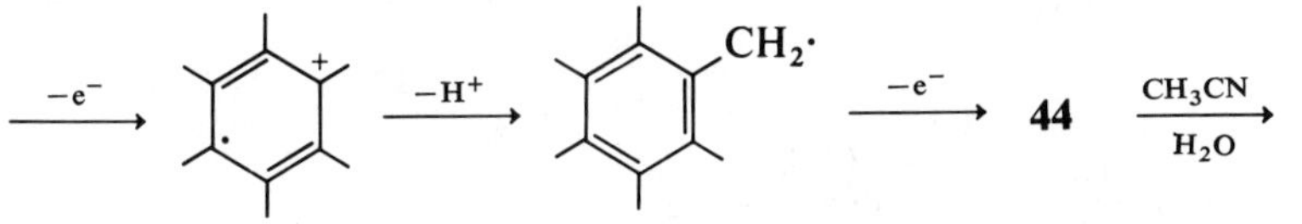

A similar mechanism can be written for the conversion of **39** to **40** (see p. 146). There had been in the literature a considerable controversy (which need not be reviewed here) over the question of 1- *vs* 2-electron oxidations in the initial step, engendered partly because of a lack of any experimental evidence for aromatic radical cations as intermediates. Almost simultaneously, however, research groups under Adams, Bard, and Peover firmly established the 1-electron nature of the oxidation by fast experimental techniques such as cyclic voltammetry and the rotated-disk electrode.[55] It was also shown that where the reactive sites of the radical cation are blocked sterically or stabilized electronically, the radical cations are stable enough for examination by spectral (esr, uv) techniques; otherwise the follow-up chemical reactions are extremely rapid, accounting for the previous difficulties in obtaining evidence for the intermediacy of the radical cation.

As we have said previously, both the structure of a radical cation and the environment in which it is generated undoubtedly influence its subsequent reactions. It is possible to account for the various reaction pathways available to radical cations in terms of the effects of these two variables. A general mechanistic scheme that permits such an analysis is shown below as Scheme I. Toluene (**45**) is chosen only for illustrative purposes, of course.

SCHEME I

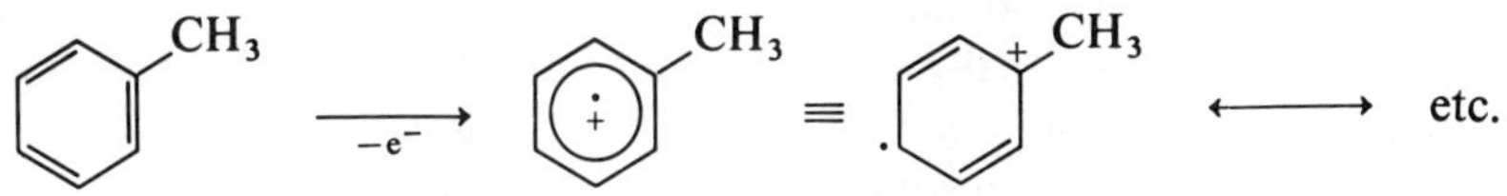

$$\underset{\mathbf{45}}{C_6H_5CH_3} \xrightarrow{-e^-} \underset{\mathbf{46}}{[C_6H_5CH_3]^{+\cdot}} \equiv \cdot C_6H_5{}^{+}CH_3 \longleftrightarrow \text{etc.}$$

Path *A* (Side-Chain Substitution)

$$\mathbf{46} \xrightarrow{-H^+} C_6H_5\dot{C}H_2 \xrightarrow{-e^-} C_6H_5CH_2^{\;+} \xrightarrow{Nu:} C_6H_5CH_2Nu$$

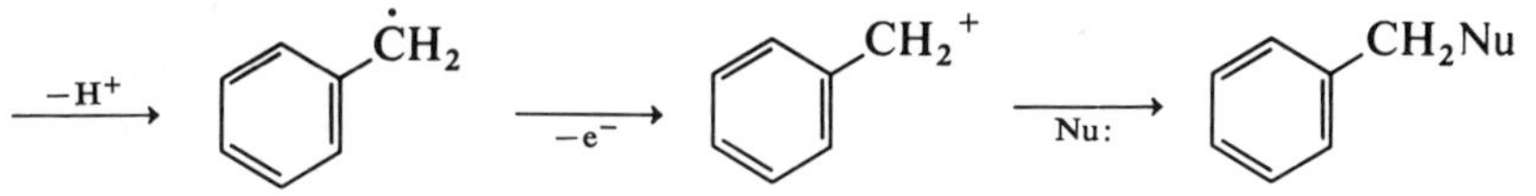

(Nu:=a nucleophile)

Path *B* (Nuclear Substitution and Addition)

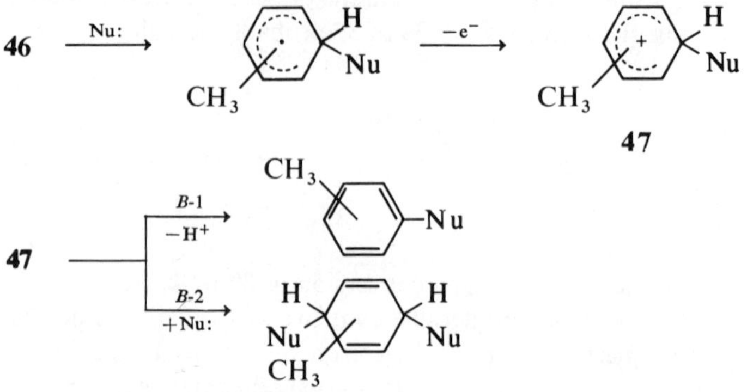

46 $\xrightarrow{\text{Nu:}}$ [radical: CH_3, H, Nu] $\xrightarrow{-e^-}$ [cation: CH_3, H, Nu] **47**

47 $\xrightarrow[-H^+]{B\text{-}1}$ [CH_3, Nu aromatic]

47 $\xrightarrow[+\text{Nu:}]{B\text{-}2}$ [H, Nu, CH_3, H, Nu]

Path *C* (Dimerization)

45 + **46** $\longrightarrow$ [CH_3, CH_3 radical cation dimer] $\xrightarrow[-2H^+]{-e^-}$ [CH_3, CH_3 biphenyl]

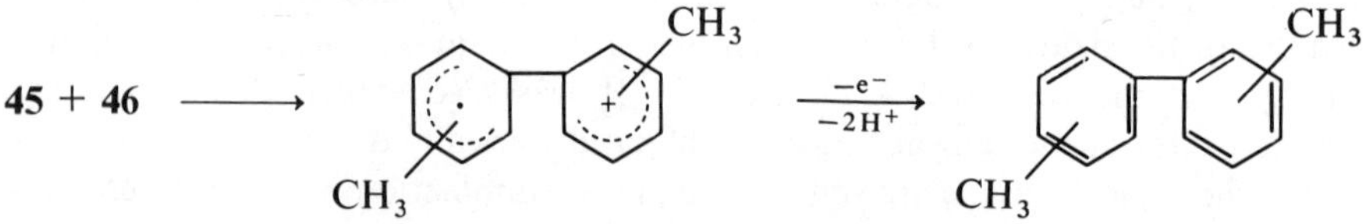

Within this mechanistic framework, it is possible to make predictions concerning the most likely course of a given reaction according to how the structure of the radical cation and the nature of the solvent and other components of the medium should cause a preference for one or another of the various paths. For example, one would expect that path *A* should be preferred over path *B* when there is no strong nucleophile present in the medium. Path *A* should also be preferred over path *B* when the starting hydrocarbon is highly alkylated, since in this case the alkyl substituents would both enhance the stability of the benzyl radical formed by deprotonation of the radical cation and slow (sterically) nucleophilic attack upon the aromatic ring. These predictions are found to be true experimentally: oxidation of alkylaromatics in acetic acid results in clean side-chain substitution (path *A*) when a nonnucleophilic salt is used as supporting electrolyte; if acetate ion is present in the solution as electrolyte, a mixture of side-chain and nuclear-substitution products is obtained[56]:

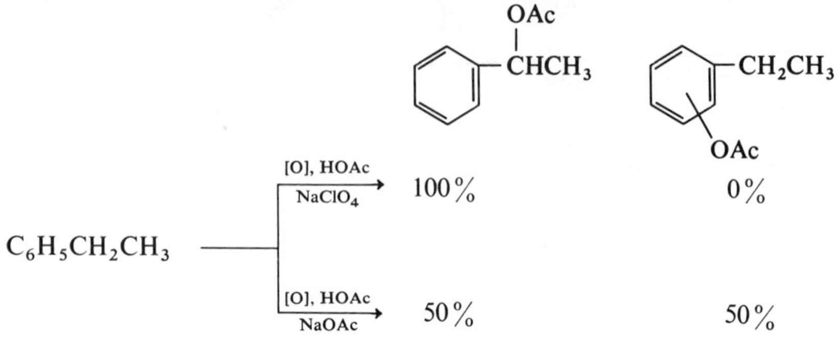

		$C_6H_5CH(OAc)CH_3$	OAc-C_6H_4-CH_2CH_3
$C_6H_5CH_2CH_3$	[O], HOAc / $NaClO_4$ $\longrightarrow$	100%	0%
	[O], HOAc / NaOAc $\longrightarrow$	50%	50%

The highly alkylated aromatics durene and hexamethylbenzene afford only side-chain substituted products even in the presence of acetate ion[57]:

$\xrightarrow[\text{NaOAc}]{\text{[O] HOAc}}$ OAc

In the presence of good nucleophiles and where the ring is not highly alkylated, one would expect the radical to undergo nucleophilic attack preferentially. Where the nucleophile is another component of the mixture, paths *B*-1 or *B*-2 would be initiated by the nucleophile. Sometimes, however, the starting hydrocarbon will itself be a good nucleophile, as a consequence of the presence of one or more strong electron-supplying substituents such as alkoxyl or amino groups. In this case, particularly in the absence of any other good nucleophiles in the solution, the starting material may attack the radical cation to generate a dimeric product (path *C*). The requisite conditions for dimerization, that is, a very nucleophilic starting material and the absence of an added strong nucleophile, are met in the oxidative coupling of **41** to **42** (p. 287) in acetonitrile containing tetraethylammonium perchlorate as electrolyte.[52]

Not only do structure and environment control which of the three main reaction paths is chosen by a radical cation, they may also exert an influence when several subordinate reactions are possible *within* one of the three main categories. Path *A*, for example, involves eventual formation of a benzylic carbonium ion. If such a carbonium ion is generated by oxidation of a hydrocarbon in wet acetonitrile, with which nucleophile will it react, the better nucleophile (water) or that which is in greater excess (acetonitrile)? The structure of the carbonium ion will play a role in this choice, since highly reactive carbonium ions should react indiscriminately, that is, with the nucleophile that is present in excess, and, conversely, less active carbonium ions should choose the better nucleophile.[52] (This reasoning is, of course, merely a restatement of the *Selectivity Relationship.*[58]) This argument appears to be correct: oxidation of *p*-methylanisole in wet acetonitrile affords the amide **49**, while under the same conditions pentamethylanisole is converted to the alcohol **51**.[52]

CH_3O–C$_6$H$_4$–CH_3 $\xrightarrow{\text{[O]}}$ CH_3O–C$_6$H$_4$–CH_2^+

48

$\xrightarrow[H_2O]{CH_3CN}$ CH_3O–C$_6$H$_4$–$CH_2NHCOCH_3$

49

CH_3O [O] → CH_3O CH_2^+ **50** $\xrightarrow[H_2O]{CH_3CN}$ CH_3O CH_2OH **51**

This dichotomy can be rationalized as being due to the steric and electronic stabilization provided by the methyl groups in cation **50**, which render it more selective and hence more likely than **48** to react with the better nucleophile, water. This is a clear example of the effect of structure on reactivity of the benzyl cations generated in path *A*. The effect of the other variable, environment, is demonstrated dramatically by Nyberg's study of the electrochemical oxidation of hexamethylbenzene in acetonitrile–acetic acid mixtures. He found that the ratio of amide **37** to ester **52a** depends not only on the ratio of acetonitrile to acetic acid but also on the nature of the supporting electrolyte (Table 8.1).[59]

Table 8.1 Product dependence on experimental conditions in the electrochemical oxidation of hexamethylbenzene

Supporting Electrolyte	Molar Ratio CH_3CN:HOAc	Molar Ratio **37**:**52a**
Sodium perchlorate	50:50	53:47
Sodium perchlorate	95:5	76:24
Sodium perchlorate	99:1	84:16
Tetrabutylammonium perchlorate	99:1	78:22
Tetrabutylammonium tetrafluoroborate	99:1	19:81

Source: K. Nyberg, *Chem. Commun.*, 774 (1969). (Reprinted by permission of The Chemical Society.)

It was suggested that **52a** predominates over **37** in the presence of tetrafluoroborate ion because the latter is solvated preferentially by acetic acid and is present in excess at the electrode surface at anodic potentials, these two phenomena causing the local concentration of acetic acid at the electrode surface to be higher than in bulk solution. A second example of this anion effect was found by Eberson and Olofsson, who observed that the ratio of **37** to **52b** in acetonitrile–water (9:1) is 95:5 with sodium perchlorate as the supporting electrolyte and 5:95 when the electrolyte is tetrabutylammonium tetrafluoroborate.[49a]

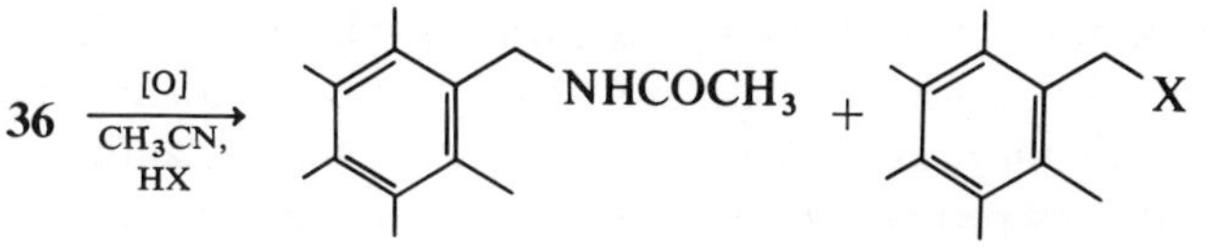

36 $\xrightarrow[\text{CH}_3\text{CN, HX}]{[O]}$ NHCOCH$_3$ + X

37 52a, X=OAc
52b, X=OH

The factors governing which of the two alternative reactions under path *B*, that is, nuclear substitution (*B*-1) or addition (*B*-2), is actually chosen in a specific case are not yet completely understood. It appears that path *B*-1 is almost always preferred. This is not surprising, since loss of a proton from **47** restores the aromatic system and hence should have a low activation energy. A similar situation arises in electrophilic aromatic substitution where an intermediate (**53**) similar to **47** is generated. This intermediate can also either lose a proton to regenerate the aromatic system or react with a nucleophile (Y^-) to

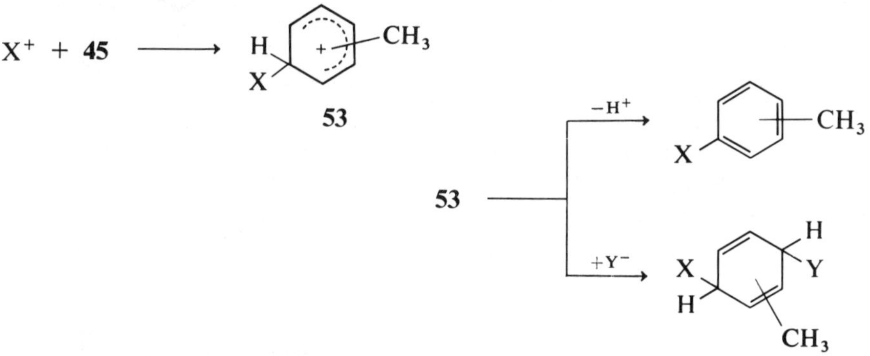

X^+ + **45** ⟶ H, X, CH$_3$

53

53 —($-H^+$)⟶ X, CH$_3$
—($+Y^-$)⟶ X, H, H, Y, CH$_3$

give an addition compound,[60] and in most cases the first of these two paths is observed. Certain hydrocarbons do, however, form addition compounds, as, for example, in the reaction of anthracene with bromine.[61] It is observed that such compounds also generally give addition compounds (path *B*-2) upon electrochemical oxidation in the presence of nucleophiles; for example,

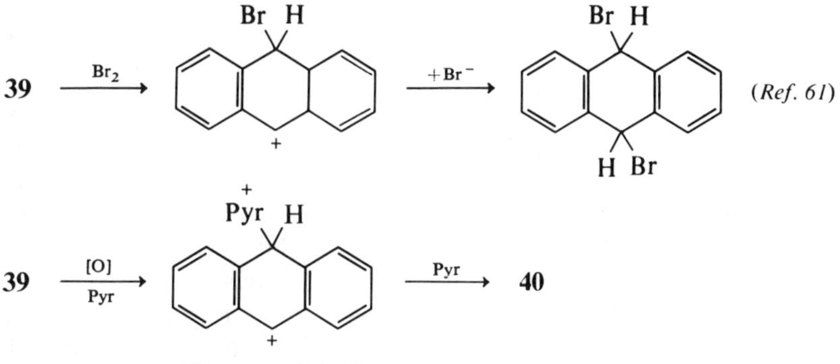

39 $\xrightarrow{Br_2}$ Br H, + $\xrightarrow{+Br^-}$ Br H, H Br (*Ref. 61*)

39 $\xrightarrow[\text{Pyr}]{[O]}$ $\overset{+}{\text{Pyr}}$ H, + $\xrightarrow{\text{Pyr}}$ **40**

(Pyr=pyridine)

Experiments performed by Parker and Eberson illustrate, however, the real difficulty of predicting whether path *B*-1 or *B*-2 will be followed in any specific case. In one study they carried out the oxidation of 9-phenylanthracene (**54**) in acetonitrile–lithium perchlorate containing various dimethylpyridines (lutidines).[62] The rates of the *B*-1 (substitution) and *B*-2 (addition) processes may be symbolized by k_H and k_C, respectively. It was found that $k_C \gg k_H$ for the

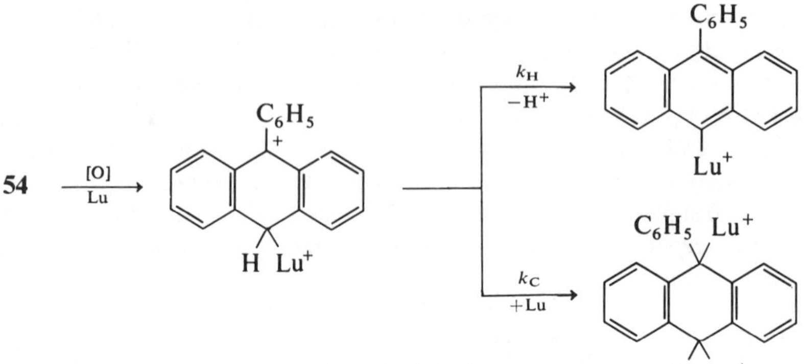

(Lu=lutidine)

unhindered nucleophile 3,5-lutidine. With the more hindered nucleophiles 2,5- and 2,6-lutidine, $k_H \gg k_C$, since the latter path (*B*-2) involves attack at a position sterically hindered by the phenyl substituent. Even more complex behavior was observed with 9,10-dibromoanthracene (**55**).[63] It was found that

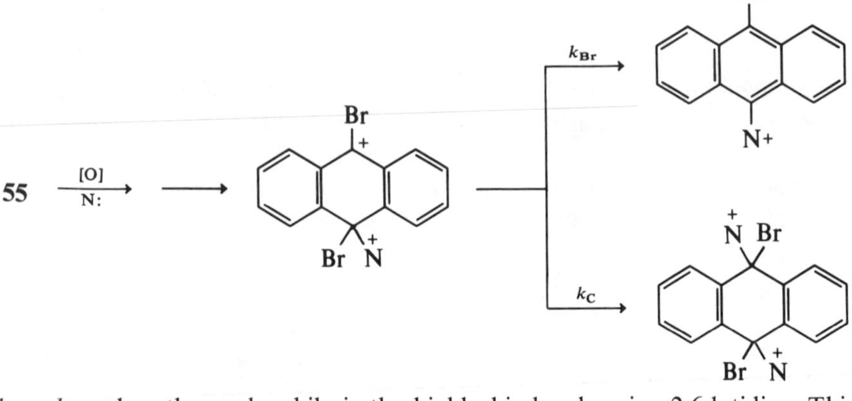

$k_C \approx k_{Br}$ when the nucleophile is the highly hindered amine 2,6-lutidine. This experiment demonstrates that attack at bromine (k_{Br}) and attack at carbon (k_C) have comparable steric requirements. Surprisingly, $k_{Br} \gg k_C$ for the unhindered amines, pyridine and 3,5-lutidine, and $k_C \gg k_{Br}$ for 2,5-lutidine. In view of the results with 2,6-lutidine, the latter differences can not be steric in origin; they have yet to be explained. Much more needs to be learned about the factors governing the balance between paths *B*-1 and *B*-2, although, as has already been noted, the former path seems to be preferred in simple benzenoid systems.

Synthetic Applications

Side-chain oxidation of alkylaromatics is potentially a synthetic technique of great value, since it both functionalizes an otherwise inert alkyl substituent and permits the introduction of a variety of substituents into the benzylic position of the hydrocarbon. It follows from the mechanistic discussions of the preceding section that the principal experimental requirement for successful side-chain oxidation is the absence of good nucleophiles in the medium, although this requirement may be relaxed with highly alkylated hydrocarbons. Other specific requirements are listed below under the individual reactions.

Side-chain acetoxylation can be effected in good yield with a wide variety of aromatic compounds.[56,57] Preferred conditions consist of controlled-potential oxidation of the aromatic hydrocarbon in glacial acetic acid containing a non-nucleophilic electrolyte, such as a tetraalkylammonium or alkali metal tosylate, perchlorate, or tetrafluoroborate.

Side-chain acetamidation is also a reaction of considerable generality. The corresponding benzylic amines are available by hydrolysis of the amides. It appears that best yields are obtained using acetonitrile dried over a molecular

$$\mathrm{ArCHR_2} \xrightarrow[\mathrm{H_2O}]{[\mathrm{O}]\mathrm{CH_3CN}} \underset{\mathrm{NHAc}}{\mathrm{ArCR_2}} \xrightarrow{\mathrm{H_2O}} \underset{\mathrm{NH_2}}{\mathrm{ArCR_2}}$$

sieve (the solvent still contains enough water for the Ritter reaction after this treatment) and a nonnucleophilic electrolyte such as sodium or tetraalkylammonium perchlorate or possibly tosylate (but *not* tetrafluoroborate).[49a] There is evidence suggesting that yields are better at a carbon anode than at platinum.[49a]

Side-chain hydroxylation of alkylarenes can be carried out in aqueous acetonitrile containing tetrabutylammonium tetrafluoroborate as electrolyte.[49,50] Yields are only fair, however, since oxidation of the resulting benzylic alcohols to carbonyl compounds does occur slowly under the reaction conditions. A much better method of hydroxylation would consist of side-chain acetoxylation followed by hydrolysis of the resulting benzylic acetate. The acetates are considerably more resistant to oxidation than the corresponding alcohols.

Other side-chain substitution reactions are known or could be developed. For example, oxidation of alkylaromatics in methanol containing cyanide ion has been shown to afford side-chain methoxylated products in poor-to-fair yields, accompanied by products of nuclear substitution by cyanide ion.[64] As the mechanistic analysis of the preceding section indicates, however, this is actually a poor way to carry out side-chain methoxylation, since cyanide ion is a good nucleophile. Oxidation of the hydrocarbon in an anhydrous alcohol containing a nonnucleophilic electrolyte, or, for solid or expensive alcohols, in acetonitrile containing the alcohol and a tetrafluoroborate salt,[59] should result in much improved yields of the alkoxylated compound. Replacement of the alcohol by other weak nucleophiles would result in other new side-chain functionalization reactions. If one oxidizes in the complete absence of *any* nucleophile, it is also possible to carry out a Friedel–Crafts alkylation of the starting hydrocarbon by the benzyl cation generated in path *A*[65]:

[O] $\longrightarrow$ CH_2^+ $\longrightarrow$

Path *C*, leading to a biphenyl, should of course compete with this process when the starting material is quite nucleophilic, and this is exactly what is observed experimentally[65]: diphenylmethane derivatives are formed from alkylaromatics, and biphenyls from oxygenated aromatics (*cf.* the conversion of **41** to **42**).

Anodic nuclear substitutions by a wide variety of substituents, such as acetoxy, cyano, cyanato, bromo, fluoro, and carbomethoxy, have been recorded in the literature.[57,66,67] While the many studies of these reactions have been important in establishing mechanisms of organic anode processes, nuclear substitution (path *B*-1) does not appear, except for a few exceptions to be mentioned below, to be particularly promising from the synthetic point of view. There are a number of reasons for this: mixtures of isomers are generally formed, yields are not high, and often the initial substitution product undergoes further oxidation. It is true that anodic fluorination of organic compounds[66] is of considerable commercial interest, but this is due in part to the facts that substitution is unselective and that polyfluorination is usually observed. These are, on the other hand, undesirable features for most synthetic purposes. There do happen to be a few exceptions, however, to the generalization that anodic nuclear substitution is not of real value. One reaction of possible synthetic interest is the anodic displacement of an alkoxy substituent by cyanide[53b]:

OCH_3 / OCH_3 $\xrightarrow[CH_3OH,\ CN^-]{[O]}$ OCH_3 / CN

(95%)

OCH_3 / OCH_3 / OCH_3 $\xrightarrow[CH_3OH,\ CN^-]{[O]}$ OCH_3 / CN / OCH_3

(86%)

It has been suggested that this reaction, which is limited to polyalkoxy aromatics, proceeds as follows:

$$p\text{-}CH_3OC_6H_4OCH_3 \xrightarrow[\text{(path } B)]{-2e^-,\ +CN^-} [\text{cyclohexadienyl cation with } C(CN)(OCH_3)]^+ \xrightarrow{CN^-} p\text{-}CH_3OC_6H_4CN + CH_2{=}O + HCN$$

The final step may be generalized as:

$$B : H{-}CH_2{-}O{-}X^+ \longrightarrow BH + CH_2{=}O + X$$

It is conceivable that other nucleophiles, such as fluoride, chloride, cyanate, or pyridine, could replace cyanide in this substitution reaction.

An excellent anodic nuclear substitution reaction that promises to be of considerable synthetic utility has been reported by Miller and coworkers.[68] They carried out iodination of aromatic hydrocarbons through oxidation of a mixture of the hydrocarbon and iodine at a platinum anode in acetonitrile–lithium perchlorate:

$$ArH + I_2 \xrightarrow[LiClO_4]{[O]CH_3CN} ArI$$

Unlike most of the reactions discussed heretofore in this chapter, this reaction does *not* proceed via initial oxidation of the aromatic hydrocarbon. Rather, it is molecular iodine that is oxidized at the anode to one or more active iodinating agents, suggested by Miller to be N-iodoacetamide and N-iodoacetonitrilium ion ($CH_3C{\equiv}N^+I$). The best yields of halogenated product are, in fact, obtained by oxidation of a solution of iodine in acetonitrile, followed by addition of the hydrocarbon. Yields (percent) of monoiodoaromatic from benzene, *p*-xylene, toluene, and anisole were 96, 100, 94 (47 *para*, 47 *ortho*), and 80 (56 *para*, 24 *ortho*), respectively. The iodinating agent is very potent: iodination of *p*-xylene was complete within a few seconds, and any number of iodine atoms up to four could be added to the *p*-xylene nucleus merely by passage of the proper number of faradays through the iodine–acetonitrile solution before addition of the hydrocarbon. Miller has pointed out the merits of this as opposed to other iodination methods: it is fast, proceeds in high yield, and eliminates the need for acid catalysts, heat, or halogenating agents which happen at the same time to be strong oxidizing agents.

8.2.2 Phenols and Phenol Derivatives

Differences in oxidative behavior between aromatic hydrocarbons and phenols or their derivatives are mainly in degree rather than in kind. Indeed, in order to discuss mechanisms of electrochemical oxidation of aromatic hydrocarbons in an earlier section, it was necessary to consider at the same time oxidations of oxygenated aromatics. There are, however, some anodic reactions more or less unique to phenols and their derivatives, and these warrant separate discussion.

Much of the difference between oxidative behavior of simple aromatic hydrocarbons and phenols or their derivatives is due to the powerful stabilizing effect of oxygen upon cationic intermediates. For example, the radical cation of *p*-xylene is capable of only fleeting existence, while the radical cation of the hydroquinone ether **56a** is moderately stable in the absence of nucleophiles.[69] Of course, oxidation in the presence of a good nucleophile should initiate path *B*

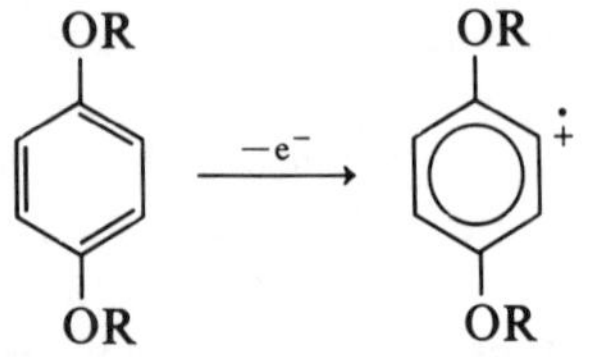

OR OR

$\xrightarrow{-e^-}$

OR OR

56a, R$=CH_3$ **57**
56b, R$=$Ac
56c, R$=$H

of Scheme I (p. 290) by nucleophilic attack on **57a**. This is observed experimentally: oxidation of **56a** in methanol containing sodium methoxide produces the otherwise synthetically inaccessible quinone ketal **58**.[70]

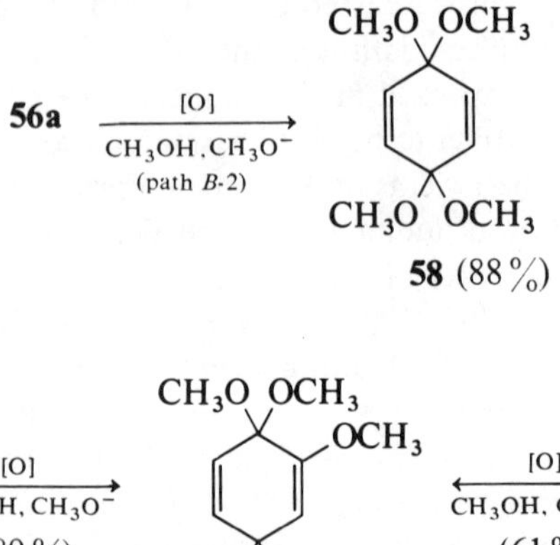

CH_3O OCH_3

56a $\xrightarrow[\text{CH}_3\text{OH}, \text{CH}_3\text{O}^-]{[\text{O}]}$ (path *B*-2)

CH_3O OCH_3

58 (88%)

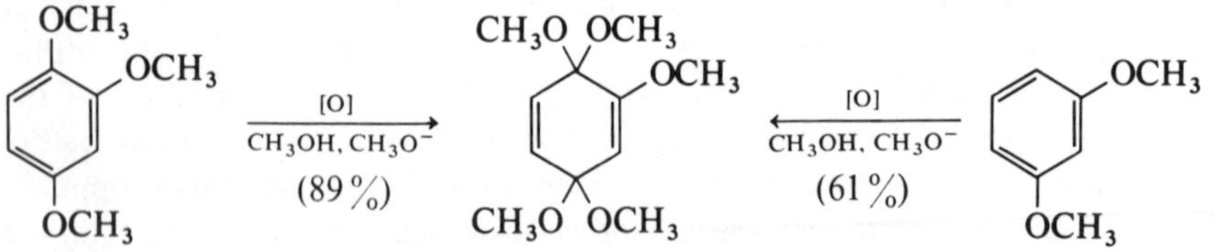

OCH_3 OCH_3 OCH_3 — $\xrightarrow[\text{CH}_3\text{OH}, \text{CH}_3\text{O}^-]{[\text{O}]}$ (89%) — CH_3O OCH_3 OCH_3 CH_3O OCH_3 — $\xleftarrow[\text{CH}_3\text{OH}, \text{CH}_3\text{O}^-]{[\text{O}]}$ (61%) — OCH_3 OCH_3

59 **60**

58 $\xrightarrow{BF_3}$ **59** +

OCH_3 OCH_3
CH_3O
OCH_3 OCH_3

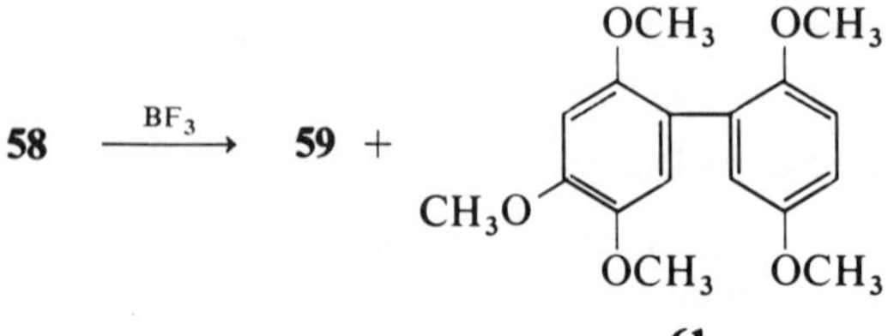

61

A number of other examples of this reaction, some shown here, are known.[5] The reaction constitutes a synthetic route not only to quinone ketals but also to polyalkoxyaromatics and polymethoxybiphenyls such as **61**, since, for example, by appropriate choice of reaction conditions either **59** or **61** can be made to predominate upon treatment of **58** with acid.[70] (The conversion of **58** to **61** presumably proceeds by a sequence involving electrophilic attack upon **59** by cation **62**.)

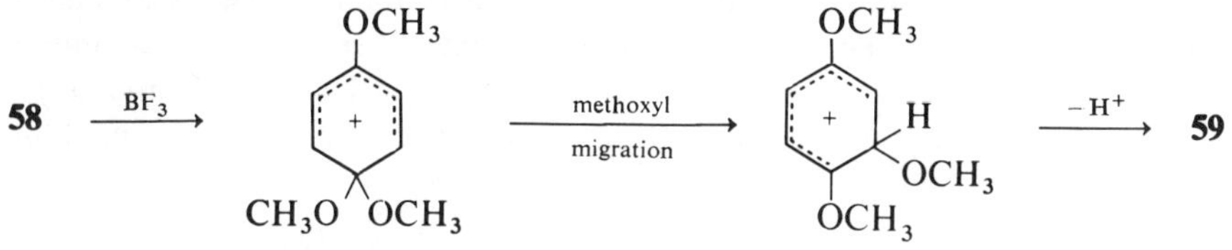

58 $\xrightarrow{BF_3}$ [OCH_3 / + / CH_3O OCH_3] **62** $\xrightarrow[\text{migration}]{\text{methoxyl}}$ [OCH_3 / + / H, OCH_3 / OCH_3] $\xrightarrow{-H^+}$ **59**

59 + **62** ⟶ ⟶ **61**

Notice that in these oxidations **56a**, **59**, and **60** possess no aliphatic side chain, thus precluding path *A*. Path *B* is presumably favored over path *C* because of the presence of methoxide ion. If, however, the electrolysis were to be carried out in the absence of an external nucleophile, the radical cations should react with starting material by path *C*, leading to biphenyls. This is observed; *cf.* the conversion of **41** to **42**. Oxidation of hydroquinone itself, that is, **56c**, or of hydroquinone monoethers or monoesters,[71] proceeds to benzoquinone irrespective of the presence of other nucleophiles, no doubt because **57c** and related species can readily undergo proton loss, that is, path *A*:

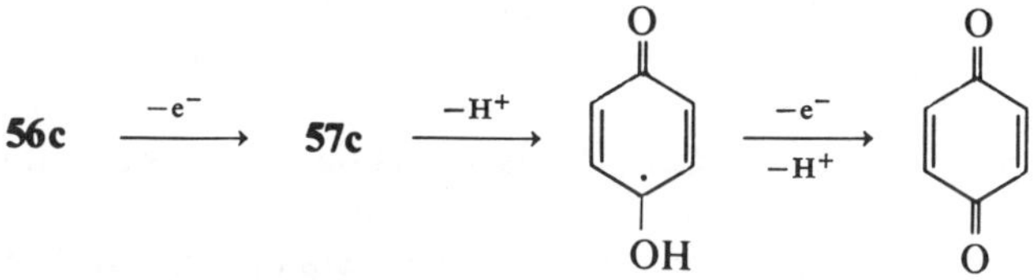

56c $\xrightarrow{-e^-}$ **57c** $\xrightarrow{-H^+}$ [O / · / OH] $\xrightarrow[-H^+]{-e^-}$ [O / O]

There is an interesting effect, not understood at this time, of alkyl substituents upon radical cations in the hydroquinone series. As has already been mentioned, **57a** is moderately stable in the absence of nucleophiles. In contrast, however, the radical cation of the durohydroquinone ether **63a** is so shortlived

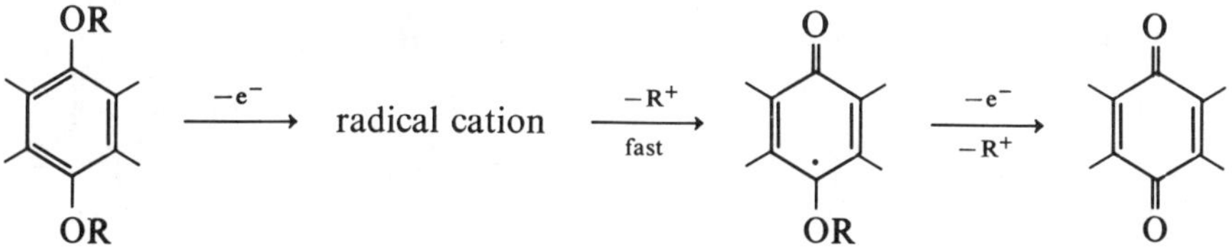

[OR / OR] $\xrightarrow{-e^-}$ radical cation $\xrightarrow[\text{fast}]{-R^+}$ [O / · / OR] $\xrightarrow[-R^+]{-e^-}$ [O / O]

63a, R=CH_3
63b, R=Ac

64

in acetonitrile–lithium perchlorate that it cannot be detected by cyclic voltammetry, even at high scan rates.[72] Apparently, the radical cation undergoes rapid methyl transfer, since duroquinone (**64**) is isolated in quantitative yield from an oxidation at controlled potential. Similar behavior was observed upon oxidation of the diacetate **63b**.[72] The fate of the methyl and acetyl groups in these oxidations was not investigated. The oxidation of compounds such as **63** may well be of interest as a mild, rapid, neutral alkylation or acylation procedure. From another point of view, it may be a mild neutral method of effecting oxidative cleavage of appropriately substituted hydroquinone ethers. Related examples of this phenomenon include conversion of **65** (a vitamin E model compound) to **66**,[73] and oxidation of the quinol phosphate **67** to 2-methyl-1,4-naphthoquinone.[74]

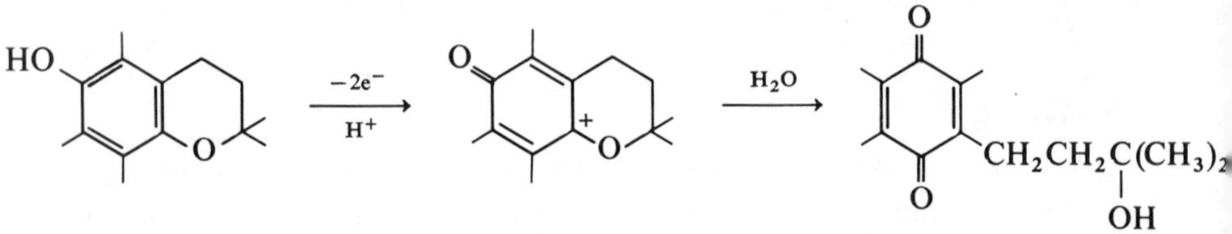

65 **66**

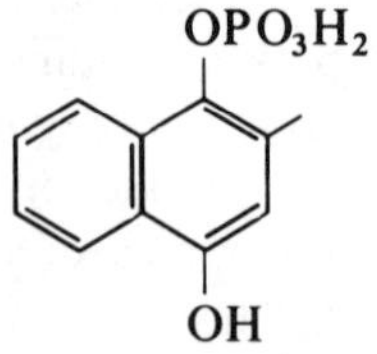

67

Adams and coworkers have described a very interesting procedure for hydroxylation of aromatic rings.[75] It is based on the electrochemical generation of quinones under experimental conditions where they undergo immediate nucleophilic attack by the solvent (water). The overall process is exemplified by conversion of 2-carboxyhydroquinone (**68**, R = CO_2H) to 2-carboxy-3-hydroxyquinone (**71**, R = CO_2H) upon oxidation of **68** at a carbon anode in 2-*M* perchloric acid. The initial hydroxylation product, triol **70**, is of course oxidized to **71** immediately upon formation. If it were desired to isolate **70** rather than **71**,

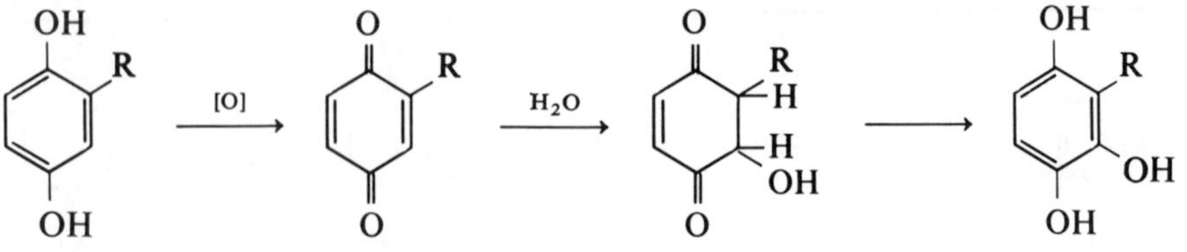

68 **69** **70**

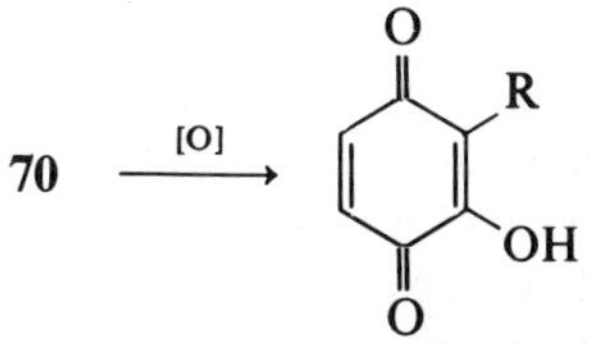

70 $\xrightarrow{[O]}$ 71

the solution containing **71** could be reduced electrochemically as soon as oxidation was complete. Naturally, quinone **69** undergoes nucleophilic attack by water only when R is strongly electron withdrawing (R = NO_2, CO_2H, CHO); when R is methyl, chlorine, hydroxyl, or hydrogen, **69** is the product isolated. Adams and coworkers also reported a related hydroxylation reaction that utilizes the same general principle. Controlled-potential oxidation of 1,5-dihydroxynaphthalene (**72**) in 2-*M* perchloric acid yields juglone (**73**), presumably formed by a variant of path *B* (p. 290). Dihydrojuglone (**74**) must be an intermediate in the formation of **73**, but it could be shown by a control experiment that **74** is easier to oxidize than **72** and hence does not survive the oxidation conditions. It can be obtained, if desired, by reduction of **73**. These hydroxylation reactions are among the very few known useful anodic nuclear substitutions, presumably because oxidation of the intermediates **70** and **74** produces a quinone that is stable to further oxidation.

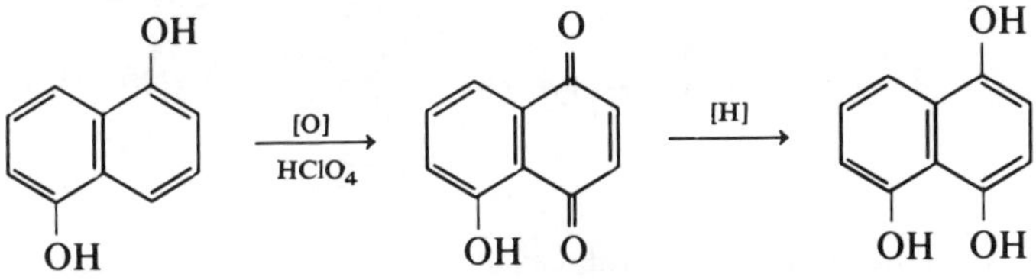

72 $\xrightarrow[HClO_4]{[O]}$ 73 $\xrightarrow{[H]}$ 74

8.2.3 Amines

The electrochemical oxidation of aliphatic amines does not appear at present to be of any synthetic interest, in view of the complex product mixtures that are generally obtained.[76] The oxidation of aromatic amines, on the other hand, should be of some synthetic utility, although this has not been the primary goal of most studies in this area to date. The initial step in the oxidation is, as with most aromatic hydrocarbons, removal of an electron to generate a radical cation. The radical cations from most aromatic amines appear to be extremely reactive, however, and so one generally observes only products of one or even more ECE sequences initiated by the radical cation. The dimerization reaction (path *C* of Scheme I) is naturally very important with these compounds, in view of the high degree of nucleophilicity of the aromatic ring, resulting from the electron-donating power of the nitrogen substituents. The anodic chemistry of aromatic amines will be discussed here by first examining the most highly arylated species, triphenylamine and its derivatives, and then pointing out changes in behavior as the degree of aryl substitution is decreased.

Triphenylamines in which all three of the *para*-positions bear a substituent other than hydrogen lead to relatively stable radical cations, as shown by both cyclic voltammetry and esr spectroscopy.[77] When one of the *para*-positions is unsubstituted, however, more or less rapid coupling (path *C*) takes place:

$$\underset{\mathbf{75}}{(C_6H_5)_3N} \xrightarrow{-e^-} \underset{\mathbf{76}}{(C_6H_5)_3N^{\dot{+}}}$$

$$\mathbf{75} + \mathbf{76} \xrightarrow[\text{path } C]{} \underset{\mathbf{77}}{(C_6H_5)_2N\text{-}C_6H_4\text{-}C_6H_4\text{-}N(C_6H_5)_2}$$

$$\mathbf{77} \xrightarrow{-2e^-} \underset{\mathbf{78}}{(C_6H_5)_2\overset{+}{N}{=}C_6H_4{=}C_6H_4{=}\overset{+}{N}(C_6H_5)_2}$$

Since the benzidine **77** is, in general, easier to oxidize than the starting amine, it is oxidized to the dication **78** immediately upon formation. Benzidine **77** can be obtained by subsequent electrolytic reduction of **78**. Under some experimental conditions, **78** undergoes hydrolysis to benzoquinone and diphenylamine. The net effect of the oxidation–hydrolysis sequence is therefore degradation of a triarylamine to a diarylamine, a procedure of possible synthetic interest. By a similar sequence, some diarylamines have been degraded to anilines.[78a]

N,N-Dialkylanilines are also converted to benzidines upon electrochemical oxidation under most conditions, unless the *para*-position is blocked, in which case a moderately stable radical cation is generated. Weinberg and coworkers have shown, however, that the oxidation products are pH dependent.[79] Controlled-potential oxidation of N,N-dimethylaniline (**79**) in methanol containing ammonium nitrate as electrolyte does afford tetramethylbenzidine (**80**). In

$$\underset{\mathbf{80}}{Me_2N\text{-}C_6H_4\text{-}C_6H_4\text{-}NMe_2} \qquad \underset{\mathbf{81}}{C_6H_5\text{-}N(CH_3)CH_2OCH_3}$$

methanol containing potassium hydroxide as electrolyte, on the other hand, the product is N-methoxymethyl-N-methylaniline (**81**). Weinberg has suggested that both reactions proceed via the dimethylaniline cation radical (**82**), adsorbed on the electrode surface.[79a] In alkaline media, **82** should be deprotonated to a neutral radical (**83**), which would then be converted to **81** via oxidation and reaction with solvent. In the absence of added base, **82** would undergo coupling with starting material to yield **80**, that is, a path *C* sequence. This mechanistic argu-

$$\mathbf{79} \xrightarrow{-e^-} \underset{\mathbf{82}}{[C_6H_5\overset{+\cdot}{N}(CH_3)_2]_{ads}} \xrightarrow{^-OH} \underset{\mathbf{83}}{[C_6H_5N(CH_3)CH_2\cdot]_{ads}}$$

$$\mathbf{83} \xrightarrow[\text{fast}]{-e^-} [C_6H_5N(CH_3)CH_2^+]_{ads} \xrightarrow{CH_3OH} \mathbf{81}$$

ment does not of itself require that **82** be adsorbed and react on the electrode surface. Data from electrochemical kinetic studies,[79a] and also certain anomalies in the anodic methoxylation of other amines in methanol–potassium hydroxide,[79b,c] do, however, support a mechanism involving oxidation of an adsorbed species. Under these conditions, for example, benzyldimethylamine (**84**) affords a 1:4 mixture of **87** and **88**, the *latter* predominating.[80] Radical **85** should be substantially more stable than its isomer, **86**, so the benzylic ether **87** should be the major product. Weinberg argued that if the radical cation of **84** is adsorbed such that the aromatic ring is lying flat on the electrode surface, removal of one of the methyl protons should be sterically preferred (it is already statistically preferred) to removal of a benzylic proton.

$$\mathbf{84} \xrightarrow[-H^+]{-e^-} \underset{\mathbf{85}}{C_6H_5\dot{C}HN(CH_3)_2} + \underset{\mathbf{86}}{C_6H_5CH_2N(CH_3)CH_2\cdot}$$

$$\xrightarrow[CH_3OH]{-e^-} \underset{\mathbf{87}}{C_6H_5CH(OCH_3)N(CH_3)_2} + \underset{\mathbf{88}}{C_6H_5CH_2N(CH_3)CH_2OCH_3}$$

Alkylanilines and diarylamines with at least one free *para*-position are converted anodically to benzidine derivatives via shortlived radical cations.[4,78b,c] Diarylamines with both *para*-positions blocked are converted to 5,10-diaryl-5,10-dihydrophenazines upon electrochemical oxidation.[78c] N-Alkyldiarylamines in which both *para*-positions are blocked undergo 1-electron oxidation to stable radical cations.[78c] At more positive potentials, these radical cations are oxidized further to dications, which then undergo an interesting cyclization to afford carbazole derivatives; for example, N-methyl-di-*p*-tolylamine is converted to 3,6,9-trimethylcarbazole.[78c]

The anodic behavior of aryl hydrazines in aprotic media has been investigated by Cauquis and coworkers.[80] Phenylhydrazine itself is converted to phenyldiimide.[80a] 1,1-Diphenylhydrazine undergoes oxidation to a moderately stable nitrenium ion (**89**). Interesting heterocyclic adducts can be obtained from reaction of **89** with added olefins (and sometimes also the solvent, acetonitrile)[80b,c]:

$$(C_6H_5)_2NNH_2 \xrightarrow[-H^+]{-2e^-} (C_6H_5)_2\overset{+}{N}{=}NH$$

89

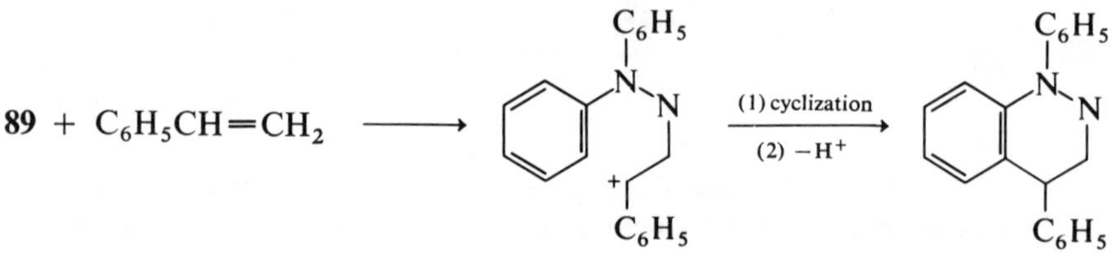

$$\mathbf{89} + C_6H_5CH{=}CH_2 \longrightarrow$$

C6H5 N N + C6H5 (1) cyclization (2) −H+ C6H5 N N C6H5

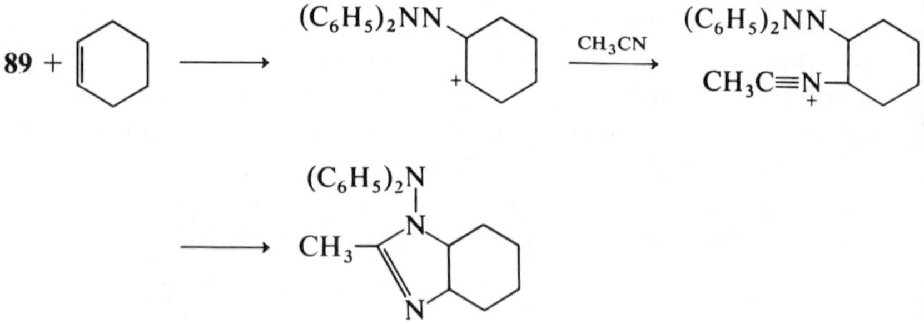

89 + ⟶ (C6H5)2NN + CH3CN (C6H5)2NN CH3C≡N+

⟶ (C6H5)2N N CH3 N

It is very likely that a wide variety of heterocyclic systems could be prepared from **89** by reaction of it with other species such as acetylenes, carbonyl compounds, or oximes.

8.3 OTHER SUBSTRATES

The anodic behavior of a number of other compound types is described in the comprehensive review of organic oxidations by Weinberg and Weinberg.[5] No useful purpose would be served by attempting the same depth of coverage here. In the two preceding sections of this chapter, particularly the second, the principal mechanisms of electrochemical oxidation have already been encountered. Application of these principles to other systems will normally be straightforward, but it may be helpful here to describe the anodic behavior of a few functional groups not already encountered, in order to provide some idea of the scope of electrochemical oxidations.

8.3.1 Alkanes and Alkenes

Although of unquestioned interest in fuel-cell studies, the electrochemical oxidation of alkanes appears at present to be of relatively little synthetic interest (but see reference 85b). The initial electrochemical step is best described as[81]

$$RR' \longrightarrow (RR')_{\text{adsorbed}} \xrightarrow[-R']{-2e^-} R^+ \longrightarrow \text{products}$$

$$(R' = \text{H or alkyl})$$

In most cases, the reaction products of the electrochemically generated carbonium ion are themselves oxidizable at the potential necessary for the primary electrochemical step, and so the hydrocarbon is systematically degraded (needless to say, this is true only under powerfully oxidizing conditions). One area where this reaction may be of interest, however, is in cases where the carbonium ion is relatively stable by virtue of delocalization of the positive charge, as in the following examples:

$$\xrightarrow{-2e^-} \quad + \quad \xleftarrow[-H^+]{-2e^-}$$ (*Ref. 82*)

$$[\;\;]_2 \xrightarrow[CH_3CN]{-2e^-} 2 \quad O^+$$ (*Ref. 83*)

$$\left[R_1R_2C{-}NR'_2 \right]_2 \xrightarrow[CH_3CN]{-2e^-} 2\; R_1R_2C{=}\overset{+}{N}R'_2$$ (*Ref. 84*)

Reports of the direct electrochemical oxidation of simple olefins are rare. Fleischmann and Pletcher have reported, however, that extremely anodic potentials can be attained in acetonitrile containing tetraalkylammonium tetrafluoroborate or hexafluorophosphate salts as electrolytes, and that under these conditions it is possible to carry out the oxidation of olefins.[85] There is other evidence that electrochemical oxidation of olefins is of synthetic interest. Oxidation of norbornene in methanol containing methoxide ion afforded ethers

$$\xrightarrow[CH_3OH,\; CH_3O^-]{[O]} \quad OCH_3 \quad + \quad OCH_3,\; OCH_3$$

90 **91**

90 and **91**.[86] This reaction might, in principle, be initiated by electrochemical oxidation of either norbornene or methoxide. One could determine which of these pathways was involved through voltammetric techniques and controlled-potential electrolysis, but this was not done. In either case, a likely intermediate is the radical **92**, which could afford **90** by hydrogen atom abstraction from the solvent; oxidation of **92** followed by Wagner–Meerwein rearrangement would afford **91**.

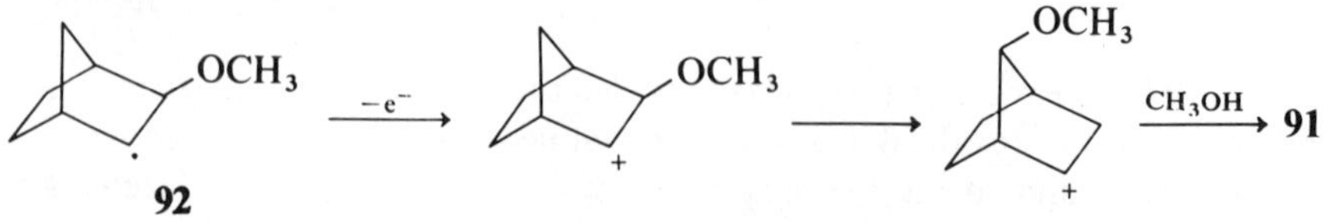

OCH_3 $\xrightarrow{-e^-}$ OCH_3 ⟶ OCH_3 $\xrightarrow{CH_3OH}$ **91**

92

Although it is unclear whether or not norbornene is the electroactive species in the above reaction, there is no doubt that the first step in electrochemical oxidation of aryl olefins is electron removal from the pi system to generate a radical cation. This species may undergo dimerization (a reaction akin to path *C* of p. 290), may react with nucleophiles in the medium, that is, path *B*, or may be oxidized further to a dication. Examples of each of these kinds of behavior are shown below.

Dimerization

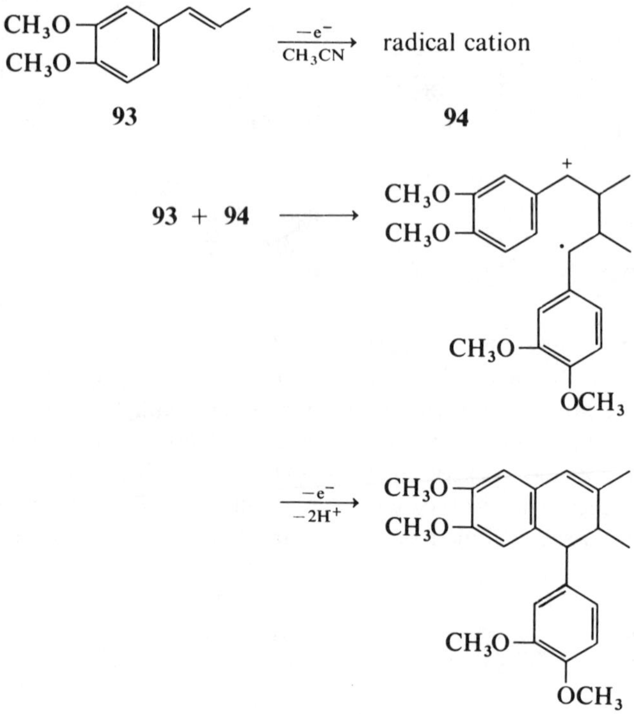

CH_3O, CH_3O $\xrightarrow[CH_3CN]{-e^-}$ radical cation

93 **94**

93 + **94** ⟶ CH_3O, CH_3O, CH_3O, OCH_3

$\xrightarrow[-2H^+]{-e^-}$ CH_3O, CH_3O, CH_3O, OCH_3

(*Ref. 87*)

Reaction with Nucleophiles

$$\underset{\mathbf{95}}{C_6H_5CH{=}CHC_6H_5} \xrightarrow[CH_3OH,\ CH_3O^-]{-e^-} \underset{\mathbf{96}}{C_6H_5\dot{C}H\overset{+}{C}HC_6H_5}$$

$$\mathbf{96} \xrightarrow{+CH_3O^-} C_6H_5\dot{C}HCH(OCH_3)C_6H_5 \xrightarrow[+CH_3O^-]{-e^-} \underset{\mathbf{97}}{C_6H_5CH(OCH_3)CH(OCH_3)C_6H_5}$$

(*Ref.* 88*a*)

$$\mathbf{93} \xrightarrow[Pyr]{[O]} (CH_3O)_2C_6H_3CH(\overset{+}{Pyr})CH(\overset{+}{Pyr})CH_3$$

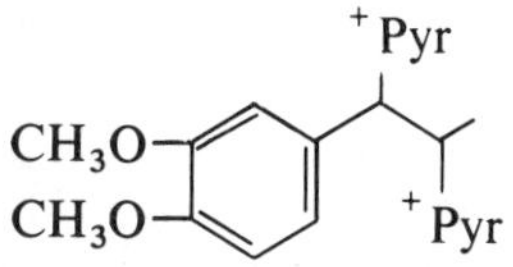

(*Ref.* 87)

(Pyr = pyridine)

Oxidation to Dication

$$ArCH{=}CHAr \xrightarrow[CH_3CN,\ HOAc,\ H_2O]{-2e^-} Ar\overset{+}{C}H\overset{+}{C}HAr \longrightarrow ArCH(OAc)CH(OH)Ar$$

(*Ref.* 89)

(Ar = *p*-anisyl)

Although the product in the last reaction could conceivably have arisen by an ECE pathway, the reaction does actually involve the dication, since it is found that another reaction (dimerization) occurs when the electrolysis is carried out at a less anodic potential, such that oxidation proceeds only to the radical cation.[89]

It is interesting to note that the *bis*-methoxylation of stilbene (**95**) is stereoselective: *dl*-**97** predominates in the product from oxidation of *trans*-stilbene, and *meso*-**97** is the major product when *cis*-stilbene is oxidized under the same conditions.[88a] (On the other hand, see reference 88d.) These stereochemical results amount to overall *cis*-addition of the two methoxyl substituents. As with the apparently anomalous substitution pattern observed by Weinberg in the electrochemical oxidation of **84**,[80] this stereochemistry is best explained as arising via adsorption of the molecule with the pi system lying flat on the electrode surface. Oxidation of **95** in methanol containing halide ion results in formation of a halohydrin[88b]; this and a similar reaction observed with cinnamic acid[88c] (and also oxidations of several olefins in acetonitrile containing halide ion[88d]) are probably initiated by oxidation of the halide, not the olefin. Olefins activated by electron-donating substituents, such as nitrogen,[90] sulfur,[93,94] or oxygen,[93] are also readily oxidized. For example, anodic behavior of a variety of nitrogen-substituted olefins such as 1,1-*bis*-[dimethylamino]-ethylene (**98**) has been examined.[90,95] Olefin **98** undergoes clean anodic dimerization,[90a] presumably by a mechanism similar to the formation of **77** from **75** or **80** from **79**. The diene **99** is, of course, easier to oxidize than **98**, and is converted to a dication immediately upon formation. It is possible to isolate **99** by electrochemical reduction of the dication.

$$((CH_3)_2N)_2C{=}CH_2 \xrightarrow{[O]} ((CH_3)_2N)_2C{=}CH{-}CH{=}C(N(CH_3)_2)_2$$

98 **99**

8.3.2 Carbanions

Electrochemical oxidation of carbanions and related species often leads to dimeric products via intermediate free radicals. Enolates of β-dicarbonyl compounds, for example, afford dehydrodimers in fair-to-good yield[5]:

$$(RCO)_2CH^- \xrightarrow{-e^-} (RCO)_2CH\cdot \longrightarrow (RCO)_2CHCH(COR)_2$$

The reaction has been used to synthesize *vic*-dinitro compounds, that is,[5]

$$R_2\bar{C}NO_2 \xrightarrow{[O]} R_2C(NO_2){-}C(NO_2)R_2$$

Oxidation of the anions of aliphatic nitro compounds in the presence of other anions affords *gem*-substituted derivatives of the original nitro compound[5]:

$$R_2\bar{C}NO_2 \xrightarrow[X^-]{-e^-} R_2C(X)NO_2 \qquad X = NO_2^-, N_3^-, \text{etc.}$$

Finally, oxidation of Grignard reagents affords dimeric hydrocarbons:

$$RMgX \xrightarrow{-e^-} RR$$

This reaction may be preferable to the Kolbe reaction where the alkyl halide RX is more readily available than the acid RCO_2H, or where the Kolbe cannot be used, such as where R is aryl or vinyl. In an interesting variation on this reaction, Schäfer and Küntzel intercepted the radicals formed by oxidation of Grignard reagents with olefins to generate new radicals, which then dimerized; for example,[91]

$$\mathrm{BuMgX} \xrightarrow{[O]} \mathrm{Bu\cdot} \xrightarrow{\text{styrene}} \mathrm{C_6H_5\dot{C}H(CH_2)_4CH_3} \longrightarrow \mathrm{CH_3(CH_2)_4\underset{C_6H_5}{\underset{|}{C}}H{-}\underset{C_6H_5}{\underset{|}{C}}H(CH_2)_4CH_3}$$

Magnesium is formed at the cathode, but addition of alkyl halide to the medium to react with the magnesium deposit, generating more Grignard reagent, enables the reaction to be operated continuously in a flow cell. In a related reaction, addition of azide radicals to olefins was shown to lead ultimately to 1,4-diazides, which could be reduced to 1,4-diamines[92]:

$$\mathrm{N_3^-} \xrightarrow{[O]} \mathrm{N_3\cdot} \xrightarrow{\text{styrene}} \mathrm{N_3CH_2\dot{C}HC_6H_5} \longrightarrow \mathrm{N_3CH_2\overset{C_6H_5}{\overset{|}{C}}H\underset{C_6H_5}{\underset{|}{C}}HCH_2N_3} \xrightarrow{[H]} \mathrm{H_2NCH_2\overset{C_6H_5}{\overset{|}{C}}H\underset{C_6H_5}{\underset{|}{C}}HCH_2NH_2}$$

References

[1]B. C. L. Weedon, *Advan. Org. Chem.*, **1**, 1 (1960).
[2]A. K. Vijh and B. E. Conway, *Chem. Rev.*, **67**, 623 (1967).
[3]L. Eberson, in S. Patai, ed., *Chemistry of the Carboxyl Group*, New York, Wiley, 1969, chap. 6.
[4]R. N. Adams, *Accounts Chem. Res.*, **2**, 175 (1969).
[5]N. L. Weinberg and H. R. Weinberg, *Chem. Rev.*, **68**, 449 (1968).
[6]H. Kolbe, *Ann.*, **69**, 257 (1849).
[7a]M. Finkelstein and R. C. Peterson, *J. Org. Chem.*, **25**, 136 (1960); [b]L. Rand and A. F. Mohar, *J. Org. Chem.*, **30**, 3156, 3885 (1965).
[8]W. A. Pryor, *Free Radicals*, New York, McGraw-Hill, 1966, p. 316.
[9]L. Eberson, *Acta Chem. Scand.*, **17**, 2004 (1963).
[10]W. J. Koehl, Jr., *J. Org. Chem.*, **32**, 614 (1967).
[11]W. A. Bonner and F. D. Mango, *J. Org. Chem.*, **29**, 430 (1964).
[12]H. Breederveld and E. C. Kooyman, *Rec. Trav. Chim.*, **76**, 297 (1957).
[13]G. W. Kenner, M. A. Murray, and C. M. B. Tylor, *Tetrahedron*, **1**, 259 (1957).
[14]P. J. Bunyan and D. H. Hey, *J. Chem. Soc.*, 324, 2771 (1962).
[15a]L. Eberson, *Acta Chem. Scand.*, **17**, 1196 (1963); [b]L. Eberson and B. Sandberg, *Acta Chem. Scand.*, **20**, 739 (1966); [c]but see L. Eberson and S. Nilsson, *Acta Chem. Scand.*, **22**, 2453 (1968).
[16a]F. Fichter and R. E. Myer, *Helv. Chim. Acta*, **17**, 535 (1934); [b]F. Fichter and H. Stenzl, *Helv. Chim. Acta*, **22**, 970 (1939).
[17a]J. Petersen, *Z. Elektrochem.*, **18**, 710 (1912); [b]P. Karrer and M. Stoll, *Helv. Chim. Acta*, **14**, 1189 (1931).

[18]R. F. Garwood, C. J. Scott, and B. C. L. Weedon, *Chem. Commun.*, 14 (1965).
[19]G. Cauquis and B. Haemmerle, *Bull. Soc. Chim. Fr.*, 182 (1970).
[20]F. Fichter and S. Lurie, *Helv. Chim. Acta*, **16**, 885 (1933).
[21]H. A. Offe, *Z. Naturforsch.*, **2b**, 182, 185 (1947).
[22]D. G. Bounds, R. P. Linstead, and B. C. L. Weedon, *J. Chem. Soc.*, 2393 (1953).
[23a]A. Wurtz, *Ann. Chim. et Phys.*, **44**, 291 (1855); [b]A. Wurtz, *Jahresberichte*, 575 (1855).
[24]H. C. Brown, *Hydroboration*, Menlo Park, Calif., Benjamin, 1962, pp. 77–80.
[25]W. von Miller and H. Hofer, *Chem. Ber.*, **28**, 2427 (1895).
[26]H. Hunsdiecker, *Chem. Ber.*, **75**, 1197 (1942).
[27]S. Stallberg-Stenhagen, *Arkiv. Kemi*, **3**, 273 (1951).
[28]D. G. Bounds, R. P. Linstead, and B. C. L. Weedon, *J. Chem. Soc.*, 448 (1954).
[29]R. P. Linstead, J. C. Lunt, and B. C. L. Weedon, *J. Chem. Soc.*, 3331 (1950).
[30]F. Sorm, M. Streibl, V. Jarolim, L. Novotny, L. Dolejs, and V. Herout, *Chem. Listy*, **48**, 575 (1954); *Chem. Ind.*, 252 (1954).
[31a]E. J. Corey and J. Casanova, Jr., *J. Amer. Chem. Soc.*, **85**, 165 (1963); [b]E. J. Corey, N. L. Bauld, R. T. Lalonde, J. Casanova, Jr., and E. T. Kaiser, *J. Amer. Chem. Soc.*, **82**, 2645 (1960).
[32]L. Rand and C. S. Rao, *J. Org. Chem.*, **33**, 2704 (1968).
[33]E. J. Corey and R. R. Sauers, *J. Amer. Chem. Soc.*, **81**, 1739, 1743 (1959).
[34]P. S. Wharton, G. A. Hiegel, and R. V. Coombs, *J. Org. Chem.*, **28**, 3217 (1963).
[35a]H. H. Westberg and H. J. Dauben, Jr., *Tetrahedron Lett.*, 5123 (1968); [b]P. Radlick, R. Klem, S. Spurlock, J. Sims, E. E. van Tamelen, and T. Whitesides, *Tetrahedron Lett.*, 5117 (1968); [c]N. E. Rowland, F. Sondheimer, G. A. Bullock, E. LeGoff, and K. Grohmann, *Tetrahedron Lett.*, 4769 (1970); [d]H. Plieninger and W. Lehnert, *Chem. Ber.*, **100**, 2427 (1967).
[36]E. E. van Tamelen and D. Carty, *J. Amer. Chem. Soc.*, **89**, 3922 (1967).
[37]R. I. Reed, *Ion Production by Electron Impact*, London, Academic, 1962, p. 28ff.
[38]M. Finkelstein and R. C. Peterson, *J. Org. Chem.*, **25**, 136 (1960).
[39]B. Wladislaw and H. Viertler, *J. Chem. Soc.* (B), 576 (1968).
[40a]W. Koehl, Jr., *J. Amer. Chem. Soc.*, **86**, 4686 (1964); [b]V. D. Parker, *Chem. Commun.*, 1164 (1968).
[41a]L. Eberson and K. Nyberg, *Acta Chem. Scand.*, **18**, 1567 (1964); [b]J.-M. Kornprobst, A. Laurent, and E. Laurent-Dieuziede, *Bull. Soc. Chim. Fr.*, 3657 (1968).
[42]B. Wladislaw and A. M. J. Ayres, *J. Org. Chem.*, **27**, 281 (1962).
[43a]J. G. Traynham and J. S. Dehn, *J. Amer. Chem. Soc.*, **89**, 2139, 2799 (1967); [b]J. G. Traynham, E. E. Green, and R. L. Frye, *J. Org. Chem.*, **35**, 3611 (1970).
[44]J. T. Keating and P. S. Skell, *J. Org. Chem.*, **34**, 1479 (1969).
[45]J. T. Keating and P. S. Skell, *J. Amer. Chem. Soc.*, **91**, 695 (1969).
[46a]F. Fichter and H. Reeb, *Helv. Chim. Acta*, **6**, 450 (1923); [b]T. D. Binns, R. Brettle, and G. B. Cox, *J. Chem. Soc.* (C), 584 (1968).
[47]T. Shono, T. Nishiguchi, S. Yamane, and R. Oda, *Tetrahedron Lett.*, 1965 (1969).
[48]R. B. Woodward and R. Hoffmann, *Conservation of Orbital Symmetry*, Weinheim/Bergstrasse, Verlag Chemie, 1970, chap. 5.
[49a]L. Eberson and B. Olofsson, *Acta Chem. Scand.*, **23**, 2355 (1969); [b]L. Eberson and K. Nyberg, *Tetrahedron Lett.*, 2389 (1966).
[50a]R. P. Linstead, B. R. Shephard, and B. C. L. Weedon, *J. Chem. Soc.*, 3624 (1952); [b]M. Leung, J. Herz, and H. W. Salzberg, *J. Org. Chem.*, **30**, 310 (1965).
[51]H. Lund, *Acta Chem. Scand.*, **11**, 1323 (1957).
[52]V. D. Parker and R. N. Adams, *Tetrahedron Lett.*, 1721 (1969).
[53a]V. D. Parker and B. E. Burgert, *Tetrahedron Lett.*, 4065 (1965); [b]S. Andreades and E. W. Zahnow, *J. Amer. Chem. Soc.*, **91**, 4181 (1969).

[54a]S. D. Ross, M. Finkelstein, and R. C. Peterson, *J. Amer. Chem. Soc.*, **89**, 4088 (1967); [b]K. Nyberg, *Acta Chem. Scand.*, **24**, 473 (1970).
[55a]L. S. Marcoux, J. M. Fritsch, and R. N. Adams, *J. Amer. Chem. Soc.*, **89**, 5766 (1967); [b]G. Manning, V. D. Parker, and R. N. Adams, *J. Amer. Chem. Soc.*, **91**, 4584 (1969); [c]J. Phelps, K. S. V. Santhanam, and A. J. Bard, *J. Amer. Chem. Soc.*, **89**, 1752 (1967); [d]M. E. Peover and B. S. White, *J. Electroanal. Chem.*, **13**, 93 (1967).
[56]L. Eberson, *J. Amer. Chem. Soc.*, **89**, 4669 (1967).
[57]L. Eberson and K. Nyberg, *J. Amer. Chem. Soc.*, **88**, 1686 (1966).
[58]L. M. Stock and H. C. Brown, *Advan. Phys. Org. Chem.*, **1**, 35 (1963).
[59]K. Nyberg, *Chem. Commun.*, 774 (1969).
[60]R. O. C. Norman and R. Taylor, *Electrophilic Substitution in Benzenoid Compounds*, New York, American Elsevier, 1965.
[61]E. de B. Barnett and J. W. Cook, *J. Chem. Soc.*, **125**, 1084 (1924).
[62a]V. D. Parker and L. Eberson, *Tetrahedron Lett.*, 2843 (1969); [b]V. D. Parker, *Acta Chem. Scand.*, **24**, 3455 (1970).
[63]V. D. Parker and L. Eberson, *Chem. Commun.*, 973 (1969).
[64a]V. D. Parker and B. E. Burgert, *Tetrahedron Lett.*, 2415 (1968); [b]K. Koyama, T. Susuki, and S. Tsutsumi, *Tetrahedron*, **23**, 2675 (1966); [c]K. Yoshida and T. Fueno, *Chem. Commun.*, 711 (1970).
[65]K. Nyberg, *Acta Chem. Scand.*, **24**, 1609 (1970).
[66]M. Stacey, J. L. Tatlow, and A. G. Sharpe, *Advan. Fluorine Chem.*, **1**, 129 (1960).
[67]V. D. Parker, *Chem. Ind.*, 1363 (1968).
[68]L. L. Miller, E. P. Kujawa, and C. B. Campbell, *J. Amer. Chem. Soc.*, **92**, 2821 (1970).
[69]A. Zweig, W. G. Hodgson, and W. H. Jura, *J. Amer. Chem. Soc.*, **86**, 4124 (1964).
[70]B. Belleau and N. L. Weinberg, *J. Amer. Chem. Soc.*, **85**, 2525 (1963).
[71a]V. M. Clark, M. R. Eraut, and D. W. Hutchinson, *J. Chem. Soc.*, (C), 79 (1969); [b]E. Adler and R. Magnuson, *Acta Chem. Scand.*, **13**, 505 (1959); [c]M. D. Hawley and R. N. Adams, *J. Electroanal. Chem.*, **8**, 163 (1964).
[72]V. D. Parker, *Chem. Commun.*, 610 (1969).
[73a]V. D. Parker, *J. Amer. Chem. Soc.*, **91**, 5380 (1969); [b]M. F. Marcus and M. D. Hawley, *Biochim. Biophys. Acta*, **201**, 1 (1970); [c]M. F. Marcus and M. D. Hawley, *J. Org. Chem.*, **35**, 2185 (1970).
[74]C. A. Chambers and J. Q. Chambers, *J. Amer. Chem. Soc.*, **88**, 2922 (1966).
[75]L. Papouchado, G. Petrie, J. H. Sharp, and R. N. Adams, *J. Amer. Chem. Soc.*, **90**, 5620 (1968).
[76]P. J. Smith and C. K. Mann, *J. Org. Chem.*, **34**, 1821 (1969).
[77a]R. F. Nelson and R. N. Adams, *J. Amer. Chem. Soc.*, **90**, 3925 (1968); [b]E. T. Seo, R. F. Nelson, J. M. Fritsch, L. S. Marcoux, D. W. Leedy, and R. N. Adams, *J. Amer. Chem. Soc.*, **88**, 3498 (1966).
[78a]D. W. Leedy and R. N. Adams, *J. Amer. Chem. Soc.*, **92**, 1646 (1970); [b]R. F. Nelson and R. N. Adams, *J. Amer. Chem. Soc.*, **90**, 3925 (1968); [c]R. F. Nelson, private communication.
[79a]N. L. Weinberg and T. B. Reddy, *J. Amer. Chem. Soc.*, **90**, 91 (1968); [b]N. L. Weinberg and E. A. Brown, *J. Org. Chem.*, **31**, 4058 (1966); [c]N. L. Weinberg, *J. Org. Chem.*, **33**, 4326 (1968).
[80a]G. Cauquis and M. Genies, *Tetrahedron Lett.*, 3537 (1968); [b]G. Cauquis and M. Genies, *Tetrahedron Lett.*, 3403 (1970).
[81]J. O'M. Bockris, E. Gileadi, and G. E. Stoner, *J. Phys. Chem.*, **73**, 427 (1969).
[82]D. H. Geske, *J. Amer. Chem. Soc.*, **81**, 4145 (1959).
[83]A. T. Balaban, C. Bratu, and C. N. Rentea, *Tetrahedron*, **20**, 265 (1964).
[84]C. P. Andrieux and J.-M. Saveant, *Bull. Soc. Chim. Fr.*, **35**, 4671 (1968).

[85a]M. Fleischmann and D. Pletcher, *Tetrahedron Lett.*, 6255 (1968); [b]J. Bertram, M. Fleischmann, and D. Pletcher, *Tetrahedron Lett.*, 349 (1971).
[86]T. Inoue, K. Koyama, and S. Tsutsumi, *Bull. Chem. Soc. Japan*, **40**, 162 (1967).
[87]J. J. O'Connor and I. A. Pearl, *J. Electrochem. Soc.*, **111**, 335 (1964).
[88a]T. Inoue, K. Koyama, T. Matsuoka, and S. Tsutsumi, *Tetrahedron Lett.*, 1409 (1963); [b]T. Inoue, K. Koyama, and S. Tsutsumi, *Bull. Chem. Soc. Japan*, **40**, 162 (1967); [c]R. Leininger and L. A. Pasuit, *Trans. Electrochem. Soc.*, **88**, 73 (1945); [d]N. L. Weinberg and A. K. Hoffmann, *Can. J. Chem.*, **49**, 740 (1971); [e]K. Koyama, T. Ebara, T. Tani, and S. Tsutsumi, *Can. J. Chem.*, **47**, 2484 (1969).
[89a]V. D. Parker and L. Eberson, *Chem. Commun.*, 340 (1969); [b]L. Eberson and V. D. Parker, *Acta Chem. Scand.*, **24**, 3553 (1970).
[90a]J. M. Fritsch, H. Weingarten, and J. D. Wilson, *J. Amer. Chem. Soc.*, **92**, 4038 (1970); [b]C. P. Andrieux and J.-M. Saveant, *J. Electroanal. Chem.*, **28**, 339 (1970).
[91]H. Schäfer and H. Küntzel, *Tetrahedron Lett.*, 3333 (1970).
[92]H. Schafer and E. Steckhan, *Tetrahedron Lett.*, 3835 (1970); [b]H. Schafer, *Angew. Chem.*, **82**, 134 (1970).
[93]R. W. Hoffmann, *Angew. Chem. Int. Ed. Engl.*, **7**, 754 (1968).
[94]D. L. Coffen, J. Q. Chambers, D. R. Williams, P. E. Garrett, and N. D. Canfield, *J. Amer. Chem. Soc.*, **93**, 2258 (1971).
[95]N. Wiberg, *Angew. Chem. Int. Ed. Engl.*, **7**, 766 (1968).

9 EQUIPMENT

A number of decisions relating to electrochemical power sources, cell design, solvents, and so forth, must be made at the outset of any investigation in organic electrochemistry. Certain of these, such as choice of solvent or whether to use constant-current or controlled-potential electrolysis, must be made from chemical experience and application of the principles outlined in earlier chapters of this book. Solutions to other problems, more technical in nature, such as the choice of specific instruments for preparative-scale electrolysis or for voltammetry, depend more or less on the state of the instrumental art at the time. The present chapter, which should by no means be considered exhaustive, attempts to describe certain equipment choices available to the beginning investigator. Rather than recommendations concerning specific instruments (which may quickly become outdated), the heaviest emphasis is on features to be desired in individual pieces of equipment. Some specifics are discussed for purposes of clarity, however. A number of references are included to extensive discussions elsewhere of specific subjects not treated in detail here.

As with other more or less specialized research areas, initiation of experimentation in organic electrochemistry requires an investment of money and time at the outset. This initial investment is surprisingly low, however, considering the very wide range of synthetic and mechanistic applications to which the apparatus may be put. It may help here to place the total expense on a scale familiar to the organic chemist by noting that equipment for polarography, cyclic voltammetry, controlled-potential electrolysis, and coulometry can be acquired commercially for less than the cost of many commercial gas chromatographs, and can be constructed for far less.

It may be helpful to the reader to point out at this time the general orientation of this chapter. It is felt by the author that an extended discussion of experimental methods, including instrumental design, comparison of specific instruments, operating practices, and so on, valuable though such a discussion might be, is not appropriate to the aim of this book, which is to present an overview of the field of organic electrochemistry, with emphasis on synthetic applications. As a practical point, furthermore, far more space would be required for such a discussion than is possible within the space limitations of this book. Rather, the intent in this chapter is to point out the kinds of questions and problems with which one is confronted when beginning experimentation in organic electrochemistry. Once the reader understands, for example, the necessary and desirable features in a potentiostat or wave generator, he may then consult the chemical and manufacturers' literature and/or an electronics technician, and begin to make choices among the various options. A further advantage of this emphasis on principles rather than specific instruments is that it is unlikely to become outdated by advances in technology and the market place.

9.1 POWER SOURCES

9.1.1 Constant-Current Electrolysis

Apparatus for constant-current electrolysis is electrically very simple. One wishes to apply a constant known current between anode and cathode. This can be done in a number of ways.[1] One method is to apply the voltage output of a laboratory autotransformer (Variac) to a simple current rectifier; the rectified current can then be passed through the cell and an ammeter in series with the cell. Manual control of the electrolysis current is effected by adjustment of the Variac output. This method, or variations on it, such as use of a constant-voltage dc power supply (in series with a rheostat) as the variable-current source, is an inexpensive method of effecting constant-current electrolysis. It does require more or less constant attention and manual adjustment of the current during electrolysis, because of changes in cell resistance, temperature, and so on. A better approach, where extensive experimentation in constant-current electrolysis is contemplated, would involve substitution of a constant-current regulated power supply for the manually operated power sources. The advantages of this procedure are that electrolysis currents are maintained constant automatically, and hence improved reproducibility is achieved in difficult experiments. The specifications required for such a supply need not be extravagant: current capability adjustable from a few milliamperes to an ampere or more, with current control not required to be better than ± 5–10%. Many relatively inexpensive commercially available power supplies would suffice.

9.1.2 Controlled-Potential Electrolysis

Recall from Chapter 3 that a controlled-potential electrolysis requires a third electrode in the solution in addition to the anode and cathode. This reference electrode is positioned near the working electrode and monitors the potential difference between the working and reference electrodes by an external voltage-measuring circuit. In the simplest form of controlled-potential electrolysis, the voltage-measuring circuit is independent of the power circuit (Figure 9.1).[2] A controlled-potential electrolysis can then be carried out by manually adjusting the output voltage (ΔV) of the power supply at a level sufficient to maintain the potential difference between the working and reference electrodes at the desired constant value (ΔE). Equipment for such an experiment is simple and readily assembled: one needs only a high-impedance laboratory voltmeter and a

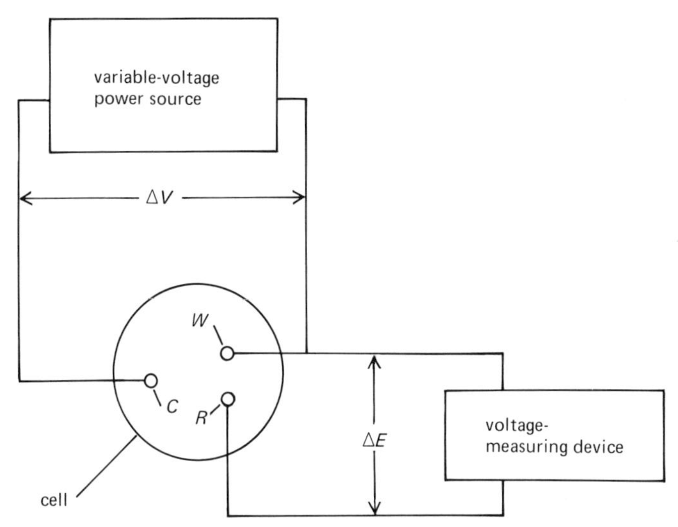

Figure 9.1 Schematic representation of a controlled-potential electrolysis experiment. *C*, counter electrode; *W*, working electrode; *R*, reference electrode.

variable-voltage power supply, such as a Variac–rectifier combination or an electronic regulated power supply. This arrangement has the disadvantages that constant adjustment of ΔV is necessary to maintain ΔE at the desired level and that the system is not capable of very precise control of working electrode potentials; hence, it would normally be utilized only where no more than an occasional electrolysis is planned. For improved operation, it is possible to incorporate into the system, with relatively little additional expense or effort, an electronic feedback circuit coupling the voltage-*measuring* (ΔE) and voltage-*controlling* (ΔV) circuits. This is simple enough that the resulting instrument for automatic control of working electrode potentials, the potentiostat, is now used almost universally for controlled-potential electrolysis.

Although there are a number of other features that the "ideal" potentiostat ought to possess, those that are of particular interest in a potentiostat to be used for preparative-scale controlled-potential electrolysis are high output voltage capacity, high current capability, sensitive and stable control of the working electrode potential, rapid response to changes in the working electrode potential, and a wide range of control potentials. Some desirable features of potentiostat design, such as high current capacity and fast response time, are to a certain extent mutually incompatible, so one normally chooses a compromise design that is best for the type of experiment contemplated. (Thus, a potentiostat to be used in studies of electrode kinetics should have fast response times, but need not supply large currents.) There is little doubt that the most desirable features in a potentiostat to be used for preparative-scale electrosynthesis of organic compounds are high output voltage and high current capacity. These have been discussed at some length in Chapter 3; simply recall that a high output voltage must be applied across the anode and cathode in order to pass substantial currents through high-resistance organic solvents, and that high currents are necessary for reduction of appreciably large amounts of material in a reasonably short time. A potentiostat for organic electrosynthesis should have an output voltage capability of 100 V or more, and an output current of 0.5–2.0 A or more, depending on the scale on which electrolysis is intended. Although potentiostats of high current capacity have been commercially available for a number of years, it is only relatively recently that instruments have appeared on the market that are capable of high output voltages as well. The latter is very often the more important of the two, especially in aprotic solvents, since a high solution resistance can prevent utilization of the full current capability of the potentiostat. With modern potentiostats, slow response times are rarely a problem in preparative experiments; likewise, precise potential control is not as important as in more sophisticated experiments. Most potentiostats can easily maintain working electrode potentials within $\pm$1–5 mV of the desired value, and it is doubtful that control is really necessary at a level better than $\pm$10 mV for most preparative electrolyses. The range of working electrode potentials useful for organic experimentation is on the order of $\pm$3–4 V (relative to the reference one is using), and again most potentiostats have this capability.

The decision whether to buy a commercially available potentiostat or to build one in one's own laboratory requires consideration of a number of factors: availability of an instrument with the necessary features, one's financial resources and electronics capability, time necessary to build the instrument, and so on. Generally speaking, commercial instruments will be more expensive than those built in one's own laboratory, but will have had the inevitable "bugs" worked out of them. A few years ago, the choices among commercial potentiostats were limited, but now there is a wide variety of useful instruments available.[3b] The interested reader may examine manufacturers' literature,* keeping in mind the preceding discussion of desirable features of a potentiostat for preparative electrolysis. One point should be mentioned here with respect to commercial potentiostats: certain of these have built-in, or obtainable as accessories, highly accurate (0.1–0.2%) coulometers designed primarily for quantitative electroanalytical purposes. One electrolyzes at a potential where the material to be determined (usually present in trace amounts) is reduced (or oxidized) and measures the (small) amount of current required for electrolysis, then relates this quantity to the amount of substance present. The accuracy of coulometry as required in mechanistic investigations, such as determination of the number of electrons consumed per mole of electroactive material in a preparative electrol-

* At this writing, a number of manufacturers (e.g., Wenking, Tacussel, Chemical Electronics, Hickling, Princeton Applied Research, Amel (Italy), and Anotrol) supply potentiostats that should serve acceptably for preparative organic electrolyses. Others will no doubt enter this growing field. Manufacturers often advertise in the professional journals such as *Analytical Chemistry* or *Journal of the Electrochemical Society*.

ysis, is of a lower order (1–3% is adequate), hence the added expense of a very accurate coulometer is not justified. If one does wish to do very accurate coulometry on a preparative scale, the coulometer used should have a capacity of at least 500 C. (For example, notice that 965 C would be consumed during electrolysis of only 0.5 g of a material of molecular weight 100 undergoing a 2-electron process.)

Many investigators have constructed potentiostats in their own laboratories. The advice of a competent electronics technician can be valuable here when one's electronics background is limited. There are some advantages associated with construction of a potentiostat as opposed to purchase of one. Apart from the substantial saving in cost, this choice allows for incorporation of only those features required for one's own purposes; at the same time, it is sometimes easier to modify such an instrument to meet changing needs than it is to modify a commercial instrument.

When considering the possibility of constructing a potentiostat, it is useful to recall the way in which a potentiostat operates (Chapter 3). It must sense fairly small changes in the working electrode potential (relative to the reference electrode) and use these changes as an error signal to adjust the output voltage of the potentiostat so as to restore the working electrode potential to its original value. In 1958, Lingane described a number of potentiostat circuits that could perform this function adequately.[3a] This review is very considerably out of date, however, because of subsequent advances in electronics, and better potentiostat designs are now available.[3b] One possibility utilized by a number of investigators is the use of a so-called *programmable power supply*.[4] A programmable power supply differs from conventional regulated power supplies by virtue of the fact that it contains a feedback circuit designed to detect changes in the potential of one of its outputs (+ or −) relative to a reference potential and to vary the voltage output of the power supply to restore that output to its original potential. This is, of course, exactly the desired function of a potentiostat, and hence very little effort is required to convert a programmable power supply into a functioning potentiostat. The specifications (output voltage and current capabilities, response times, etc.) required of the power supply can be met by one of the many commercially available programmable power supplies; an electronics technician's advice can be helpful with this choice.* One attractive feature resulting from the use of programmable power supplies is the fact that two or more can be coupled together when the voltage or current capability of one is inadequate. Thus, two 60-V, 1-A supplies might be operated in series to produce a potentiostat of 120-V, 1-A capacity for use in high-resistance solvents,[4a] or they could be operated in parallel to produce a 60-V, 2-A potentiostat for larger-scale electrolyses.

Although convenient, it is not necessary to use the internal feedback

* One manufacturer, Kepco, Inc., has explicitly recognized and discussed the use of programmable power supplies as potentiostats.[5] A potentiostat based on a Kepco programmable power supply has given satisfactory service in the author's laboratory for several years.

circuitry of a programmable power supply to control power-supply output voltage. Instead, one can build an external feedback circuit to control the output of an ordinary regulated power supply. A convenient way of doing this was suggested by Bottei and Boczkowski, who used operational amplifiers as sensing and control devices to govern the output voltage of a simple power supply.[6] These authors constructed a potentiostat of high output-voltage capability for use with high-resistance organic solvents by assembling several readily available and relatively inexpensive commercial components.

Potentiostats possessing the virtues of rapid response and sensitive potential control can be constructed quite easily using an operational amplifier not as a control device but as the potentiostat itself.[7] Because of the low current capabilities of most operational amplifiers (1–5 mA), such potentiostats are not adequate for electrolysis of gram amounts of organic compounds (some operational amplifiers with higher current capability are, however, commercially available) but are adequate for electrolyses on the milligram scale, for example, when products are to be analyzed by gas chromatography. Aside from preparative applications, one real use that potentiostats based on operational amplifiers do have in electroorganic chemistry is in cyclic voltammetry and polarography, since the currents passed in these techniques are less than 1 mA, and these techniques require accurate and sensitive control of potentials as well as fast response times, features that operational amplifier potentiostats do possess. This application will be discussed in Section 9.3.

9.2 COULOMETRY

Recall from Section 3.3 that the role played by coulometry in mechanistic organic electrochemistry is (1) in determinations of n, that is, the number of electrons consumed per molecule of starting material and, (2) in obtaining information about the nature of coupled chemical reactions from analysis of current-time behavior during the electrolysis. These applications should be kept in mind when deciding how to obtain coulometric data. A variety of coulometric methods have been described in the literature, but many of these have been designed for use in quantitative analysis, often of what may be only trace components of mixtures.[8] Consequently, such applications have called for development of very accurate and sensitive coulometers. As was noted previously, the application we are discussing here does not require such accuracy: while coulometers designed for trace electroanalysis can measure quite precisely quantities of electricity in the 1–50 C range, one is interested for our purposes in a coulometer with capacity of 500–5000 C and modest accuracy ($\pm$1–3%). The following discussion is limited to a few relatively simple methods of gathering coulometric data at this level of accuracy. Extended discussions of coulometry may be found elsewhere.[8]

A very useful general method of measuring the quantity of electricity consumed in an electrolysis is to follow the voltage drop across a standard resistor in series with the cell, using a strip-chart recorder. It is useful to have available a number of different standard resistors, selectable by a multiposition switch, so that sensitivity and accuracy can be improved by choosing the proper resistor to produce a full-scale deflection of the recorder pen. The area under the current–time curve produced by the recorder is integrated to obtain the total quantity of electricity passed in the electrolysis. Integration could be performed by any of the standard methods: planimetry, cutting and weighing, or preferably, mechanically (i.e., a Disc integrator) or by simultaneous electronic integration. The procedure used in the author's laboratory involves recording the current–time curve on a 1-mV recorder equipped with a Disc integrator and its digital printer accessory. The latter is activated at the beginning of the electrolysis; at the end it prints out the total number of counts measured by the integrator during the electrolysis. This system has the convenient feature that on a given current range one integrator count is equal to a fixed number of faradays, and that the total number of counts recorded by the printer need only be multiplied by the appropriate factor* to obtain the total number of faradays consumed in the electrolysis. The use of a 1-mV recorder makes accurate current measurements difficult above 100 mA, but it has the advantage that since most commercial gas chromatographs have a 1-mV output, the recorder–integrator combination can be used for either coulometry or quantitative gas chromatography by incorporation of a simple switch. For coulometry at higher current levels, a 10- or 100-mV recorder (or plug-in conversion unit for the 1-mV recorder) can be used.

The graphical record of current *vs* time obtained using this method will, on occasion, suggest mechanistic possibilities such as the formation and decay of intermediates in the electrolysis (Section 3.3). Such information is lost in those methods, such as charging of a capacitor, voltage-to-frequency conversion, digital integration, and so on, that afford only the total amount of electricity consumed in the electrolysis, but otherwise they are quite satisfactory for determination of n values.

9.3 CYCLIC VOLTAMMETRY

There are four components to a cyclic voltammetric instrument: triangular wave generator, potentiostat, recorder, and cell. Design of cells will be dealt with separately in Section 9.4. Of the remaining three, let us first briefly discuss the recorder. Depending on the voltage sweep rate, the recorder will be either an

* This factor is readily computed from the characteristics of the recorder, integrator, and standard resistor. For example, for a Disc integrator with full-scale output of 6000 counts/min, a 1-mV recorder, and a standard resistor of 0.01 Ω (i.e., full-scale current of 100 mA; this low resistance is obtained by a voltage-dividing network), one count is equal to 10^{-3} C, or $1.036 \times 10^{-8}\ \mathscr{F}$.

oscilloscope or a mechanical X–Y recorder. In either case, the voltage output of the wave generator is applied to the X axis of the recording device; the electrolysis current is recorded on the Y axis by following the voltage drop across a precision resistor in series with the cell. (It is possible, but less convenient, to use a strip-chart, i.e., Y-time, recorder for slow cyclic voltammetry where an X–Y recorder is not available.)

The two components of special interest here are the triangular wave generator and the potentiostat. These are discussed separately below, even though in practice the two are often combined in a single chassis.

Triangular wave generators are widely available commercially, and several designs have been described in the literature. One's choice of a specific design must be guided by a number of considerations. Characteristics of wave generators vary, depending on the applications for which they are intended. A good wave generator for cyclic voltammetry should provide for a wide range (amplitude) of output potentials, sensitive setting of anodic and cathodic potential limits, a wide range of voltage sweep rates, and both single- and multiple-cycle operation. The first of these requirements is easily met by most wave generators: a range of working electrode potentials of ± 4 V is adequate for almost all common electrochemical operations, and most generators have at least this wide an amplitude. A related feature, the option of setting both the cathodic and anodic scan limits independently, is not always available, however. That is, some wave generators permit one to set only a single limit; the other is then 0 V (relative to one's reference electrode). When investigating a cathodic process occurring at -1.5 V with a wave generator of this type, one would set the cathodic limit at, say -2 V; the wave generator would then cycle the cathode potential between 0 and -2 V and back to 0 V again (and then repeat this if set for multiple cycle operation). The ability to set both anodic and cathodic limits, so that, for example, one could cycle the potential between -1 and -2 V in the above example, provides an extra measure of flexibility, particularly in wave-clipping experiments (Section 3.2). If not built into the wave generator, this feature can be supplied separately via an external voltage offset applied to the working electrode.

The very great utility of cyclic voltammetry in mechanistic studies arises from the fact that by observing the electrochemical behavior of a substance over a wide range of potential scan rates one can obtain a much clearer insight into the course of the reaction than is possible from measurements taken at only a single scan rate. Hence, a useful wave generator must of necessity produce a very wide range of scan rates, roughly from 10 mV/sec to 10^3 V/sec. (The lower limit is determined by the point at which natural convection sets in, and above the upper limit the contribution to the total current from the charging current becomes excessive.) When examining specifications of wave generators, it will be noted that many have outputs expressed in cycles per second, while the operational parameter in cyclic voltammetry is the voltage sweep rate, expressed in volts per second. The potential sweep rate will be a function of the size of the cycle in those generators where the output is expressed in cycles per second rather than volts per second. Thus, a cyclic scan from 0 to -2 V and back is equal to 40 V/sec at 10 cycles/sec, but when scanning from 0 to -1 V and back, the sweep rate is only 20 V/sec at 10 cycles/sec. This point is of importance where the width of a cycle is changed during an experiment, such as during wave-clipping experiments.

It is generally observed that the shape of a cyclic voltammogram is slightly different on the second and succeeding cycles from that traced on the first cycle, even in the case of a simple reversible process. This arises from the fact that slight changes take place in the concentration gradients of oxidized and reduced species at the electrode surface during successive cycles; the steady state is usually reached after a few cycles.[9] For qualitative purposes the steady-state voltammogram is often adequate, but for quantitative and semiquantitative experiments the cyclic voltammogram measured on the first cycle is usually required. For example, the mechanistic criteria of Nicholson and Shain (Chapter 3) were calculated for the first cycle.[9] However, it is also often important to be able to record the second and later cycles in mechanistic investigations, since peaks may appear in the second cycle that were not present in the first [*cf.* the study of aniline by Adams and Bacon (p. 85)[10]], hence the desirability of having a wave generator that is capable of both single- and multiple-cycle operation. A final feature of some utility is the capacity to produce asymmetric triangular waves, that is, those in which the sweep rate on the forward sweep is either lesser or greater than that of the backward sweep.

The cost of a commercial triangular wave generator incorporating all of the above features would be several hundred dollars. Shain and Myers have, however, published construction details for a wave generator including all of these features and more.[11] The instrument is based on solid-state digital circuitry and is compact, inexpensive, and readily assembled. Since the unit was specially designed with electrochemical applications in mind, it also possesses some features not available in some very expensive commercial wave generators. For example, the starting, anodic, and cathodic potential scan limits may all be set independently. This feature could be of real value in mechanistic studies. Other features include a range of sweep rates from 2 mV/sec up to 1000 V/sec, single- or multiple-sweep operation, a range of asymmetry ratios of cathodic : anodic scan rates up to 500:1, and scan limit setting to an accuracy of ca. 1mV. Clearly, the construction of an instrument of this type should be considered in preference to purchase of a commercial instrument. The paper by Shain and Myers, incidentally, also illustrates a model circuit for a potentiostat to be used with the wave generator if so desired. We now turn our attention to this question of the type of potentiostat that is useful in cyclic voltammetry.

If the potentiostat used for preparative electrolyses has a fast enough response time, it can also be used for cyclic voltammetry. Often, however, it will not be fast enough; furthermore, experimental efficiency and productivity are enhanced by the use of separate potentiostats for cyclic voltammetry and preparative electrolysis, so that the two kinds of experiment can be carried out independently in the laboratory. The degree of sophistication required in a potentiostat for preparative electrolysis is not necessary in a potentiostat to be used for cyclic voltammetry, because of the low currents and short times characteristic of the latter technique. In fact, Shain and Schwarz have pointed out that, in principle, a potentiostat can be constructed using a single operational amplifier.[7] (This paper, two by Reilley,[12a] and the books by Bair[12b] and Malmstadt and Enke[12c] will serve as an introduction to the nature and use of operational amplifiers.) Potentiostats for preparative electrolysis usually incorporate additional operational amplifiers, both to stabilize the system against voltage drifts during operation and to minimize errors in measured electrode potentials as a result of either current flow through the reference electrode or voltage drop across the current-measuring device.[7,13] Drift will, on the other hand, be essentially negligible on the time scale (seconds) necessary to measure a cyclic voltammogram, so drift stabilization is not as important a consideration as it would be on the preparative time scale. A number of investigators have actually described potentiostats for cyclic voltammetry that consist of only one operational amplifier.[14] However, in view of the low cost of high quality solid-state operational amplifiers, the increased sophistication of a stabilized potentiostat, such as the design described by Shain,[11] would be worth the additional construction effort.

9.4 CELLS

The electrochemical cell must perform several distinct functions, which are not too different whether one is interested in a quantitative voltammetric experiment at the microampere level or a large-scale preparative electrolysis. Explict recognition of these functions may help in design of cells for given experiments. The cell must: (1) isolate the electrolysis solution from deleterious influences, (2) afford firm control over those experimental variables that are liable to affect the electrode reaction, and (3) ensure that the proper type of mass transport is operative in the experiment.

Note, for example, the necessity for isolation of the solution from possibly destructive agents such as atmospheric oxygen and water, and also more insidious sources of interference. Prominent among the latter are reaction of products formed at the working electrode with reagents generated at the counter electrode, or diffusion of the products themselves to the counter electrode where they might undergo undesired electrochemical reactions. For all of these reasons, the electrolysis solution is usually isolated from contact with the counter electrode and atmosphere. The latter is readily excluded by a preliminary nitrogen purge followed by a careful sealing of the cell, but in really precise work it may be necessary to effect electrolysis on a vacuum line[15] or in a dry box. Isolation of the electrolysis solution from the counter electrode is somewhat harder to arrange. What is required is a porous membrane between the working and counter electrode compartments that is permeable to ions of the supporting electrolyte but not to the electrolysis substrate and products. Most often, the membrane separating the working and counter electrode compartments is a sintered glass disk, although other substances have been used.[16] Unfortunately, such disks, especially when of fine porosity, exhibit a high electrical resistance that, especially when in combination with high-resistance organic solvents, may seriously limit the total amount of current that can be passed in a preparative electrolysis. On the other hand, too coarse a frit will not impede the flow of solution through the frit. For these reasons, a medium-porosity fritted disk is often used, as a compromise. Elving has recently described a method of preparing low-resistance membranes for use in experiments where DMF is the solvent.[17] This method depends on the fact that a hot, concentrated solution of methylcellulose in DMF containing a dissolved electrolyte sets to a clear gel upon cooling. This gel has low electrical resistance (because of its electrolyte content), but at the same time it exerts considerable physical resistance to solution flow. The hot methylcellulose solution is poured onto the frit separating the counter and working electrode compartments, and allowed to cool. Since the frit serves only as a physical support for the gel, a very coarse, and hence low electrical resistance, frit suffices. This gel has been used routinely for several years at Wesleyan University. Unfortunately, methylcellulose does not form a gel in all solvents (1,2-dimethoxyethane and acetonitrile have been tried),[18] so the generality of this procedure is limited. It happens, however, that gels made up in DMF are stable for hours when in contact with solvents such as 1,2-dimethoxyethane and acetonitrile.[18] Therefore, the gel can be made up in DMF, excess DMF decanted away, and a different solvent added. DMF is slowly leached from the gel during use in this manner.

In reactions where products formed at the working electrode do not react at the counter electrode or with substances formed at the counter electrode, as in, for example, many Kolbe reactions, anode and cathode need not be in separate compartments. A one-compartment cell[19] is simpler and has lower internal resistance than a cell with divided anode and cathode compartments, and may be of use in large-scale reactions where the above restrictions are met. In careful work, such as when one is examining the effects of experimental conditions on an electrode reaction, a divided cell is always necessary.

The second cell function mentioned above, that of ensuring firm control over experimental parameters, includes a variety of factors such as provisions for adequate temperature control, introduction of reagents and withdrawal of samples, degassing of the solution, and so on. For accurate voltammetric measurements, cells should be immersed in a bath thermostatted to $\pm 1°C$. Considerable heat may be evolved in some preparative electrolyses; this may call for external or internal cooling of the cell, particularly when chemical reactions are coupled to the electron transfer. In the latter case, it is possible that the course of the overall electrochemical process could actually be changed by overheating. Under this category, it may be relevant to note that proper placement of the reference electrode can be important in a controlled-potential electrolysis involving electrode processes whose potentials are closely spaced. Harrar and Shain showed that in such a situation it is best to locate the reference electrode between the counter and working electrodes and at the point of highest current density.[20] The reference electrode should also be positioned as closely as possible to the working electrode, to minimize the *iR* drop between the two.

Obtaining mass transfer of the desired type could mean efficient stirring of the solution in a preparative electrolysis, or conversely, elimination of convection in an experiment whose success depends on mass transfer by diffusion only, such as polarography or cyclic voltammetry. This would require careful thermostatting of the cell and elimination of other sources of convection, such as vibration. This category would also include problems related to the planarity of the electrode where linear diffusion is to be assumed.

The preceding principles can be reduced to practice in many different ways, and the reader should consult the literature for the many designs that others have used.[21] No attempt will be made to reproduce cells here that are already adequately described elsewhere. Rather, for the purpose of illustrating certain common features of cell design, two cells used in the author's laboratory are shown in Figures 9.2 and 9.3.

The larger cell, shown in Figure 9.2, is designed for electrolysis of materials on the gram scale (quantities of solution from 25 to 100 ml), but could be adapted to the electrolysis of substantially larger amounts of material if desired. The cell is constructed from a standard-taper 50/50 female joint. It utilizes a mercury pool cathode, to which electrical contact is made via a platinum wire inserted through the capillary sidearm. Four standard-taper side-arms (three 14/20 and one 24/40) provide for introduction of accessories, including a variable-speed stirring motor equipped with standard-taper joint. The anode compartment consists of a length of glass tubing ending in a coarse-porosity fritted disk and fitted into a Teflon sleeve, which is machined to take the anode compartment and to fit tightly into the cell. Some comments on this arrangement may be helpful. In general, the best placement for the anode is parallel to the cathode surface, and the areas of the anode and cathode should also be similar, to ensure that potential gradients between the two are as uniform as possible. In a cell of this size, these considerations dictate that the anode be inserted from the top, hence the use of a stirrer entrance on the side. In a larger cell there would be room to insert both stirrer and anode (and even the other utilities) through the top of the cell. The solution could also be stirred magnetically if so desired. The reference electrode is connected to the cell by an adjustable Teflon adapter.[21q] A rubber serum cap over one 14/20 port permits injection of reagents and withdrawal of samples using precision syringes, and also degassing with nitrogen

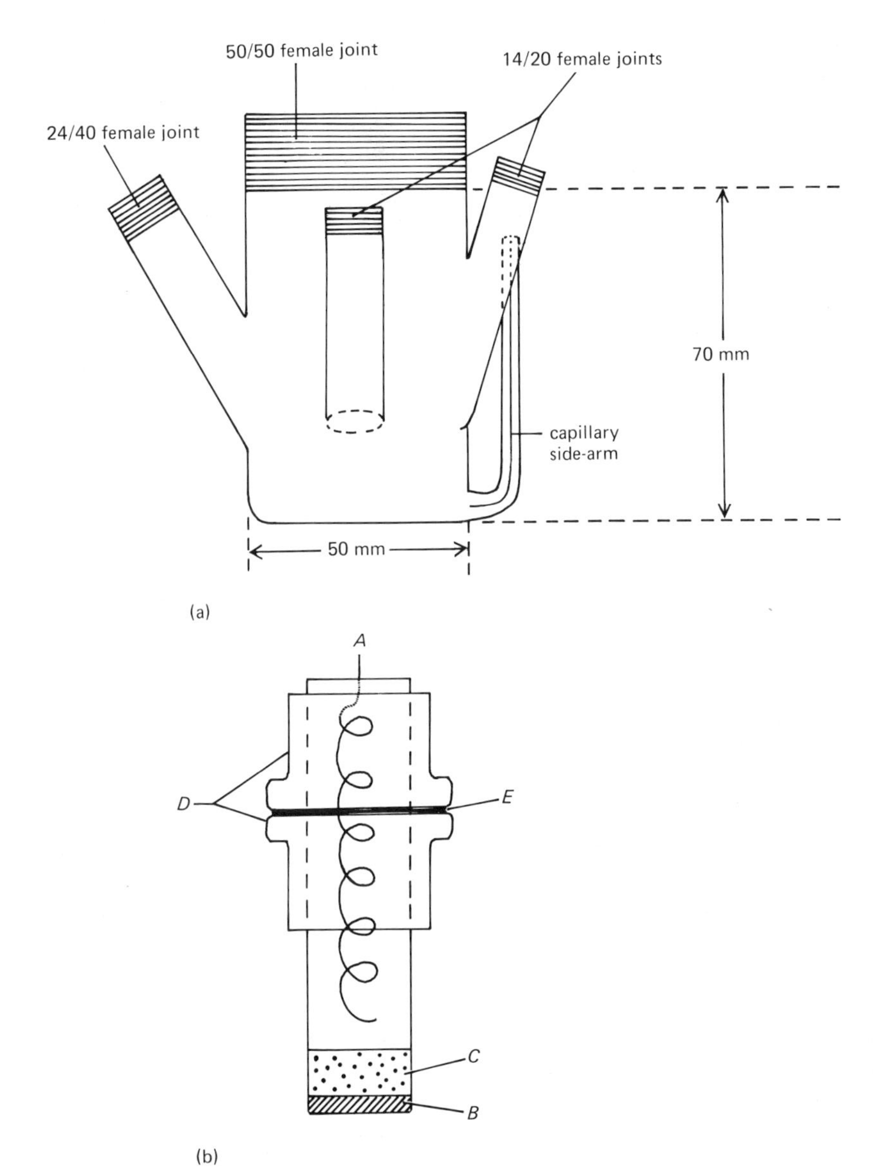

Figure 9.2 Electrochemical cell. (a) Working electrode compartment. A third port with 14/20 female joint is hidden from view. (b) Counter electrode compartment. *A*, counter electrode; *B*, fritted glass disk (coarse); *C*, methylcellulose gel; *D*, Teflon plug, machined to fit the working electrode compartment; *E*, rubber O-ring. The two halves of the Teflon plug screw together.

before electrolysis, with the aid of two syringe needles piercing the serum cap, one needle attached to the nitrogen source and the other serving as a pressure vent. Gaseous reagents would be added in the same manner. The cell has also been used with solid metal cathodes (electrical contact is made through the serum cap).

The small cell (Figure 9.3) is designed for electrolysis of small amounts

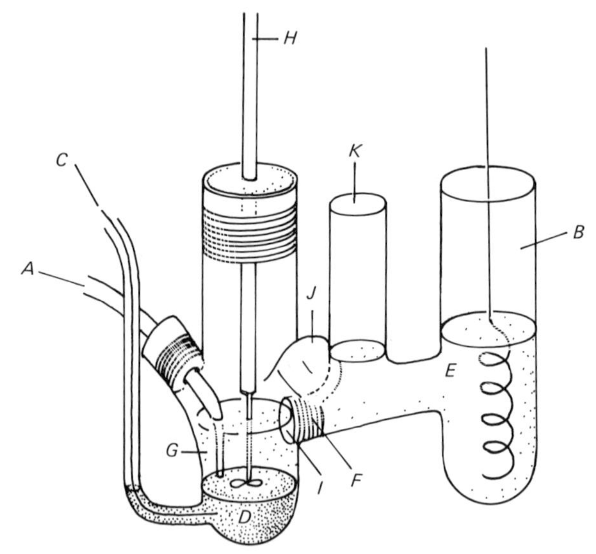

Figure 9.3 Cell for small-scale electrolysis. *A*, reference electrode probe; *B*, anode; *C*, capillary for electrical contact to cathode; *D*, mercury pool cathode; *E*, anode compartment; *F*, methylcellulose gel; *G*, cathode compartment; *H*, stirrer; *I*, fritted disk (coarse); *J*, serum cap for degassing, addition of reagents, and withdrawal of samples; *K*, port for addition of methylcellulose gel.

(20–100 μmole) of material, especially when the electrolysate is to be analyzed directly by gas chromatography. The volume of solution electrolyzed is 5–7 ml, and electrolyses are complete in 7–10 min. A methylcellulose gel is added through port *K* and allowed to set before using the cell. The stirrer *H* can be replaced by a cap bearing a dropping mercury electrode and stationary electrode, so that polarography and cyclic voltammetry can be carried out on a single sample of solution. Because of the small size of the cell, a separate nitrogen inlet is not incorporated; instead, degassing is carried out with the aid of syringe needles piercing the serum cap, as in the cell discussed in the previous paragraph. As in that cell, electrical contact to the mercury is made through the capillary sidearm by a fine-gauge platinum wire. Others have used instead a platinum wire sealed into the bottom of the cell, but cleaning the platinum becomes difficult.

9.5 MATERIALS

9.5.1 Electrode Materials

A mercury-pool electrode is perfectly adequate for most preparative-scale cathode reactions. Mercury has a number of advantages: a clean reproducible surface can readily be obtained, it has a high overvoltage for hydrogen evolution, and potentials measured polarographically can be compared directly with those in the preparative electrolysis. The toxicity of mercury must not be forgotten, but Christian has shown that no real hazard exists when mercury safety regulations are obeyed.[22] Mercury does have the disadvantage that, as a liquid of considerable surface tension, it does not lend itself readily to the construction of planar electrodes, at which linear diffusion may be assumed. This is unimportant in preparative electrolysis and even in routine polarography, but for some quantitative and semiquantitative applications, such as use of the Nicholson–Shain mechanistic criteria in cyclic voltammetry, nonplanarity of the electrode may introduce serious complications (see p. 93).[23] Several solutions to the problem of producing a planar mercury electrode have been reported. One method is to coat a planar platinum electrode with a mercury film.[24] Alternately, one may take advantage of the fact that mercury wets platinum; thus, the height of a platinum ring surrounding a mercury pool can be adjusted to a point where the mercury surface is flat, as a result of a balance between its inherent tendency to be convex and its countertendency to be concave when wetting platinum.[25]

Qualitative analysis of the electrode behavior of a substance by cyclic voltammetry does not, on the other hand, require the use of a planar electrode. For experiments of this sort the experimentally simple hanging mercury drop electrode (HMDE) may be employed. As the name implies, this is simply a drop of mercury used as a cathode.[26] One may construct a HMDE in several ways. A common practice, which takes advantage of the adhesion of mercury onto platinum, involves suspending a drop of mercury from a short length of platinum wire. Shain has described a useful HMDE arrangement, in which a dropping mercury electrode produces the drops for the HMDE.[27] Another kind of HMDE (available from Brinkmann Instruments) consists of a mercury-filled microburet, with which drop sizes are readily reproducible.[28]

Platinum is perhaps the most commonly used material for organic oxidations, although a number of other substances, including carbon, are also employed. It should be recognized that the surfaces of most solid electrodes may contain films consisting of variable amounts of oxides, adsorbed organic materials, hydrogen, and so on, depending on the history of the electrode. Since surface conditions can affect electrochemical behavior, such films are a potential source of nonreproducibility. It is a wise procedure, therefore, to pretreat solid electrodes in a standard way before each use so that, even if surface impurities are not completely removed, they are at least relatively constant throughout a series of experiments. Adams has described at length the electrochemical characteristics of a wide variety of solid materials commonly used as anodes, including questions of surface preparation, reproducibility, and potential limits.[29] Swann has discussed the subject briefly from a more empirical point of view.[30]

9.5.2 Solvents

Mann has recently published an extensive review of the physical and electrochemical properties, and purification procedures, for solvents commonly used in organic electrochemistry.[31] This and other recent reviews[32] make any detailed discussion of the subject unnecessary here. Recall, however, that the ideal solvent should not only have low electrical resistance, but also be chemically inert, cheap, volatile, easily purified, and a good solvent for electrolytes. No solvent possesses all these virtues and a compromise must be made, depending on one's needs. For example, generation of aromatic radical anions requires the use of aprotic media, at the expense of the high electrical resistance associated with such solvents. A good solvent for oxidations might not be good for reductions, and vice versa. Thus, nitrobenzene[33] and dichloromethane[34] are rather good solvents for studying anodic processes, despite the fact that both are readily reduced electrochemically. The latter solvent has the disadvantage that it has a high electrical resistance and is a poor solvent for electrolytes. (Baizer has, however, found that methyltriphenylphosphonium tosylate is freely soluble in dichloromethane.[35])

Of course, the reason for using organic solvents, in spite of their high electrical resistance, is the fact that organic compounds are poorly soluble in aqueous media. A solvent of high water content (and hence high conductivity) that would also dissolve large amounts of organic solutes would be of obvious utility in organic electrochemistry. An interesting class of solvents possessing these properties does, in fact, exist. It is not widely known that concentrated aqueous solutions of certain salts are quite good solvents for organic compounds. Such solutions are known as *hydrotropes*, and a surprisingly wide variety of salts possess this property.[36] The most common hydrotropes are ammonium or alkali metal arenesulfonates, whose hydrotropic properties were discovered and used in electrochemical applications by McKee many years ago,[37] but even some inorganic salts such as sodium thiocyanate exhibit hydrotropism. Baizer has developed a valuable commercial process for the electrochemical hydrodimerization of acrylonitrile to adiponitrile (Chapter 7), which is based on the hydrotropic properties of arenesulfonate salts.[38] It is important to note that solvent properties are normally observed only in concentrated solutions of the hydrotropes. Hydrotropic solutions are highly structured; although apparently homogeneous on a macroscopic scale, they contain a high proportion of micelles (hydrophobic subregions) into which the organic solute may be incorporated.[39] The necessary micellar solvent structure is apparently constructed only in concentrated solution. (McKee developed a useful procedure, presently used in many parts of the world, for extracting cellulose from wood, which is based on the fact that dilution of hydrotropic solutions with water destroys their solvent properties.[36]) Hydrotropes will undoubtedly become more important in organic electrosynthesis in the future. A related class of compounds, the low-melting tetraalkylammonium salts studied by Gordon,[40a,b] which are rather good organic solvents and have high conductivity, have been little used in electrochemical applications but appear to be very promising. Tetrahexylammonium benzoate, a liquid at room temperature, has indeed been found useful for electrochemical experimentation.[40c]

A final point of information: when operating in protic solvents, one can often derive a great deal of mechanistic information from the pH dependence of the electrochemical behavior of a substance (*cf.* Chapters 4, 6, and 7). In investigations of this type, it is generally necessary to buffer the solution against electrochemically generated acid or base. Useful buffers for use over a wide range of pH values in mixed aqueous–organic solvents have been described by Bates[41] and others.[42]

9.6 POLAROGRAPHY

The book by Meites[43] is an extremely useful guide to experimentation in polarography, and hence polarographic practice need not be discussed here. A word about instrumentation may, however, be helpful. In principle, the voltage ramp required for a polarographic experiment can be supplied by the same wave generator used for cyclic voltammetry. Operational amplifier potentiostats of the sort discussed in Section 9.3 can also be used for polarography, but may need to be stabilized against oscillations due to loss of potential control, which sometimes happens when the drop falls from the dropping mercury electrode. On the other hand, circuits specifically for transistorized polarographs have been described in the literature.[44] It is a virtual certainty that any chemistry department or research facility of any size will have a polarograph already available, but where a commercial instrument is to be used it should be of the 3-electrode type; many are 2-electrode instruments, in which the anode also serves as the reference electrode. In such instruments, recorded potentials will be in error by an amount equal to iR, where i is the polarographic current and R is the solution resistance between anode and cathode. The iR term can be substantial in organic solvents; worse, it varies over the height of the polarographic wave (as i is increasing) and makes the wave appear more drawn out (irreversible) than it actually is. Some commercial 2-electrode polarographs can be adapted to 3-electrode operation by the addition of optional iR compensators sold for this purpose.

9.7 REFERENCE ELECTRODES

It ought to be abundantly clear to the reader by this time that a stable reference electrode whose potential is accurately known is a *sine qua non* for careful, reproducible electrochemical experimentation. There are certain problems that are peculiar to the use of reference electrodes in nonaqueous solvents that do not arise in aqueous systems, but they are well recognized, and procedures for dealing with them are adequately developed. There are a number of good compilations of information and data relevant to the use, fabrication, reproducibility, and so on, of reference electrodes in organic solvents, to which the reader may refer.[45]

References

[1a]S. Swann, Jr., in A. Weissberger, ed., *Technique of Organic Chemistry*, 2nd ed., New York, Wiley, 1956, vol. 2, pp. 398–399; [b]B. C. L. Weedon, *Advan. Org. Chem.*, **1**, 1 (1960); [c]J. J. Lingane, *Electroanalytical Chemistry*, 2nd ed., New York, Wiley, 1958, pp. 499–511.

[2a]L. Meites, in A. Weissberger, ed., *Technique of Organic Chemistry*, 3rd ed., New York, Wiley, 1959, vol. 1, pp. 3292–3305; [b]Lingane, *op. cit.*, pp. 296–391; [c]G. A. Rechnitz, *Controlled-Potential Analysis*, New York, Macmillan, 1963, pp. 19–25; [d]M. J. Allen, *Organic Electrode Processes*, London, Chapman and Hall, 1958, pp. 20–46.

[3a]Lingane, *op. cit.*, pp. 308–339; [b]A. J. Bard and K. S. V. Santhanam, in A. J. Bard, ed., *Electroanalytical Chemistry*, New York, Marcel Dekker, 1970, vol. 4, pp. 289–297.

[4a]R. E. Dessy and R. L. Pohl, *J. Amer. Chem. Soc.*, **90**, 1995, 2005 (1968); [b]A. J. Fry and R. G. Reed, *J. Amer. Chem. Soc.*, **91**, 6448 (1969).

[5]P. Birman, *Power Supply Handbook*, Flushing, N.Y., Kepco, Inc., 1965, p. 129.

[6a]R. S. Bottei and R. J. Boczkowski, *J. Chem. Educ.*, **47**, 312 (1970); [b]K. Lowe, *J. Chem. Educ.*, **47**, 846 (1970).

[7]W. M. Schwarz and I. Shain, *Anal. Chem.*, **35**, 1770 (1963).

[8a]G. W. C. Milner and G. Phillips, *Coulometry in Analytical Chemistry*, New York, Pergamon, 1967; [b]Meites, *op. cit.*, pp. 3319–3328; [c]Rechnitz, *op. cit.*, pp. 25–32; [d]D. D. DeFord and J. W. Miller, in I. M. Kolthoff and P. J. Elving, eds., *Treatise on Analytical Chemistry*, New York, Wiley, 1963, part I, vol. 4, pp. 2475–2531.

[9]R. S. Nicholson and I. Shain, *Anal. Chem.*, **36**, 706 (1964).

[10]J. Bacon and R. N. Adams, *J. Amer. Chem. Soc.*, **90**, 6596 (1968).

[11]R. L. Myers and I. Shain, *Chem. Instrumentation*, **2**, 203 (1969).

[12a]C. N. Reilley, in "Topics in Chemical Instrumentation," *J. Chem. Educ.*, **39**, Nov. and Dec. (1962); [b]E. J. Bair, *Introduction to Chemical Instrumentation*, New York, McGraw-Hill, 1962; [c]H. V. Malmstadt and C. G. Enke, *Electronics for Scientists*, Menlo Park, Calif., Benjamin, 1962.

[13]W. L. Underkofler and I. Shain, *Anal. Chem.*, **35**, 1778 (1963).

[14a]W. M. Schwarz and I. Shain, *J. Phys. Chem.*, **69**, 30 (1965); [b]S. P. Perone, *Anal. Chem.*, **38**, 1158 (1966); [c]R. S. Nicholson, *Anal. Chem.*, **37**, 1351 (1965); [d]I. Shain, J. E. Harrar, and G. L. Booman, *Anal. Chem.*, **37**, 1768 (1965).

[15]K. S. V. Santhanam and A. J. Bard, *J. Amer. Chem. Soc.*, **88**, 2669 (1966).

[16a]Allen, *op. cit.*, pp. 31–33; [b]Swann, *op. cit.*, pp. 405–406; [c]R. G. Clem, F. Jakob, and D. Anderberg, *Anal. Chem.*, **43**, 292 (1971).

[17]G. Dryhurst and P. J. Elving, *Anal. Chem.*, **39**, 607 (1967).

[18]W. E. Britton and R. Scoggins, unpublished observations.

[19a]Weedon, *op. cit.*, pp. 24–25; [b]Meites, *op. cit.*, pp. 3294–3298; [c]Swann, *op. cit.*, pp. 400–401.

[20]J. E. Harrar and I. Shain, *Anal. Chem.*, **38**, 1148 (1966).

[21a]Allen, *op. cit.*, pp. 33–40; [b]R. N. Adams, *Electrochemistry at Solid Electrodes*, New York, Marcel Dekker, 1969, pp. 267–270; [c]Meites, *op. cit.*, pp. 3294–3302; [d]Rechnitz, *op. cit.*, pp. 32–39; [e]R. E. Dessy, W. Kitching, and T. Chivers, *J. Amer. Chem. Soc.*, **88**, 453 (1966); [f]K. B. Wiberg and T. P. Lewis, *J. Amer. Chem. Soc.*, **92**, 7154 (1970); [g]S. Andreades and E. W. Zahnow, *J. Amer. Chem. Soc.*, **91**, 4181 (1969); [h]R. E. Dessy and E. L. Pohl, *J. Amer. Chem. Soc.*, **90**, 1996, 2009 (1968); [i]J. E. Hickey, M. S. Spritzer, and P. J. Elving, *Anal. Chim. Acta*, **35**, 277 (1966); [j]R. Pasternak, *Helv. Chim. Acta*, **31**, 753 (1948); [k]N. Dinh-Nguyen, *Acta Chem. Scand.*, **12**, 585 (1958); [l]B. Mooney and H. I. Stonehill, *J. Chem. Soc.*, (A), 1 (1967); [m]M. M. Baizer, *J. Electrochem. Soc.*, **111**, 215 (1964); [n]H. W. Sternberg, R. E. Markby, I. Wender, and D. M. Mohilner, *J. Electrochem. Soc.*, **113**, 1060 (1966); [o]A. J. Fry, M. A. Mitnick, and R. G. Reed, *J. Org. Chem.*, **35**, 1232 (1970); [p]G. S. Alberts and I. Shain, *Anal. Chem.*, **35**, 1859 (1963); [q]R. C. Buchta and D. H. Evans, *Anal. Chem.*, **40**, 2181 (1968); [r]V. J. Puglisi, G. L. Clapper, and D. H. Evans, *Anal. Chem.*, **41**, 279 (1969); [s]R. E. Sioda, *Electrochim. Acta*, **13**, 375 (1968).

[22]G. D. Christian, *J. Electroanal. Chem.*, **23**, 172 (1969).

[23]R. N. Adams, in I. M. Kolthoff and P. J. Elving, eds., *Treatise on Analytical Chemistry*, New York, Wiley, 1963, part I, vol. 4, pp. 2388–2390.

[24]L. Ramaley, R. L. Brubaker, and C. G. Enke, *Anal. Chem.*, **35**, 1088 (1963).

[25]J. R. Kuempel and W. B. Schaap, *Anal. Chem.*, **38**, 664 (1966).

[26]W. Kemula and Z. Kublik, *Advan. Anal. Chem. Instrumentation*, **2**, 123 (1963).

[27a]I. Shain, in I. M. Kolthoff and P. J. Elving, eds., *Treatise on Analytical Chemistry*, New York, Wiley, 1963, part I, vol. 4, pp. 2560–2562; [b]G. S. Alberts and I. Shain, *Anal. Chem.*, **35**, 1859 (1963).

[28]S. P. Perone, D. O. Jones, and W. F. Gutknecht, *Anal. Chem.*, **41**, 1154 (1969).

[29]Adams, *op. cit.*, pp. 19–42 and 187–209.

[30]Swann, *op. cit.*, pp. 399–404.

[31]C. K. Mann, in A. J. Bard, ed., *Electroanalytical Chemistry*, New York, Marcel Dekker, 1969, vol. 3, pp. 57–127.

[32a]D. S. Reid and C. A. Vincent, *J. Electroanal. Chem.*, **18**, 427 (1968); [b]G. Le Guillanton, *Bull. Soc. Chim. Fr.*, 2359 (1963).

[33]L. Marcoux, P. Malachesky, and R. N. Adams, *J. Amer. Chem. Soc.*, **89**, 5766 (1967).

[34a]J. Phelps, K. S. V. Santhanam, and A. J. Bard, *J. Amer. Chem. Soc.*, **89**, 1752 (1967); [b]Mann, *op. cit.*, pp. 122–123.

[35]M. M. Baizer, *J. Electrochem. Soc.*, **111**, 215 (1964).

[36a]R. H. McKee, *Ind. Eng. Chem.*, **38**, 382 (1946); [b]H. S. Booth and H. E. Everson, *Ind. Eng. Chem.*, **40**, 1491 (1948); [c]P. S. Putanov and J. D. Babin, *Chim. et Ind.*, **85**, 263 (1961).

[37a]R. H. McKee and B. G. Gerapostolon, *Trans. Electrochem. Soc.*, **68**, 329 (1965); [b]J. Chodkowski and H. Czajka-Gluchowska, *Rocz. Chem.*, **31**, 1303 (1957); see *Chem. Abstr.*, **52**, 9816 (1958).

[38]M. M. Baizer, *J. Electrochem. Soc.*, **111**, 215 (1964).

[39]P. A. Winsor, *Trans. Faraday Soc.*, **44**, 376 (1948); *Trans. Faraday Soc.*, **46**, 762 (1950).

[40a]J. E. Gordon, *J. Amer. Chem. Soc.*, **87**, 1499, 4347 (1965); [b]J. E. Gordon, *J. Org. Chem.*, **30**, 2760 (1965); *J. Org. Chem.*, **31**, 1925 (1966); [c]C. G. Swain, A. Ohno, D. K. Roe, R. Brown, and T. Maugh, II, *J. Amer. Chem. Soc.*, **89**, 2648 (1967).

[41]R. G. Bates, *Determination of pH*, New York, Wiley, 1964, pp. 222–229.

[42a]M. Runner and F. G. Wagner, *J. Amer. Chem. Soc.*, **74**, 2529 (1952); [b]J. Volke and A. M. Kardos, *Coll. Czech. Chem. Commun.*, **33**, 2560 (1968).

[43]L. Meites, *Polarographic Techniques*, 2nd ed., New York, Wiley, 1965.

[44a]H. C. Jones, W. L. Belew, R. W. Stelzner, T. R. Mueller, and D. J. Fisher, *Anal. Chem.*, **41**, 772 (1969); [b]R. Bezman and P. S. McKinney, *Anal. Chem.*, **41**, 1560 (1969); [c]D. Hume, *J. Electroanal. Chem.*, **7**, 245 (1969).

[45a]J. N. Butler, in P. Delahay and C. A. Tobias, eds., *Advances in Electrochemistry and Electrochemical Engineering*, New York, Wiley, 1970, vol. 7, pp. 77–175; [b]C. K. Mann and K. K. Barnes, *Electrochemical Reactions in Nonaqueous Systems*, New York, Marcel Dekker, 1970, pp. 17–28; [c]Adams, *op. cit.*, pp. 288–291; [d]G. J. Hills, in D. J. G. Ives and G. J. Janz, eds., *Reference Electrodes, Theory and Practice*, New York, Academic, 1961, pp. 433–463.

INDEX